U0027547

Forensic Examination of Glass and Paint

Taylor & Francis Forensic Science Series

Edited by James Robertson

Forensic Sciences Division, Australian Federal Police

Firearms, the Law and Forensic Ballistics

T A Warlow

ISBN 0 7484 0432 5

1996

Scientific Examination of Documents: methods & techniques, 2nd edition

D Ellen

ISBN 0 7484 0580 1

1997

Forensic Investigation of Explosions

A Beveridge

ISBN 0 7484 0565 8

1998

Forensic Examination of Human Hair

J Robertson

ISBN 0 7484 0567 4

1999

Forensic Examination of Fibres, 2nd edition

J Robertson & M Grieve

ISBN 0 7484 0816 9

1999

Forensic Examination of Glass and Paint

Analysis and Interpretation

Edited by

BRIAN CADDY

London and New York

First published 2001 by Taylor & Francis
11 New Fetter Lane, London EC4P 4EE

Simultaneously published in the USA and Canada
by Taylor & Francis Inc,
29 West 35th Street, New York, NY 10001

Taylor & Francis is an imprint of the Taylor & Francis Group

Typeset in Times by Wearset, Boldon, Tyne and Wear
Printed and bound in Great Britain by TJ International Ltd, Padstow, Cornwall

Every effort has been made to ensure that the advice and information in this book is true and accurate at the time of going to press. However, neither the publisher nor the authors can accept any legal responsibility or liability for any errors or omissions that may be made. In the case of drug administration, any medical procedure or the use of technical equipment mentioned within this book, you are strongly advised to consult the manufacturer's guidelines.

British Library Cataloguing in Publication Data
A catalogue record for this book is available from the British Library

Library of Congress Cataloging in Publication Data
Forensic examination of glass and paint: analysis and interpretation / edited by Brian Caddy.
p. cm. – (Taylor & Francis forensic science series)
Includes bibliographical references and index.
1. Criminal investigation. 2. Glass–Identification. 3. Paint–Identification. I. Caddy, Brian. II. Series.

HV8077.5.G55 F67 2001
363.25'62–dc21

2001027208

ISBN 0-7484-0579-8

Contents

List of contributors

José R. Almirall

Metro-Dade Police Department Crime Laboratory, Miami, FL 33172, USA.

John Bentley

Taplow, Maidenhead, Berkshire SL6 0LD, UK.

Alexander Beveridge

Royal Canadian Mounted Police, Forensic Laboratory, 501 Heather Street, Vancouver, BC, Canada V5Z 3L7.

John M. Challinor

Forensic Science Laboratory, Chemistry Centre (WA), 125 Hay Street, E. Perth, Western Australia.

Geoffrey J. Copley

British Glass, Sheffield S10 2UA, UK.

Niamh Nic Daéid

Forensic Science Unit, University of Strathclyde, Glasgow G1 1XW, UK.

Peter R. De Forest

John Jay College, New York, NY, USA.

Tony Fung

Royal Canadian Mounted Police, Forensic Laboratory, 501 Heather Street, Vancouver, BC, Canada V5Z 3L7.

Pamela S. Hamer

Forensic Alliance Ltd, Culham Science Park, Abingdon, Oxfordshire OX14 4ED, UK.

M. Lynn Henson

United States Army Central Investigation Laboratory (USACIL), Fort Gillem, Georgia, USA.

Tammy A. Jergovich

Georgia Bureau of Investigation, Division of Forensic Sciences (GBI/DOFS), Decatur, Georgia, USA.

Donald MacDougall

Royal Canadian Mounted Police, Central Forensic Laboratory, 1200 Vanier Parkway, Ottawa, ON, Canada K1G 3M8.

John McCullough

Forensic Science Laboratory, Department of Justice, Equality & Law Reform, Garda HQ, Phoenix Park, Dublin 8, Ireland.

Seán McDermott

Forensic Science Laboratory, Department of Justice, Equality & Law Reform, Garda HQ, Phoenix Park, Dublin 8, Ireland.

Wilfried Stoecklein

Bundeskriminalamt, Wiesbaden, Germany.

John I. Thornton

Forensic Analytical Specialties, Inc., Hayward, CA 94545, USA.

Sheila Willis

Forensic Science Laboratory, Department of Justice, Equality & Law Reform, Garda HQ, Phoenix Park, Dublin 8, Ireland.

Preface

This volume has taken much longer in its preparation than one would have wished but this is what might have been anticipated if the top people in their fields are asked to write relevant chapters. All the authors are exceptionally busy people so it is to their credit that they have now produced what I believe to be a valuable volume that will support the practising forensic scientist. Inevitably with a book of this nature there will be developments in the forensic sciences which it has not been possible to include and some of the authors who were able to return their manuscript early may have wished that such inclusions could have been made but to have done so would have extended the preparation time. Continuous revision means that books never get published. As the title suggests, this volume covers trace evidence as it relates to glass and paint and follows a regular pattern of defining the material describing the analytical processes and discussing how these may be applied to glass and paint. Finally each area describes how the analytical results may be interpreted for the courts. I ask the authors to forgive me for the times I had to resort to bullying tactics but I hope that they feel the final result has been worth it. I trust all find it good and valuable reading.

I would like to acknowledge all the hard work put in by the individual authors and in particular the tolerance many showed at the slow progress sometimes exhibited in producing such a volume. I have also to thank my wife for the patience she has exhibited over the last few years when I have been entrenched in my study trying my best to deliver the completed manuscripts. This volume has seen the passing of my faithful companion 'Barney' and his delightful replacement 'Elsa'. All three of these have kept me sane during this endeavour.

Brian Caddy
Glasgow

1

What is trace evidence?

PETER R. DE FOREST

1.1 Preliminary definitions

1.1.1 Trace evidence and physical evidence

The term *trace evidence* refers to a very broad category of physical evidence. The trace evidence category covers a larger variety of *physical evidence* types than any other division [1–3]. A discussion of the attributes of physical evidence that make it trace evidence will follow in a later section of this chapter. Anything or any material can be physical evidence. This term has both a legal and a scientific meaning. In the legal context it refers to tangible objects that are brought into court and formally entered into evidence as exhibits (or productions). Obviously, such objects need not be entered into evidence in a legal proceeding to constitute physical evidence from the scientific point of view. Any objects and materials from a crime scene (or suspect, victim, etc.) that may shed light on an investigation are physical evidence from the scientific perspective. They become physical evidence as soon as they are recognised as having possible significance. The process of evidence recognition is more complex and challenging than is generally appreciated. It requires considerable scientific knowledge and skill as well as experience. Evidence recognition, both at the crime scene and in the laboratory, needs to receive more attention from scientists in cooperation with non-scientist investigators.

1.1.2 Forensic science and criminalistics

Forensic science can be defined as the application of the sciences to matters of the law. That branch of forensic science which focuses on physical evidence is known in many quarters as criminalistics. Although these definitions are not universally agreed upon, forensic science is clearly the more general term. In addition to criminalistics, disciplines such as forensic pathology and toxicology fall within the more general field of forensic science. The American Academy of Forensic Sciences (AAFS) has about ten sections, the largest of which is criminalistics with about 800 members. For reasons that are primarily economic, only a fraction of the eligible criminalists in the United States are members of this section of the American Academy of Forensic Scientists. Many more are members of about eight regional associations of forensic scientists whose meetings are geographically convenient. A similar pattern holds on the international scene. There are many more scientists, whether they are called criminalists or not, who are employed by forensic science laboratories to analyse physical evidence from crime scenes than in any other branch of the field.

1.2 Concepts and the production of physical evidence

A fundamental principle of forensic science or criminalistics is that a criminal act, or more generally a human-initiated event, produces a record of itself. The record, however imperfect, is the result of human actor(s) and the events they set in motion producing interactions that result in changes in the immediate environment. These changes can be expected to follow well-established laws of nature. As a consequence of this, they can be studied scientifically. Objects get moved or broken, marks are made, and materials are changed or transferred. In the preface to his classic 1953 textbook, *Crime Investigation*, Paul L. Kirk [4] had this to say about physical evidence:

> Wherever he steps, whatever he touches, whatever he leaves, even unconsciously, will serve as silent evidence against him. Not only his fingerprints or his footprints, but his hair, the fibres from his clothes, the glass he breaks, the tool mark he leaves, the paint he scratches, the blood or semen he deposits or collects – all these and more bear mute witness against him. This is evidence that does not forget. It is not confused by the excitement of the moment. It is not absent because human witnesses are. *It is factual evidence.* Physical evidence cannot be wrong; it cannot perjure itself; it cannot be wholly absent. Only its interpretation can err. Only human failure to find it, study and understand it, can diminish its value. (The italics are in the original.)

The term *trace evidence* is difficult to circumscribe and define. In some ways the adoption and well-established use of this term is unfortunate. However, it is difficult to come up with a completely satisfactory alternative term. Two very different notions contribute to concept of trace evidence. One notion is related to size. Small amounts of material may be referred to as *traces*, and evidence of concern in forensic investigations is often very small. Here the notion of trace seems appropriate, but size is not the defining characteristic. On occasion, relatively large amounts of material may be trace evidence. The word trace also has the connotation of a *vestige* or of something remaining behind after something else has been removed, as in the phrase ‘he vanished without a trace’. This is closer to the other dimension of trace evidence which embodies the idea of an interaction which results in an exchange of material or the production of a pattern or both. This aspect is reflected in the term *transfer evidence* which is sometimes used. Despite the extensive and broadly inclusive realm of trace evidence, clearly not all physical evidence is trace evidence. As noted above and further illustrated in the quotation from Dr Kirk, physical evidence can be anything. The field of trace evidence has only a somewhat narrower purview. To sum up, it is the consequences of marks being made and material being transferred that are of concern in the area of trace evidence investigations. They are useful in evaluating possible associations and in reconstructing events.

Logically, physiological stain evidence such as blood and body fluid stains can also be regarded as trace evidence but are normally not discussed in references on trace evidence, because this subject is given specialised treatment in other sources. Conceptually, however, it is useful to consider these stains as a category of trace evidence. Physiological stains are set apart in common practice only because a different, and a more-or-less self-contained, set of analytical techniques is used for their characterisation. In forensic science casework it is essential to avoid focusing on techniques to the neglect of a larger view which encompasses all of the evidence and the context in which it occurs. This point should always be borne in mind.

1.3 Uses of trace evidence

Trace evidence has three distinct uses in forensic science. These are as *investigative aids*, as *associative evidence*, and in *reconstructions*. The least exploited of these uses is the employment of trace evidence to assist in giving direction to an investigation in its early stages. This can also be referred to as investigative intelligence. Unfortunately, historically scientific work on trace evidence in cases has been peripheral to investigations rather than being integrated with them.

Investigative aids

Although the skilful examination and the knowledgeable interpretation of trace evidence has a great inherent potential for contributing much useful information to investigations, this potential is rarely realised in practice for a number of reasons. Principal among these is the dearth of forensic scientists skilled in the examination and interpretation of trace evidence and the general lack of early involvement of forensic scientists in investigations in most jurisdictions. Examples of investigative aids or intelligence supplied by trace evidence interpretations would include suggesting a geographic provenance for an unknown sample, determining the origin of a manufactured material, determining the approximate date of manufacture of a material, providing information to eliminate suspects and thus limit the number of candidate suspects, as well as helping to develop suspects.

Associative evidence

In current practice trace evidence is most commonly used as evidence of association or as *associative evidence*. In this way it is used to investigate the existence of a link between individuals or between individuals and scenes after suspects have been developed, thus providing information of value at the adjudicative phases of a case.

Reconstruction

Trace evidence may contribute information to the understanding of how an event of concern took place. This point can be illustrated by considering situations commonly arising in the investigation of motor vehicle accidents. In addition to their value as associative evidence in linking the vehicle to the victim or the scene, paint transfers resulting from the contact of a motor vehicle with some object or another motor vehicle can be studied to provide geometric details concerning the manner in which the contact took place. This is particularly true where there have been multiple contacts and two-way exchanges of different kinds of complementary trace evidence. Beyond the motor vehicle example, there are many examples of trace evidence being used in reconstructions. At times the associative evidence, combined with the reconstruction its geometry affords, can greatly strengthen the association. In some of these cases the association can be so strong that there is no credible alternative, and the association becomes absolute even though the individual associations, provided by each component trace evidence comparison, are only suggestive. An example of this involving paint and other trace evidence will be given later in the chapter.

1.4 The history of trace evidence

Like the early history of forensic science in general, the beginnings of trace evidence examination are rather murky. Anecdotal accounts concerning the utilisation of trace evidence may date back millennia. Suspicious wives have undoubtedly noticed hairs and other traces on a husband's clothing that in their minds seemed to defy innocent explanation. It is all but certain that similar ad hoc recognitions of trace evidence occurred in early criminal investigations as well. However, trace evidence examinations as we know them had to await the development of modern science and the refinement of the microscope over the course of nearly two centuries by Anton van Leeuwenhoek, Robert Hooke, Ernst Abbe and others. The first articulated general recognition of trace evidence would appear to have occurred in the fictional exploits of Sherlock Holmes in the Sir Arthur Conan Doyle stories. Shortly after the appearance of the Holmes stories, the Austrian Magistrate Hans Gross published a book in which he advocated the use of trace evidence in criminal investigations [5]. Later Edward Oscar Heinrich in the United States, Edmund Locard in France, and Georg Popp in Germany were among the first to make use of trace evidence in casework. Interesting details of the work of some of these pioneers have been discussed by Palenik [6–8]. The details of this history will not be presented here, but the history, such as that recounted in these articles, is well worth reading. The articles present trace evidence insights as well as a historical perspective.

1.5 The scope of trace evidence

1.5.1 Introduction

For trace evidence produced by contact, two general kinds of evidence are produced, *pattern evidence* and *transferred material*. The material transferred can be important evidence from two principal perspectives. From one of these, it is the *transferred material* itself that becomes the evidence. This consideration is applicable where the comparison of the transferred material with its suspected source is used as the evidence of association or common origin between the transferred material and the donor surface. In the second way that trace evidence resulting from contact can be used, it is the *pattern* of the transferred matter on the recipient surface or the pattern of matter removed from the donor surface that is important. This pattern may be compared to one on the corresponding surface to study the question of commonality of origin, or it may be used to shed light on the way in which the interaction took place between the objects bearing the two interacting surfaces. The latter is an issue involving reconstruction. The former can be considered part of the domain of trace evidence comparisons but will not be the subject of this book.

The focus in this volume will be on paint and glass evidence and thus on the properties and characterisation of these two types of material when they appear as transfer evidence. However, in practice the other perspective or broad class of trace evidence cannot be ignored or important information may be lost. The material and the pattern or context must be considered together. Serious dilemmas can arise with respect to decisions regarding documentation and sampling. For example, in the case of paint evidence, if flakes or chips are encountered, these need to be evaluated for their potential for physical matching to a known source. In addition, layer structure, if present, needs to be considered.

The idea of material transfer across the contact boundary is often referred to as the *Exchange Principle* in recognition of the early contributions of Dr Edmund Locard of Lyon, France, to the theory and practice of trace evidence examinations. This is discussed by Nickolls [9]. It does not address the situation where the significant evidence is not the material transferred but is instead a pattern produced as a result of the interaction between the two surfaces. A more general exposition of this idea is the one contained within the quotation from the preface of Kirk's *Crime Investigation* which was included earlier in this chapter. Here contactless transfers as well as marks are considered. More general yet is the idea of viewing the crime scene as a recording medium where human activities and human-initiated events produce a *latent record* in the form of alterations to the environment. These alterations, whether they are material transfers, markings, fractures, or mere displacements of objects, are physical evidence. The totality of the physical evidence, if recognised and properly interpreted, is the record. It should be appreciated that the record is often largely a latent one. It must be recognised, decoded and interpreted.

This section will be concerned with three major divisions of trace evidence, namely contact transfers of material where the material itself takes on importance as the evidence, contactless transfers of material of evidentially useful material, and patterns of value resulting from contact. These three broad divisions encompass the scope of trace evidence. These are subdivided and discussed in the subsections that follow.

1.5.2 Evidence consisting of contact transfers of material

Much trace evidence is the result of the transfer of material across the contact boundary as two surfaces come into contact. Evidence produced in this way comprises a class of evidence known as *contact transfer* evidence. The material is transferred from what can be called a *donor* surface to a receiving or *recipient* surface. For certain combinations of pairs of materials and areas of contact, a given surface may play both the donor and recipient roles. In this case we would have what is termed a *two-way transfer*. The nature of the materials and the nature of the interaction can influence whether or not this takes place. Two-way transfers are among some of the most evidentially significant transfers. There is a need for more research regarding the details of these interactions and the ways in which they can influence trace evidence interpretations.

1.5.3 Evidence consisting of contactless transfers of material

Contact between donor and recipient surfaces is not always necessary for there to be a transfer of trace evidence. Many situations involving glass transfer provide a good example. A would-be burglar who breaks a window may get numerous microscopic bits of glass on his clothing even though his clothing did not come into direct contact with the window. During the act of breaking the window small fragments of glass are projected from the area of impact. Those projected in a retrograde fashion with respect to the direction of the applied force may be deposited and retained as trace evidence in the hair and clothing of the burglar. Stains produced by airborne blood droplets and deposits of gunshot discharge residue provide additional examples.

1.5.4 Pattern evidence from contact transfers of material

At times the nature of the material transferred may not be as important as evidence of association or prior contact as is the pattern that results from the transfer of the material. It is the pattern that the material makes on the recipient surface that has the most value as evidence of association. This can be called *static contact transfer pattern* evidence. Material present on a donor surface may be transferred to a receptor surface in a manner that preserves some characteristic geometric features of the donor surface. The pattern left by an inked rubber stamp is a good example of this. A similar situation would be the pattern left on a fingerprint card by a suitably inked fingertip or by a bloody fingertip at a crime scene. Of course in the latter case, depending on case circumstances, the blood transferred could be as important evidence as the pattern produced.

Other situations where the transferred material carries the pattern of interest are the reverse of the above. Here it is the pattern of removal of material from the donor surface compared to the geometry of the recipient surface that is utilised. The patterned rubber sole of a shoe may lift dust from a thin film that exists on a floor to form a detailed void pattern that would have value as evidence. Parts of assemblies that have remained in contact for a period of time before becoming separated present another example. A rubber seal or gasket which has pulled away from a painted surface may take an irregular pattern of paint with it. The pattern of voids on the surface from which the paint has been removed compared with the pattern of paint adhering to the rubber may provide the basis for a conclusion of a unique association at a later time. This is somewhat like evidence produced by physical separation to be discussed in the next section.

1.5.5 Pattern evidence from physical separations of material

The physical separation described above where a pattern of paint was removed from one surface by another could be viewed as a special case of pattern evidence produced by physical separation. The more general case would be where an object is cut apart or fractured producing two or more pieces. Fractures are most often random, resulting in fragments that possess detailed and distinct topology on the fracture surfaces. Such fragments have the potential of being uniquely associated with the surfaces from which they were separated.

1.5.6 Pattern evidence from contactless transfers of material

Patterns can be produced by contactless transfers of material. Unlike the *contact patterns* discussed in this section of this chapter, most of these *contactless transfer patterns* seldom have value as evidence of association. Examples of this type would include gunshot discharge residue patterns or airborne blood droplet stain patterns. These kinds of patterns more commonly have value in reconstructions rather than as associative evidence.

Those more exceptional circumstances where a contactless transfer could have value as associative evidence could occur where a patterned void is left where a characteristically shaped object has intercepted the material being deposited during the process of material deposition. The characteristically shaped object might intercept paint spray, bloodspatter, or airborne powder or dust. Here the pattern of the void itself may become evidence if there is a need to evaluate whether specific objects were present during deposition and were the cause of the void. If the object is recovered later at some other location, the possibility of its former presence can be assessed.

1.5.7 Contact pattern evidence independent of transfers of material

Marks resulting from contact between two surfaces where one surface distorts, deforms or reshapes the other comprise the evidence in this category. The marks are to some degree three-dimensional and are not due to material transfer. Material is undoubtedly transferred in the process, even if the amount transferred is insignificant and too small to be noticed. However, the critical distinction in this situation is that the transfer of material, if any, is peripheral to our present consideration. Unlike the situation discussed in Section 1.5.3, it is not the transferred material which contains the pattern information. In this circumstance, it is the pattern resulting from the alteration of one surface caused by its coming into contact with another harder one which comprises the evidence.

If there is little or no lateral movement of one surface relative to the other during the time they are in contact, a *static contact deformation pattern* is the result. This can also be referred to as *indentation evidence*. If sufficiently detailed features are present on the harder of the two surfaces, the contact may produce a unique indentation in the softer surface. Examples of these three-dimensional patterns would include footwear or tyre tread impressions in soft soil or snow; plastic fingerprint indentations in such things as soft wax, caulking, tape adhesive, and similar surfaces; as well as certain toolmarks such as firing pin and breechface indentations in cartridge primers.

Where relative lateral movement is present during contact, a *dynamic contact deformation pattern* results. These consist of three-dimensional striations. These are seen in the rifling markings on bullets and with many kinds of toolmarks resulting from cutting operations.

1.6 Associations based on physical pattern evidence

As noted in Section 1.5, contact trace evidence can be in the form of transferred material, a pattern, or both. Because of the possibility that a particular example of associative evidence may have value as pattern evidence as well as a material to be compared, both possibilities need to be considered at all times. Valuable information may be lost if one aspect is focused on at the expense of the other. Examples of this duality abound. However, for the purposes of our discussion the focus in this section will be on the pattern itself. In favourable circumstances physical pattern evidence can yield associations that are unique and indisputable. Typically, such strong associations are relatively rare when based solely on material transfers.

1.6.1 Imprints – two-dimensional static contact physical patterns

A pattern left by the deposition of a film of residue by the friction ridge skin of a fingertip or the sole of a shoe on a smooth surface is an example of a two-dimensional physical pattern. These can be referred to as *imprints*. Other examples would include tyre tracks on a garage floor and cloth imprints. If unique or individual features are transferred in the process of depositing the pattern, useful evidence with the potential for being used to demonstrate a unique association at a later time is produced.

1.6.2 Indentations – three-dimensional static contact physical patterns

When a force normal to the plane of contact between two surfaces is great enough, geometric features of the harder of the two surfaces may be impressed into the softer one. If the resulting three-dimensional pattern or indentation contains sufficient detail, a useful means of definitively associating the indentation with the object producing it at a later date has been produced. This was the *static contact deformation pattern* described in Section 1.5.5. Similar examples to those listed in Section 1.5.5 would include footwear sole tracks in snow, a friction ridge skin pattern left in soft wax, and breech face marks in a cartridge primer as representative examples of this kind of evidence.

1.6.3 Striations and wipes – dynamic contact patterns as evidence

Relative movement between two contacting surfaces can yield detailed streaks or striation patterns on one or both surfaces. These may be in the form of transferred material or they may be scratches caused by the harder of the two surfaces moving relative to the softer one. These are seen as die marks on drawn wire, extruded metals, and plastics. Cutter marks on wires and the microstriae in land impressions on bullets are additional familiar examples.

1.6.4 Fracture and physical matches

If an object is broken or torn apart, random processes are involved which result in varied and detailed topological features being produced on the fracture surface which may be unique. If the fracture surface is large enough, the detail present may be sufficient to allow a unique association to be demonstrated between the two portions. This is somewhat analogous to the fitting together of two corresponding pieces of a jigsaw puzzle, and is often called a *jigsaw fit* or a *physical match*. Physical matches can yield some of the strongest associations attainable. Under favourable circumstances, a sound theoretical basis for concluding that one object was broken from another to the exclusion of all other possibilities may be justified. Situations where there is distortion during the fracture process may limit the certainty with which positive conclusions can be drawn. In some of these situations the results of efforts to obtain a 'match' may be inconclusive. In addition to the potential for producing strong associations, good physical matches do not suffer from the need to have databases of some debatably adequate size. Statistical evaluation may not be necessary for reaching strong conclusions with physical matches and certain of the other kinds of physical pattern comparisons.

1.6.5 Matches based on continuity

The separation of materials by cutting or breaking may not produce separation surfaces suitable for the kinds of physical matching described in Section 1.6.4. Even in situations where the nature of the material could be expected to produce suitable fracture surfaces, adverse circumstances can present themselves. Sometimes a fracture surface may suffer post-fracture damage or an intervening segment or section may be missing. In such cases a direct physical match may not be possible, and it may be necessary to attempt an *indirect physical match*. Here continuity of both surface and internal features can have potential value in evaluating the possibility of an association. A prime example of such an indirect physical match would be the matching of wood grain. Numerous cases have been encountered in practice where broken or cut wooden articles have acquired importance as evidence. When direct physical matches were not possible because of a damaged end or a missing intervening piece, recourse has to be made to indirect physical matches. Many of these have been reported in articles and textbooks. Matches of this type are often necessary with pigmentation patterns in plastic bags. The cut edges of plastic bags are frequently too featureless or too distorted by the 'heat cutting' and tearing to allow a direct physical match to be attained. The under-surface of automotive paint chips may contain a cast-like replica of distinct random scratches present on the metal surface from which they have been separated. Even if what would have been the corresponding edge of one of these was damaged, the continuity of such scratch patterns could make an indirect physical match possible.

1.7 Associations based on material comparisons

1.7.1 Comparison of materials using morphological criteria

Morphological taxonomic identifications are commonly used in certain sciences, particularly biology. The sizes and shapes of biological structures are relied upon for classifications of plants and plant parts in botany. Similar processes along with crystallographic identifications are often used by geologists in identifying minerals. Rarely, if ever, is a true individualisation or unique association possible using taxonomic approaches on individual objects or particles. Taxonomic approaches, by definition, are used for identification not individualisation. However, when applied to a combination of features such as those present in a complex mixture, such as the mineral grains present in a soil sample (see Section 1.7.3), individualisation may be approached closely. Human hair comparisons present a somewhat analogous situation. An assemblage of morphological features is present in each hair. The totality of these must be relied upon to approach the goal of individualisation of human hair. Most trace evidence criminalists would agree that this goal is never reached in the case of the morphological approach to human hair comparison, although useful results can be obtained. Certainly, with adequate known reference samples, clear-cut exclusions are possible. This is a valuable result. It can eliminate a suspect from further consideration and conserve investigative resources. A failure to exclude can also have value. This 'failure' is especially valuable if it remains the case after a series of independent tests which have utilised discriminating criteria have been applied. Such a failure to exclude may not provide a numerical result to assist with determining the likelihood of association, and it does not constitute an individualisation, but it is valuable nevertheless. The significance and value of such a conclusion is often underappreciated.

1.7.2 Comparison of materials using physical properties

Physical properties of homogeneous materials are intrinsic to the material and as such are independent of the size of the sample. Mass density and optical properties of glass, such as refractive index and dispersion, present major examples that have been used for several decades. These are discussed in Chapter 3. They reflect the composition of the material. Physical properties include a number of quantifiable properties beyond those commonly used by the criminalist in characterising physical evidence. Those other than the familiar density and optical properties, i.e. absorptivity (colour) and refractive index (and dispersion), will be discussed in a separate chapter of this volume.

1.7.3 Comparison of materials composed of complex physical mixtures

Complex physical mixtures such as those composed of several different kinds of particles have considerable potential for uniqueness. One primary example would include the assortment of varied proportions of different mineral grains and other particles in soil. These other particles include pollens and plant residues such as phytoliths as well as a variety of anthropogenic particles. The potential for soil individualisation also extends to inorganic and organic chemical composition, as well as microbial and biochemical profiles. These approaches, however, are outside the scope of our discussion of the information richness of complex physical mixtures. However, dusts are complex physical mixtures and offer a similar potential. They often consist of mixtures of particles in combination with varied types and colours of fibres. Another example could be termed contrived or 'designer' trace evidence, in which specially formulated mixtures of particles are used to label or mark something that may be a target of theft. Those that have been used in the past, although not widely, have included mixtures of dyed starch grains or dyed lycopodium spores. Such unique mixtures affixed to an object can be used to identify the object when it is recovered later or they may be designed to transfer to and mark the thief, functioning as a kind of detective dye. Particle mixtures can be studied and characterised by microscopy. The process can be time-consuming and labour-intensive. In the future we can expect to see increasing automation of the particle mixture characterisation process. As the technology develops more of this will be accomplished using improved image analyser software and particle sorters.

1.7.4 Comparison of materials using refined compositional analyses

Small differences in chemical composition have long been used in attempts to individualise physical evidence. We can expect to see improvement here. These approaches depend on detecting and measuring trace chemical constituents in homogeneous materials. For fine discrimination between closely related objects or materials, it is necessary to analyse chemical species present at very low levels. The potential of this approach can be appreciated from a brief consideration of the concept of chemical purity. It should be recognised that even extremely pure materials contain large numbers of atoms or molecules of other substances. We can carry out some simplified but illustrative calculations. For example, if we consider a single microgram of a 99.9999 per cent pure material which has a molecular weight of about 100 daltons, this sample size would represent about 10^{-8} mole. Thus, it would contain about $6 \times 10^{23} \times 10^{-8}$ or about 6×10^{15} molecules. If the purity is as stated above, one part per million of these, or several billion molecules in this small sample, could be impurity species with an average molecular weight of 100 daltons. In theory, we should be able to generate a characteristic profile with over one billion molecules. In practice, a number of factors limit this potential. Past attempts at exploiting this potential have been thwarted by the limited sensitivity and quantitative reproducibility of existing analytical chemical techniques. Problems arising from contamination in evidence handling and within the laboratory are also critical. The needs of this trace evidence problem have placed stringent demands on such techniques. Although modern chemical instrumentation has been making great advances and has been adequate for many problems in chemical analysis, it has fallen short of meeting this forensic science need. However, analytical chemical techniques have been improving steadily.

Until fairly recently refined physical property measurements applied to glass samples, principally refractive index and density, provided a better means of investigating possible associations among items of glass evidence than did methods based on compositional measurements. Many methods of elemental analysis simply lacked the sensitivity, precision and quantitative reproducibility necessary to exploit subtle differences in composition for the purpose of differentiating closely related samples. Without this, similar results obtained with two samples for which the question of commonality of origin was being explored were almost meaningless. A close compositional match between two samples would not be significant, if, in reality, all of the objects in this class fell within a very narrow range that was not appreciably larger than the range spanned by the known and questioned samples.

Recent examples of research with very sensitive instrumentation for elemental analysis, applied to sample populations of glass, lead bullets and fibres, have proved useful. The most successful instrumentation for this purpose has included total reflectance X-ray fluorescence (TXRF), inductively coupled plasma (ICP) atomic emission spectroscopy (AES), and ICP mass spectroscopy (MS) [10]. These techniques have been both sensitive enough and reproducible enough to allow individual items or samples of material to be distinguished from other similar items within the sample population.

In addition to differentiating manufactured materials and objects by taking advantage of batch-to-batch differences in composition, there is the possibility of exploiting acquired compositional differences which may take place during the post-manufacturing interval. This potential has not been studied in any detail. Important here are chemical characteristics acquired between manufacturing and the separation of known and questioned material. Chemical characteristics acquired after separation can only serve to confound the significance of a comparison unless the history or nature of the post-separation exposure processes can be known.

We need to be concerned about problems with contamination. The greater the sensitivity or the more the detection limits are lowered, the more critical potential contamination problems become. There is a clear need for more attention being paid to sample handling and to intelligently selected control samples.

1.8 The role of trace evidence databases

Trace evidence databases have two more-or-less distinct uses. *Investigative databases* may be collections of reference samples or they may consist of data sets taken from such samples. They are searched when an unknown is encountered to locate a sample or data set with matching attributes. A successful search may result in the development of a suspect or in gaining insight as to how an event took place. The recent notable success of DNA databases in developing suspects during the investigative phase, although not directly analogous, serves to illustrate the potential of similar trace evidence databases. The other category of database could be referred to as *databases for interpretation* or *interpretive databases*. Such databases are used to obtain frequency of occurrence data to assist the scientist in determining the weight or value to be assigned to the correspondence of various attributes of a known and questioned sample. This second kind of database is continually evolving. It may have little value if it is not kept current. Unlike a DNA database used for interpretation, in which frequency of occurrence data for each of its populations converge on stable values as the size of the database grows (at least for systems in Hardy–Weinberg equilibrium in non-evolutionary time-scales), there is no such longer-term stability with this type of trace evidence database. It must constantly be renewed to remain relevant. Using textile fibres as an example, it can be appreciated that what may have been a rare or even non-existent fibre at one period of time could become more commonplace later. The reverse situation could also be of concern. A textile fibre, or even a whole fibre type, produced in large quantity at an earlier time may stop being used completely.

Lack of currency with a trace evidence *investigative database* may result in missed opportunities during an investigation. This may be viewed as a less serious problem than a misinterpretation stemming from the use of an out-of-date *interpretive database*.

There is a clear need for interpretive trace evidence databases. These would be very useful to help to evaluate the a priori probability of a match when one is obtained between an unknown sample and a known control sample. Too often, information about frequency of occurrence of a particular kind of trace evidence is only developed ad hoc in response to the needs of a particular major case. It would be far better if databases existed so that information helpful in evidence interpretation would be readily available for use in less-celebrated cases. Those which exist at present are few in number and limited in scope. The National Bureau of Standards (NBS) of the United States (now known as The National Institute for Standards and Technology or NIST) maintained and distributed automotive topcoat reference collections for many years. These were very useful in casework for identifying candidate vehicles from paint traces left in hit-and-run cases. The Alcohol, Tobacco and Firearms (ATF) laboratory of the United States Treasury Department developed and maintained a ball-point ink reference 'library' for use in identifying ink sources in questioned document cases. Using this collection it was often possible to show that the purported date on a document was fraudulent, because the ink used was not manufactured until years later. This is normally not considered a trace evidence problem but is illustrative of the value of databases and collections. In Japan the National Police Agency (NPA) maintains several manufactured product databases to which manufacturers supply reference samples on a regular basis.

The oldest 'trace evidence' databases would be fingerprint databases. Such databases, in the form of fingerprint card files, go back about a century. Recent advances in Automated Fingerprint Identification Systems (AFIS) have made single-digit cold searches possible during investigations. Although less well established, computerised footwear databases and firearms databases are becoming useful tools. Firearms databases, that can take advantage of both microstriation markings in the rifling on fired bullets and markings on ejected cartridge cases, are currently being put into use.

1.9 Complementary nature of trace evidence and DNA typing

In the last decade or so the methods of molecular biology have been applied to physiological stains occurring as physical evidence with great success. This development is clearly a landmark event in forensic science. Media attention has accompanied this and the abbreviation DNA has become familiar to millions among the non-scientific public as a result. The advent of forensic DNA technology and the resulting media attention may appear to have eclipsed trace evidence. However, potentially it has enhanced rather than diminished the importance of trace evidence. Properly viewed, DNA technology is another excellent tool for working with trace evidence. However, with this advance has come some negative fallout. Because of the great success of DNA analysis methods applied to blood and seminal stains, we appear to have been blinded to the true role of these techniques. We cannot forget that physical evidence occurs in a context. Focusing exclusively on applying DNA typing methods to physiological stain evidence may not produce the most useful information in a given case. Other evidence which may be complementary and even potentially more important may be overlooked.

DNA typing methods as practised are directed primarily toward the associative evidence with physiological fluid stains. These methods have potential applicability to many trace evidence problems. Such applications would include the analysis of evidence from human sources other than body fluids, materials derived from other animal species, and plant materials. We can anticipate many such applications to forensic science problems in the future.

Even with human-derived stain evidence on which DNA typing is successful, there is often a need to look beyond the results obtained to the context in which the evidence occurs. For example, DNA methods applied to physiological stains may produce only part of the link necessary to tie an item of physical evidence to a suspect. Consider, for example, a ski mask or knit cap which has bloodstains on it which can be persuasively linked to the victim by DNA typing methods. If the ski mask or cap is found discarded in a location that cannot be associated with the suspect, trace evidence found on the headgear may become crucial in investigating the completion of the link between victim and suspect. The trace evidence should not be the weakest link in the chain. With skilled scientific investigation at the crime scene and proper attention to research needs, trace evidence can complement DNA typing methods. This is too important an area to neglect. The advent of DNA typing methods has made trace evidence more important, not less important.

In nature, DNA is 'packaged' in or often associated with a discrete morphologically recognisable form. One example would be the root sheath tissue found with some hair evidence. Pooling such samples prior to DNA analysis should be avoided wherever possible. This is also true of tissue fragment samples such as those that may appear as trace evidence from shootings. The fragments may have become co-mingled with other physiological material or contaminated with fluids such as blood. Swabbing and digesting such tissue samples for DNA testing may create mixtures, confound the results, and unnecessarily complicate the interpretations. Microscopic assessment of such evidence prior to DNA analysis, even if it does not result in an effective separation in each case, is essential. At present the ability to achieve separations of DNA from different sources in a range of situations with casework samples is limited. With more research in the future it may be possible to develop methods for effecting separations of DNA from different types of sources prior to extraction and typing.

1.10 Recognition and interpretation of trace evidence

The context of trace evidence can be crucial in its evaluation and interpretation in many ways. For example, is the trace evidence affixed to a surface in such a way that it is clearly related to the event? In this or a similar situation, its significance can be remarkable. Or, in the other extreme, is the trace evidence of interest merely found in connection with a mass of accumulated debris such as fibres or hairs deposited in the interior of a poorly cared for automobile that may represent an extended, poorly defined, time-frame? It should be clear that the way in which evidence is recognised, documented and collected may be critical to the interpretation in this type of situation. An important question to ask is: does the evidence have any inherent potential value before it has been characterised?

Even where biological evidence amenable to approaches to individualisation using DNA typing is found, other kinds of trace evidence may be useful in providing the context or even authenticating it. The potentially extraordinary discriminating power of DNA typing makes defence claims of planted evidence more likely and, we hope far less frequently, might in fact tempt some unethical investigators to plant evidence. In any case this must be guarded against. In addition to protection offered by enhanced concerns about supervision, procedures and evidence documentation, other forms of trace evidence may offer some protection here as well, in the form of independent authentication of associations. As a result of the subtleties and innumerable complexities involved, fabrication of trace evidence, so that the deceit will escape the notice of an experienced trace evidence examiner, can be very difficult. When this capability is appreciated it can serve as an effective deterrent against criminal behaviour in the form of overzealousness on the part of those holding positions of public trust.

Scientific case assessment is given far too little attention. This applies to work at the crime scene and to work in the laboratory. Without this preliminary overall assessment, the risk of potentially significant evidence going unrecognised is very real. There is also the risk that effort may be spent analysing items that are ultimately shown to have no relation to the event being investigated and therefore no value. Evidence recognition is a far more challenging endeavour than is commonly realised. Experienced forensic scientists are necessary at crime scenes to deal with the recognition difficulty. Evidence recognition concerns extend to the laboratory. When the evidence from a case enters the laboratory it needs to be assessed scientifically prior to any analyses being carried out. Relevant questions need to be posed. This process defines and frames the problem. More questions may be developed as a result. The subsequent analyses need to address these questions. Trace evidence concerns and microscopy are an integral part of this case assessment.

The interpretation of physical evidence can be very complex and demanding. The context is often crucial to a proper interpretation. It can have a major impact on the value of the evidence. With associative evidence, when a positive association is found, its value is dependent not only on the expected frequency of duplication of the attributes of the questioned object in a similar object selected at random but also on factors such as where it is found and on indicators as to how long it has been there prior to recovery.

In situations where more than one association is demonstrated, additional value may be realised. However, additional care must be given to the interpretation. Evidence of multiple independent transfers would be expected to have considerably more value than a single one. Similarly, a two-way transfer would exceed the value of a one-way transfer.

Spatial relationships can also be important and enhance the value of associations. Consider two objects, each consisting of an array of several different materials in specific geometric arrangements. Were these to come into forcible contact and material transferred at each of several points of contact, a definitive association might be demonstrated even though the value of each individual one- and two-way transfer in demonstrating an association might be limited. It would be the particular configuration of each of these that would have the potential to confer a unique and therefore powerful association on this constellation of individual ones.

1.11 Hair and fibre evidence

A hair, having an axis many times longer than its width, is, by definition, a fibre. For this reason, hairs and fibres are often discussed together. However, they are dealt with separately in two dedicated volumes in this series. Operationally and organisationally, forensic science laboratories differ in how they deal with hair and fibre evidence. In some laboratories, hair is analysed in a 'biology' unit, whereas fibres are analysed in a 'chemistry' unit. Logically, both should be analysed by criminalists with specialist knowledge in microscopy and microanalytical techniques in a trace evidence laboratory unit [11].

Both hairs and fibres find value as associative evidence for use in developing proof at the adjudicative stage of a case, and are commonly thought of in this context. As a result, other uses of this kind of evidence may be overlooked. In addition to its value as associative evidence, the knowledgeable examination of hair and fibre evidence can yield information that is useful investigatively and in reconstructions. The examination of damaged hairs or fibres may reveal how they were damaged or indicate the environment in which they were exposed and something of the duration of the exposure. This may be important in reconstruction or in helping to decide the contemporaneousness of the evidence to the event under investigation and thus the relevance of the hairs and fibres to the case. Such information may help to authenticate other evidence.

1.11.1 Hair evidence

Hair is an outgrowth of specialised structures in the skin known as hair follicles. It is extruded continuously at a rate of about 3 mm per week during the active phase of a follicle's cycle. The hair shaft has three gross morphological regions. Microscopy and comparison microscopy have been used in attempts to evaluate the possibility that a given hair originated from a particular donor. Another volume in this series deals with the forensic examination of hair. Such an examination can provide very useful information if it is conducted properly. Unfortunately, at times, claims are made in particular cases that are unsupportable scientifically. Under the most ideal of circumstances it is never possible to state conclusively that a given hair came from a particular donor. However, there is a considerable range of variation among individual people with respect to hair morphology. This can even be noted on the macroscopic scale and can be appreciated by letting one's gaze scan the hair on the heads of several people standing together. Considerable differences can be observed in this way. It would not be surprising, then, that more differences could be expected on an examination of microscopic features. In fact, this is the case. Many more features can be studied. Practically speaking, this favourable circumstance is somewhat offset by the fact that the hairs on the head (or other body region) are not all the same. This intra-sample variation greatly complicates hair comparison by microscopic morphology. However, it does not invalidate it, as long as the examination is carried out by an experienced scientist. We must not underestimate the value of a 'failure to eliminate' that results from a rigorous examination and comparison of the samples. The reporting of the results must be done responsibly, with a clear explanation of the limitations that are inherent in hair comparisons generally and with specific reference to the case in hand.

In addition to head hairs, pubic hairs frequently find use as associative evidence. However, with respect to pubic hairs, there is less range of variation among individuals than is seen with head hairs. Hairs from the body regions other than the head and pubic regions are often easily identified as to somatic origin, but they are more problematic in terms of their value as associative evidence. Some show relatively little range among individuals. Studies concerning the potential evidential value of hairs from regions such as the limbs, chest and beard are needed. Methods of typing of mitochondrial DNA have excellent potential with such samples as discussed briefly below.

Mitochondrial DNA typing (mitotyping) has been found to be particularly useful with certain hair evidence. Very little nuclear DNA is found in the keratinised shaft of human hair. It appears that it is scavenged by the body before the hair becomes fully keratinised in the follicle. Only nuclear remnants can be seen in transmission electron micrographs of sections cut from the keratinised cortex. Using polymerase chain reaction (PCR) amplification techniques enough nuclear DNA is often present in the root end of both anagen and telogen hairs to allow nuclear DNA typing. There is more than ample nuclear DNA in the root sheath tissue that is often present on the proximal end of plucked anagen hairs to allow typing. The situation with the keratinised shaft is different. The most successful technique with the keratinised shaft is mitotyping. Unlike the situation with nuclear DNA, cells have multiple copies of the mitochondrial genome. This 'high copy number' is one of the principal reasons that mitotyping can be applied to small quantities of biological material. There are drawbacks. The mitochondrial genome is extremely small in comparison with that found in the cell nucleus, and the number of polymorphisms is greatly reduced with respect to nuclear DNA. Thus, the discrimination potential is much less than that of methods for nuclear DNA typing. There are additional considerations with forensic relevance. Mitochondrial DNA is inherited maternally. This can be an advantage in cases involving a missing person where no comparison sample is available. Here the known DNA type can be ascertained by obtaining and typing a sample from a relative in the same maternal line. In other cases the fact that members of the same maternal line cannot be discriminated could be a disadvantage.

1.11.2 Fibre evidence

The distinction between fibres and hairs becomes even more blurred when we consider animal hairs. Some of these are used commercially as fibres. With experience and a suitable reference collection, a trace evidence examiner can determine the species, or at least the genus, of the source of an adequate sample of animal hair. This can often be accomplished in a matter of minutes. The information can be useful investigatively, and in a general way, as associative evidence. Dyed animal hair such as rabbit may be used in knitwear and other apparel. Of course, the animal hair which finds the most extensive use as a textile fibre is wool. Natural fibres, whether hairs such as wool or vegetable fibres such as cotton, have characteristic morphologies that are useful in identifying them. Because of their ubiquitousness, undyed natural fibres are of very limited value as associative evidence. The presence of a dye in such a fibre enhances its value as evidence. The dye is often the only thing that differentiates natural fibres from others of the same generic class. Colour comparisons can be effected by comparison microscopy or microspectrophotometry. If it is necessary to compare the dyes directly, the dyes can be extracted from the fibre and compared chromatographically.

With manmade fibres, whether made from regenerated natural polymers such as cellulose or totally synthetic ones such as nylon, there are often useful comparison criteria in addition to the colour or dye comparison. These would include the cross-sectional shape and the degree of delustering. Both of these can be characterised microscopically. Many microstructural and morphological variations are possible with manmade fibres. In addition, polymers can be compared chemically by using microchemical methods such as solvent solubility and melting point testing as well as by using instrumental methods such as Fourier-transform infrared (FTIR) microscopy and pyrolysis gas chromatography (PGC). When examined this way most synthetic fibres yield little more than the generic class of the polymer. FTIR is more useful with acrylic and modacrylic fibres where subclass information can be generated in this way. Microscopy remains an important comparison tool, especially polarised light microscopy (PLM).

1.12 Glass evidence

Physical matching, where possible, of moderate-sized fragments will remain the most definitive means of proving the existence of a common origin between glass samples [12]. Individualisation attempts based on physical property measurements and refined methods of chemical analysis are unlikely to ever achieve the degree of certainty possible with physical matches. Such matches are also free of the need for reliance on databases. Physical matching of small particles might be made feasible by computerised methods capable of recognising and comparing geometric patterns. This potential will be discussed in a later section of this chapter.

With respect to glass examinations we can expect a continued move away from reliance on refined methods of physical property measurement and toward more reliance on methods of trace elemental analysis. Whereas at one time excellent discrimination among glass samples from different sources was offered by physical property measurement, the situation has changed in two respects. There is less variation among glass samples than there was formerly but techniques for reliably measuring small differences in trace elemental composition have improved by orders of magnitude. There are two major reasons for the decrease in variation among glass samples. The manufacture of glass has become increasingly concentrated in the hands of a smaller number of large-volume manufacturers. Along with this has come increased control over manufacturing tolerances so that the resulting product is more uniform.

Until recently, methods for determining chemical composition were often incapable of discriminating between small samples of glass which were easily told apart by refined measurements of the physical properties of density and refractive index. Newer methods of determining trace elemental composition on small samples have become available. These improvements in compositional analysis have reversed the picture. Elemental analysis has gained new importance. However, database development has not kept pace with the advances in analytical instrumentation. There is a need for extensive research to develop meaningful databases. The instrumentation is expensive and the collection and analysis of samples is time-consuming. Because of the increased sensitivity, contamination problems arising in the field and possibly in the laboratory have become more acute. Increased attention in preventing contamination is necessary. The lack of research funding for this and many other similar projects in other physical evidence areas is likely to continue, unless the public can be made to recognise and appreciate the disparity between the potential and the present reality with respect to trace evidence in particular and forensic science in general.

Most modern high-quality sheet glass is made by a process developed in the UK about 50 years ago. This is the *float glass process*. The molten glass is spread over a bath of molten tin. In an environment isolated from vibration the surface of the molten tin is very smooth and flat. This results in the lower surface of the forming glass sheet being nearly optically flat. The upper surface exposed to the air is also very flat. The process is a continuous one in which hot molten glass is added to one end of the bath. It cools as it progresses to the other end where the continuous sheet is cut into manageable sections.

Float glass fragments can be recognised readily in the field during forensic investigations by using a hand-held short-wave ultraviolet (UV) lamp. The surface that was in contact with the molten tin during manufacture fluoresces.

Tempered or *toughened* glass has properties that are of particular interest with respect to trace evidence. Briefly, after this glass is formed into its final shape, typically a sheet, it is heated to a temperature in the vicinity of what is known as its annealing temperature. The surface is then chilled rapidly in a controlled fashion, often with jets of compressed air. Cooled in this way the solidified glass retains the internal stress of the differential contraction induced by the uneven cooling of the outer and inner zones or layers of the glass. This results in making it very strong mechanically. However, it will fail catastrophically, fracturing into a very large number of small pieces or 'dice', when the limit of strength is exceeded. Once it is tempered, it cannot be cut without fracturing. The article must be cut to size, and any necessary holes drilled, before it is tempered. Even scoring it may cause it to fail dramatically.

Most of the many dice that form from a sheet of tempered glass when it fractures contain a portion of both original planar surfaces of the sheet. This allows a screw micrometer to be used to obtain the thickness of the original sheet from measurements taken of the individual fragments. This can be useful investigatively. Those from the immediate vicinity of a bullet impact can be much smaller than the typical 'dice'. A bit further removed from the bullet hole, the 'dice' are wedge-shaped. These can be useful in locating the point of impact and contribute to reconstructions.

Reconstructions with fractured glass can be very useful. We have already discussed physical matches. Potentially, these allow unambiguous associations to be demonstrated. Reassembling fragments can also allow the nature of the fracture process to be discerned. For example, a study of the fracture surfaces can provide information about the direction of the force producing a fracture in sheet glass. In this way one can know whether a window was broken from the outside or inside. Obviously, this can be very useful in investigations.

1.13 Paint evidence

Paint can occur as trace evidence in a variety of ways. Depending on the mechanism of contact and the nature of the contacting surfaces, paint from a painted surface may be transferred as chips or smears. When chips are encountered the possibility of a direct or an indirect physical match should always be considered. For manufactured articles in which surfaces which were formerly in contact, but which have become separated through disassembly or as a result of damage, excellent associations via physical matching of patterns of separated paint may be possible. This potential exists in situations where painting takes place both before and after the original assembly operation. Also, with chips, layer structure can be valuable. A specific sequence of several colours can be unique. Even a sequence of different white pigmented layers can be recognised and can be useful. Visible light microscopy may be supplemented by other techniques such as cathodoluminescence, scanning electron microscopy/energy dispersive X-ray spectrometry (SEM/EDS), and diffuse reflectance FTIR spectroscopy. Coloured or white, a specific sequence of several distinct layers can be considered unique and a correspondence between a questioned and a known sample in such a case can provide an individualisation. However, caution is necessary in certain contexts. For example, in an institutional setting, such as might exist on a military base, multiple locations may have similar surfaces that have received the same paint treatment over the course of many years. It would not be surprising in these circumstances for chips from such surfaces from different locations to bear the same sequence of paint layers. As always, the collection of proper control samples and a knowledge of the context of the evidence is crucial for an accurate interpretation to be rendered.

Both chips and smears may be found on tools used to force entry into a building. This is also true in automobile accidents. Sliding contact between automobiles or between an automobile and another surface commonly produces smears. However, chips are often formed and exchanged in automobile accidents when an impact results in the bending and folding of sheet metal. Wet paint can occur as physical evidence in both contact and contactless transfers, such as drips, splashes and sprays.

Paint transfers can also occur and produce valuable trace evidence even when painted surfaces are not involved directly in the contact transfer. In a triple homicide at a 'fast food' restaurant, microscopic chips of weathered paint made up part of the 'soil' in a low spot next to a water runoff drain, in the paved parking lot, near the rear exit of the restaurant. The killer stepped in this 'soil' on leaving the building after executing three employees by shooting each in the back of the head. Later, material found adhering to his boot and present on the floor of his automobile, presumably transferred there by the boot, was examined microscopically. In addition to the expected mineral grains, this was found to contain microscopic chips of bright red, bright yellow, and green paint. These colours were the same as the colours of the paint used to coat parts of the building as well as structures in the parking lot. Although this would be perhaps a garish combination of colours in other uses, these were the colours used by this restaurant chain. The reference 'soil' sample was found to contain corresponding particles of paint. The paint particles from the 'soil' and from the boot and automobile floor were found to correspond using comparison microscopy, FTIR, SEM/EDS and PGC. This proved to be very valuable evidence in the prosecution of the suspect. In many ways this evidence was more like a complex mixture of particles as discussed in Section 1.7.3 than a more typical paint comparison. In addition to the presence of numerous particles of the three different types of paint in certain ratios, other kinds of particles were also present. The totality and the context made this very persuasive evidence of association. Thus, in addition to serving as an example of a paint association and as a case of an association based on the comparison of complex physical mixtures, it serves as an illustration of the complex nature and interdependence of physical evidence.

Before selecting methods for the collection and comparison of paint samples, a careful assessment must be made. If there are chips present in the questioned sample the possibility of a physical match must be considered first. This has important implications for the collection of the known reference samples for comparison. Uninformed collection of the reference samples might make a physical match more complicated or even impossible. We should never lose sight of the fact that good physical matches can shorten the comparison process while providing the strongest comparisons. The periphery of an area on the known object from which paint has been broken needs to be studied and documented carefully before reference samples are taken.

If it is not possible to obtain a direct physical match, because the fracture surface is inherently indistinct or because it is damaged, it may be worthwhile investigating the possibility of obtaining an indirect physical match (see Section 1.6.4) by taking advantage of the possibility of the continuity of surface features before destructive chemical analysis is attempted. The reassembly of a large number of known and questioned chips may be necessary to fully assess and exploit this.

In addition to individualisations based on physical matches, methods of comparing paint samples include comparisons based on colour, texture, surface markings, layer structure, inclusions, solubility testing and chemical composition. The chemical compositional methods include SEM/EDS, XRF, XRD, FTIR and PGC and will be discussed in the different chapters of this volume. Microscopical methods are discussed in Chapter 3.

1.14 Miscellaneous types of trace evidence

A list of the types of matter that might be encountered as trace evidence in forensic science casework would be endless. This would also be true of the possibilities with the pattern aspects of trace evidence. With respect to matter, examples such as fibres, hairs, glass and paint are only the more common ones. Other kinds of trace evidence would include soil, pollens, wood particles, vegetable matter, cosmetics, polymeric materials, metal cuttings, airborne dusts, and so on. In addition, several could occur in a variety of combinations. A case solution could hinge on recognising such trace evidence. Forensic scientists need to be more aware of such possibilities. The focus in trace evidence should not be narrow. There is a need to consider a broader range of trace evidence. The philosophy of the approach to these problems is important. There is also a need to develop a knowledge of what sources of special expertise are available and how they can be accessed.

1.15 Future technology and trace evidence

The risk of error attendant with any prognostication notwithstanding, several future directions for trace evidence will be discussed. These will be reviewed briefly in this section. Some future directions and potentials which can be discerned may remain mere potentials, if not recognised and pursued in practice. If given proper attention, these should offset the apparent short-term eclipsing of trace evidence by the focus on DNA technology applied to blood and seminal stains that has been seen in recent years in many forensic science laboratories. If these present and future trace evidence potentials are not recognised, and the present trend continues, much will be lost. The loss will not be to scientists skilled in the recognition and interpretation of trace evidence but to the field of forensic science and to the effective and impartial administration of justice.

Advantages accruing from the integration of DNA technology into approaches for solving trace evidence problems were discussed in Section 1.9. This needs to be implemented more generally.

DNA technology, especially that for mitochondrial DNA typing (mitotyping), can be expected to replace microscopy of hair for ultimate comparisons, but it would be foolish if this were done without being coordinated with information derived from microscopy. Mitotyping does not approach the discrimination of some of the methods of typing nuclear DNA. In some case contexts, the fact that mitochondrial DNA is maternally inherited is very useful investigatively. In others, because it cannot distinguish among individuals in the same maternal line, its value may be limited. For these reasons, and because there is no reason to believe that the mitotype would be correlated with the microscopic morphology, mitotyping and microscopical examination can be seen as complementary approaches. In addition to the microscopic morphological comparison, information derived from a microscopical examination can include evidence of mechanical damage, environmental exposure, cosmetic treatment, other trace evidence, and/or contamination [11]. In the context of a case these can be very important. The condition of the root can also be crucial to a proper interpretation. Several examples have been discussed in a paper on hair root morphology [13].

Biological material can occur as trace evidence in ways that may represent a departure from the types of samples typically submitted to a forensic biology or forensic DNA laboratory. In addition to human physiological trace materials, trace materials could include plant and animal material where the application of DNA typing could be crucial to a case. DNA analysis methods may be very valuable with these. Nuclear material from plant tissues such as those present in grass stains on clothing represent but one example. The forensic scientist conducting the trace evidence examination may be the only one to recognise these. For this reason some knowledge of DNA technology and its capabilities and limitations can be very important to criminalists engaged in trace evidence analysis. In this way evidence may be recognised that would otherwise escape attention. It can then be referred to the appropriate specialist or laboratory.

Computer technology has much to contribute to future trace evidence analyses. Automation and the use of technicians for certain functions will not replace the need for good scientists with extensive trace evidence experience. However, computer technology can provide better tools for the scientist analysing and interpreting trace evidence. Information databases are already much more accessible. Image analysis software and automated searching devices for use with the microscope are coming on the market. So-called 'fibre finders' have been marketed which simplify the task of searching for certain target fibres among thousands of other fibres and particles. Refinements and broader applications of the technology can be expected, if a demand and profit potential is perceived by potential developers.

What might be termed 'micro-physical matching' is another area for future development that would use computer technology. As noted earlier in this chapter (Sections 1.5.5 and 1.6.4), physical matches can provide unequivocal associations. They are commonly obtained on the macro scale. They are much less frequently encountered or recognised on the micro scale. When microtopographical features such as those present on truly microscopic fracture surfaces are recognised and exploited in forensic science casework, it is on a serendipitous basis. Two reasons can be put forward for this apparent underutilisation. The first is due to the difficulty of recognising potential matches on a micro scale. Second, in some situations where there are a very large number of fragments produced in the fracture event, the sheer number of comparisons that may be necessary may make this a pragmatic impossibility. Potentially useful microtopographical features would include features such as hackle marks on glass, fractures along grain boundaries in metals, and continuity of microstriae. This approach would appear to have reasonable prospects of success with several kinds of trace evidence, including glass and paint.

Let us examine the scale of typical trace evidence, even microscopic trace evidence, in contrast with atomic scale dimensions. This provides a sound theoretical basis for evaluating the possibility of physical matches between microscopic bits of trace evidence that may have been separated from a suspected source by a random fracture process. If we consider even a relatively small particle such as one that is of a size of approximately 10 μm contrasted with atomic scale dimensions, this results in a difference of about four orders of magnitude. There is ample room over this large range of sizes for numerous random features to be present on the fracture surface of the microscopic item. Not every micro fracture would be amenable to physical matching approaches. Certainly materials exhibiting elastic deformation could prove problematic or even unsuitable and render a physical match impossible. For materials that would be suitable, we would need methods for visualisation and recording of small-scale topographic features. Techniques such as interference microscopy, SEM, scanning tunnelling microscopy and atomic force microscopy might be useful in this area. The latter two are quite esoteric but could be useful with extremely small-scale fracture comparisons. Imaging and recording the microscopic topographic features would not be enough. The required manipulations and the large number of comparisons required would render this approach impractical, unless it could be automated. We need to investigate computer technology for coding, rapidly searching, and comparing microscopic topographic features. This might build on the experience gained with systems for automated cartridge case and bullet comparisons such as the Drug Fire or IBIS systems. These technologies have been applied recently to allow the building and searching of bullet and cartridge case databases. Does this development of technology for micro-physical matching sound like a preposterously optimistic goal or 'pie in the sky'? We should remember that it wasn't so long ago, when some of us considered the possibility of trying to utilise some of the information locked within the DNA inside cell nuclei to type biological samples, that this sounded possible in theory but unlikely to be realised in practice. Look at what has been accomplished with respect to the technology of molecular biology in a relatively short period of time. There is no need to go into detail. Of course, we have to recognise that most of the advances in DNA technology applicable to forensic science were not driven primarily by the needs of this field. They were developed for other purposes and adapted for use in forensic science. Once practical results were obtained, the straightforward potential of DNA typing was clear and seized the imagination of informed lay-citizens and politicians alike, and further funding was assured. Similar public understanding accounts for much of the developmental support for automated firearms evidence and fingerprint pattern recognition and databasing systems. This kind of lay understanding of the potential with respect to other more general and complex trace evidence problems is more difficult to convey. Despite the potential, the relative obscurity of trace evidence as it relates to the process of case solutions makes it difficult to obtain the funding necessary to carry out the research and development.

The minuscule amounts of material that typically comprise transferred trace evidence raise concerns about possible contamination. We can anticipate that protective clothing will be more widely seen to be necessary and will be required at crime scenes both for the protection of personnel and to minimise contamination of the scene. Similarly, clean room technology or, perhaps more likely, 'clean bench' technology will be applied routinely in the laboratory to minimise the chances of contamination.

Technology can provide extremely useful tools for the trace evidence scientist, but it cannot solve all the problems in trace evidence. An important area for future development is the fuller integration of trace evidence with reconstructions. This will require a recognition of the potentials of this synergistic combination and an earlier involvement of experienced forensic scientists in investigations. Scientists should contribute from the outset. The whole case context must be considered. As discussed in Section 1.10 the benefits would include the development of more informative reconstructions and enhanced associations.

1.16 Summary and conclusions

Trace evidence is one of the most valuable, misunderstood, misused and underutilised forms of physical evidence. Because the approach to its analysis is challenging and demands an experienced scientist, it has never been utilised to its full potential. In some sense the term 'trace evidence' describes an approach as much as it does a broad category of physical evidence. Despite these difficulties, trace evidence and the trace evidence approach can provide powerful evidence and have a broad range of applicability. The benefits are clearly worth the effort. It needs to be integrated with the investigation. There are relatively few forensic scientists who have adequate knowledge, training and experience in microscopy and general physical evidence problems to effectively exploit the potentials of many types of trace evidence. Because of this lack of an adequate number of properly trained scientists, the trace evidence successes in casework have too often been fortuitous rather than systematic. There is a need here that must be addressed. This need cannot be satisfied by reliance on technicians and automation.

Properly recognised and utilised trace evidence has a bright future, but only if its value is fully appreciated [14]. One reason that trace evidence is not utilised more frequently is that it is harder for the lay person to understand the complex interrelationships that are typical of the context in which it occurs than to understand such things as DNA typing of bloodstains. In laboratories where submissions are based solely on assessments by non-scientists, potentially valuable evidence may not be recognised and submitted. Another reason that may explain the apparent lack of appreciation of trace evidence is that the advent of DNA typing technology has led to unrealistic expectations with respect to the generation of numbers related to discrimination potential when other kinds of evidence are encountered. Under certain circumstances the discrimination value of trace evidence as associative evidence may match or surpass that of DNA typing of biological evidence. However, even in those situations where it falls far short of the generation of astronomical numbers, it can be very valuable. This point needs wider appreciation.

The need for trace evidence databases has been recognised for some time. The building of databases requires a considerable effort. International cooperation will be necessary for these to be both most effective and to be developed most economically.

The introduction of DNA technology to forensic science has focused unprecedented attention on quality assurance and quality control (QA/QC). This is welcome and must continue, but the focus needs to be broadened to include other physical evidence types beyond those encountered routinely, including trace evidence. The QA/QC issues are as complex and important with trace evidence as they are with DNA. This asymmetry of attention needs to be recognised and rectified.

There is danger of considering one type of trace evidence without the other forms that may be present. Considerable attention has been paid to fibres, glass and paint. Relatively less attention has been paid to the myriad other types of trace evidence. This can be explained by the relative frequencies with which they are encountered in casework. These others need to receive more attention. The frequency of occurrence of certain evidence types does not necessarily correlate positively with the inherent value of the evidence within the type category. Unanticipated trace evidence may constitute the most important evidence in a particular case. Experts working with paint and glass comparisons must have an acute awareness of other forms of trace evidence. The lack of a comprehensive and integrated approach will limit the scope of case solutions unnecessarily.

Scientists must assume a more proactive role by evaluating the evidence in the context within which it occurs and by asking relevant scientific questions of it. Decisions as to what is analysed must follow a scientific assessment of the totality of the physical evidence early in the investigation. Progress in this direction will allow a move toward realisation of the potential with respect to the contributions of physical evidence to the investigative stages of a case.

1.17 References

1. De Forest P. R. (1991). Trace evidence: a holistic view and approach. *Proceedings of the International Symposium on the Forensic Aspects of Trace Evidence*, FBI Academy, Quantico, USGPO, pp. 9–15.
2. De Forest P. R. (1999). Recapturing the essence of criminalistics, Founders Lecture, California Association of Criminalists. *Science and Justice* **39**:196–208.
3. De Forest P. R. (1997). The potential for microscopic physical matches in casework. Abstract B70 in *Proceedings of the American Academy of Forensic Sciences* **3**:38.
4. Kirk P. L. (1953). *Crime Investigation*. Interscience Publishers, New York, p. 4.
5. Gross H. (1922). *Handbuch für Untersungsrichter als System der Kriminalistik* (7th edn). J. Schweitzer Verlag, Berlin.
6. Palenik S. J. (1982). Microscopic trace evidence – The overlooked clue. Part I. Albert Schneider looks at some string. *Microscope* **30**:93–100.
7. Palenik S. J. (1982). Microscopic trace evidence – The overlooked clue. Part II. Max Frei – Sherlock Holmes with a microscope. *Microscope* **30**:163–170.

8. Palenik S. J. (1982). Microscopic trace evidence – The overlooked clue. Part III. E. O. Heinrich – The 'Wizard of Berkeley' traps a left-handed lumberjack. *Microscope* **30**:281–290.
9. Nickolls L. C. (1962). Identification of stains of nonbiological origin. In *Methods of Forensic Science* (ed. F. Lundquist), Vol. 1, pp. 335–371. Interscience Publishers, London.
10. Koons R., Peters C., Rebbert P. (1991). Comparison of refractive index, energy dispersive X-ray fluorescence and inductively coupled plasma atomic emission spectrometry for forensic characterization of sheet glass fragments. *Journal of Analytical Atomic Spectrometry* **6**:451–456.
11. Robertson J. (1999). *Forensic Examination of Hair*. Taylor & Francis, London.
12. Nelson D. F. (1965). The examination of glass fragments. In *Methods of Forensic Science* (ed. A. S. Curry), Vol. 4, pp. 117–119, Interscience Publishers, London.
13. Petraco N., Fraas C., Callery F. X., De Forest P. R. (1998). The morphology and forensic significance of human hair roots. *Journal of Forensic Sciences* **33**:68–76.
14. Thorwald J. (1967). *Crime and Science*. Harcourt Brace and World, New York.

2

The composition and manufacture of glass and its domestic and industrial applications

GEOFFREY J. COPLEY

2.1 Introduction

The glass industry is divided into broad sectors for the manufacture of containers (bottles and jars), flat glass (for architecture and transport glazing), glass fibre (for reinforcement and insulation), domestic glass (kitchen and tableware) and technical glasses (for a host of scientific and industrial uses). Manufacturing processes differ from sector to sector [1, 2]. Technical and commercial developments in glass manufacture over the past, say, forty years have yielded new products, and methods of manufacture have changed dramatically in terms of speed of production, the quality of glass produced and the number of peripheral processes for treating glass. There are many different glass compositions but they fall into a limited number of types which simplifies classification [3]. A composition is developed to meet the requirements of the manufacturing process, the properties required in the use of the end product and the economics of production. Manufacturers' catalogues show a wide range of products, with compositions which vary from product to product. In the large tonnage sectors (container, flat and domestic) compositions tend to be similar within each sector but there are differences of detail. Glass is manufactured in most technically advanced countries and there is a good deal of international trade. Since glass is a highly durable material, products can remain in use for long periods of time, with church windows providing an extreme example. Samples of glass arising from a particular site or event may therefore possess an easily determined composition be it ancient or modern, domestic or foreign.

2.2 Definition of a glass

Glass is defined as a product of fusion which has cooled to a rigid state without crystallisation. Glass is therefore, by definition, amorphous or non-crystalline. Glasses are essentially supercooled liquids and they possess a unique combination of properties: transparency with or without colour, durability, electrical and thermal resistance, a range of thermal expansions, with hardness, rigidity and stability. Glasses are made by melting together, and chemical reactions between, inorganic materials, many of them naturally occurring oxide minerals, of which the principal one is silica sand. Commercial glasses are therefore silicates. Depending upon the combination of properties needed in the end product, other additives are selected to optimise manufacture and performance.

2.3 Commercial glass types

The main constituent of practically all commercial glasses is silica sand. Silica (SiO_2) is a glass former and can itself be fused to produce fused silica, but the temperature needed to do this is in excess of 1700°C. Fused silica is therefore limited to a specialised range of products where high temperature resistance is required. The addition of other chemicals (fluxing agents) to silica reduces the temperature of fusion. Sodium oxide, usually added as the carbonate, is the principal additive which gives melting temperatures between 1500 and 1650°C and which are accessible in large-scale furnaces. Glass types are categorised by their principal constituents and such mixtures form the basis of the alkali silica glasses. Silica is the glass former and the additive is designated the modifier. To give the glass durability other additives are needed and the most important of these are limestone (calcium oxide) and alumina (aluminium oxide). These additives provide two additional categories, the soda-lime-silicate and the alumino-silicate glasses. Other additives are used to confer particular properties. An important series of glass is based upon mixtures of two glass formers, boric oxide (B_2O_3) and silica. These are the boro-silicate glasses with properties of temperature and thermal shock resistance. Lead silicate glasses, based upon silica and lead oxide, form important ranges of optical, crystal and electrical glasses. The principal types of glass and their applications are summarised in Section 2.7. While the melting processes for all glasses are broadly similar (though differing in scale and glass composition), it is at the forming stage of the production process, that is at the conversion of the molten glass into the final product, that there is great diversity. Glass-forming processes are illustrated in Section 2.5. The precise behaviour of the glass as it cools from a molten fluid to a rigid solid is crucial. The prevention of crystallisation as the glass cools from the melting temperature and the rate of change of the viscosity as the glass cools and stiffens to rigidity are vital. The correct behaviour is obtained by the choice of the glass composition and the forming process. After forming, the glass product must be cooled further at a controlled rate, the annealing process, to prevent the development of stress.

2.4 Glass-manufacturing process

Glass manufacture in all sectors broadly follows the stages illustrated in Figure 2.1 (see for example reference [4]).

RAW MATERIALS →	MELTING →	FORMING →	ANNEALING →	WAREHOUSE
Storage	Refining	Shaping	Controlled	or
Weighing	Homogenising		cooling	SECONDARY
Mixing				PROCESSING

Figure 2.1 The glass-manufacturing process

2.4.1 Raw materials

Glass-making materials are specified, transported and stored to maintain high standards of consistency of composition and freedom from impurities. The raw materials are weighed and thoroughly mixed before being delivered to the furnace. All major components and minor additives are present in the charge (batch) delivered to the furnace. Waste glass (cullet) is an important raw material which assists the melting process. All factory created waste is recycled to the furnace and, in the container sector, a large quantity of glass is recovered from the consumer through the bottle-bank system for reprocessing bottles and jars.

2.4.2 Glass melting

The melting of the raw materials takes place in furnaces constructed of refractory blocks which are capable of withstanding temperatures in excess of 1500°C and the corrosive nature of molten glass. The greatest tonnage of glass is melted in tank furnaces which produce a continuous flow of glass to feed automatic glass-forming machines. In small-scale manufacture in, for example, the hand-made crystal sector where an intermittent supply of glass is needed, the glass is melted in high-purity refractory pots. The sizes of the furnaces varies from sector to sector. A large float (flat) glass furnace will have a throughput of 800 tonnes a day. Container furnaces are smaller, and throughputs of 50 to 300 tonnes a day are common. A pot will hold up to 1 tonne of glass, though several pots may be accommodated in one furnace structure. Furnaces are heated by natural gas or heavy fuel oil and heat is recovered from the waste gases as they leave the furnace and is used to preheat the air for combustion. Electrical melting is used in some cases but it is limited by economic factors to small furnaces. Electricity is used in combination with fossil fuels in container glass furnaces to boost production at times of high demand. The temperature of the raw materials is raised rapidly in the furnace and the melting and chemical reactions must be completed before the glass is formed. The glass is fluid at high temperature and molten glass behaves as a viscous liquid, with the viscosity decreasing (the liquid becoming more fluid) as the temperature is raised.

2.4.3 Refining and homogenising

Gases, principally carbon dioxide and water, are emitted as the raw materials melt and react. The molten glass must be held at a temperature which is high enough and with a viscosity which is low enough to permit all gas bubbles to rise to the surface and hence be eliminated. This removal of bubbles takes place in the refining stage as the molten glass passes through the furnace. Thermal, and in some cases mechanical, stirring ensures that the glass is fully homogenised. Small localised variations in chemical composition in the glass will lead to difficulties in forming and will cause visible faults in the product due to small variations in the refractive index. When the glass is thoroughly homogenised and refined the temperature of the molten glass is reduced as it passes down the furnace towards the forming stages. Glass has no fixed melting or freezing point but it changes gradually and continuously from a viscous liquid to a rigid solid as its viscosity increases on cooling. If glass is held for a sufficient time in the temperature range between, say, 800 and 1000°C, then crystals will start to form. This process is known as devitrification and must be avoided by using a sufficiently rapid cooling rate through the critical range. Glass ceramics are made by the controlled crystallisation of silicate glass compositions.

2.4.4 Glass forming

It is the gradual and continuous change in viscosity with temperature which permits molten glass to be converted into such a wide range of products. In the appropriate range of viscosity, molten glass can be subjected to many different processes to make products ranging from flat sheets several square metres in area to filaments with diameters less than that of a human hair. The rate of change of viscosity in the forming range must be appropriate for the forming process being used. For example, full lead crystal is made by slow hand-forming methods; therefore, the viscosity of the glass must increase slowly as the temperature falls, giving the glass a long working range. At the other extreme, the viscosity of glass for fibre manufacture must increase rapidly for the very rapid forming process. Container glass has an intermediate behaviour appropriate to high-speed mechanical forming methods. Glass-forming methods are illustrated in Section 2.5.

2.4.5 Annealing

Glass, like other materials, contracts on cooling. However, because of its low thermal conductivity, it does not cool uniformly. This can produce uncontrolled stress in the article. Badly annealed glass cannot withstand shock and is liable to break in use. Undesirable stress can be avoided by slow cooling. This is the annealing process. Annealing is carried out in kilns or continuous lehrs, through which the glass passes on a slow-moving conveyor.

2.5 Glass-forming processes

Glass-forming processes vary widely from product to product but the following examples serve to illustrate the diversity (see reference [5]).

2.5.1 Glass blowing

For two thousand years mouth blowing was the main method of forming glass articles. The last few years of the nineteenth century saw the beginnings of blowing glass with compressed air and the twentieth century brought in mechanised blowing and forming. Mouth blowing is still used in the traditional crystal glass sector. A hollow iron pipe is dipped into a pot of molten glass and the glass is gathered at the end of the pipe. It is manipulated by rolling and blowing through the blowing iron. The glass may be reheated and placed within an iron or wooden mould, which is kept wet, as the glass is blown into its final shape. After forming the article is reheated and annealed to prevent the development of stress and the cracking of the product.

2.5.2 Automatic glass container forming

Containers are made by fully automatic high-speed processes [5]. The process is outlined in Figure 2.2.

The furnace produces a fully melted, refined and homogeneous glass. The molten glass then flows, under gravity, through an orifice at a carefully controlled temperature and is sheared into precise lengths (known as gobs) which drop into the moulds of the automatic forming machine. There are two different blowing processes: the press and blow process for making wide-necked jars, and the blow and blow process for bottles. The finished articles are released from the moulds at a temperature of about 500°C when they are sufficiently rigid to bear their own weight without sagging. The articles are transferred to a conveyor belt which passes through the annealing lehr. Most glass containers undergo surface treatments as they enter the annealing lehr – hot end processes – and as they leave the lehr – cold end processes. The forming machine is made up of identical sections placed side by side. This has the advantage that each section can be stopped for adjustment or maintenance or to change the moulds without stopping the whole machine. Each section may be capable of forming one, two or three gobs simultaneously. The speed of production depends upon the size of the product being made, but machines producing over 300 containers a minute are typical.

2.5.3 Flat glass manufacture

Flat glass is made by two processes. The greatest tonnage made for high-quality products for glazing, for architecture and transport, is made by the float glass process [5]. Wired glass and patterned glass are made by rolling [5].

2.5.3.1 The float glass process

The float glass process [5] is the principal method for producing flat glass throughout the world and has replaced all other processes. Figure 2.3 shows the layout of the process.

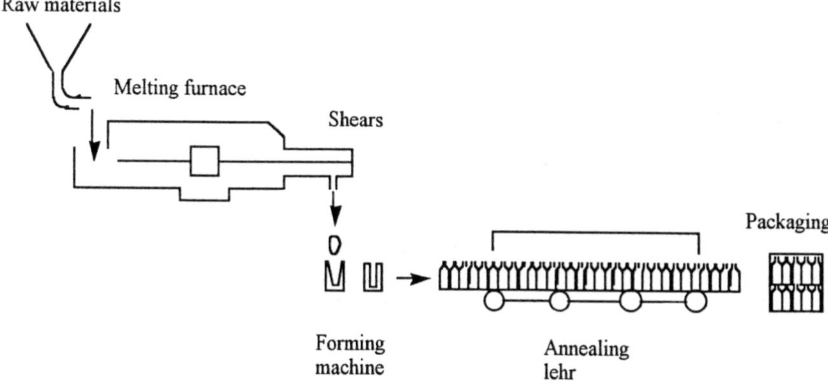

Figure 2.2 Glass container manufacture

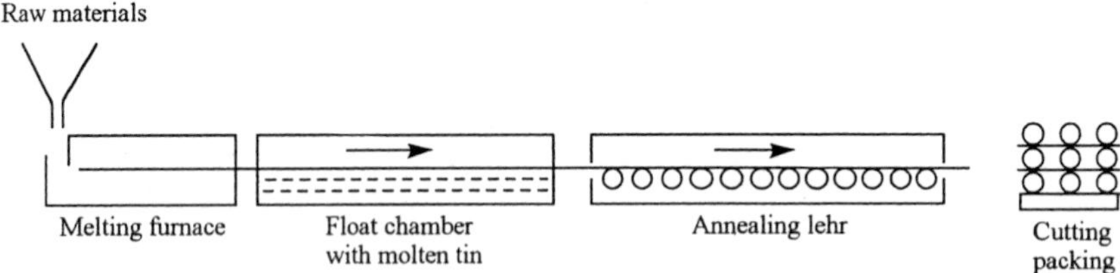

Figure 2.3 The float glass process

The fully homogenised glass emerges from the melting furnace at a controlled temperature and flows into the float chamber. The chamber contains a bath of molten tin in a chemically controlled, non-oxidising atmosphere. The glass forms a ribbon which is drawn continuously along the tin bath.

The glass is held at a high enough temperature (1000°C) for a long enough time for irregularities to flow out and for the surfaces to become flat and parallel. The thickness, in the range 2.5 to 25 mm, is controlled by stretching or constraining the ribbon as it is forming. The ribbon is cooled to about 600°C while advancing along the molten tin until the surfaces are hard enough for it to be lifted onto the take-out rollers without marking the bottom surface. The ribbon then passes through the annealing lehr to the automatic warehouse where the glass is cut automatically. The glass has uniform thickness and bright fire-polished surfaces without the need for grinding and polishing; it has a high optical quality free from defects.

2.5.3.2 The rolled glass process

The rolling process [5] is used for the manufacture of patterned or textured glass for decorative glazing, and wired glass for fire-resistant glazing. For patterned glass, the continuous stream of molten glass from the furnace is poured between water-cooled rollers. Patterned glass is made in a single-pass process in which the glass flows to the rollers at a temperature of about 1050°C. The bottom cast iron or steel roller is engraved with the negative of the pattern, the top roller being smooth. Thickness is controlled by adjustment of the gap between the rollers. The ribbon leaves the rollers at a temperature of about 850°C and is supported over a series of water-cooled steel rollers to the annealing lehr. After annealing the glass is cut to size. Wired glass is made in a double-pass process which uses two independently driven pairs of water-cooled forming rollers each fed with a separate flow of molten glass from a common melting furnace. The first pair of rollers produces a continuous ribbon of glass half the thickness of the end product. This is overlaid with a wire mesh. A second feed of glass, the same thickness as the first, is then added with the wire mesh sandwiched between the two. The sandwich of glass and wire mesh is then passed through the second pair of rollers which form the final ribbon of wired glass. After annealing, the ribbon is cut by special cutting and snapping arrangements.

2.5.4 Glass fibre manufacture

There are two broad groups of fibreglass products: (a) continuous glass filament which is used for the manufacture of composite materials, that is for reinforcement of plastics, rubber and gypsum, and (b) glass wool, which is used for thermal insulation. The fibre-forming processes are shown schematically in Figure 2.4 (see reference [5]).

2.5.4.1 Continuous glass fibre

Continuous glass fibre is a continuous strand made up of a large number of individual filaments of glass. Molten glass (Figure 2.4(a)) is fed from the furnace through a channel to a series of bushings which contain over 1600 accurately dimensioned holes or forming tips in its base. Filaments of glass are drawn mechanically downwards from the bushing tips at a speed of several thousand metres per minute, giving a filament diameter which may be as small as 9 μm or one-tenth the diameter of a human hair. From the bushing the filaments run to a common collecting point where a coating of size is applied and they are subsequently brought together as a bundle or strands on a high-speed winder.

2.5.4.2 Glass wool process

Glass wool is made in the process shown in Figure 2.4(b). From the furnace a stream of glass flows by gravity from a bushing into a rapidly rotating alloy steel dish or crown which has several hundred fine holes around the periphery. The molten glass is thrown out through the holes by centrifugal force to form filaments which are further extended into fine fibres by a high-velocity blast of hot gas. After being sprayed with a suitable bonding

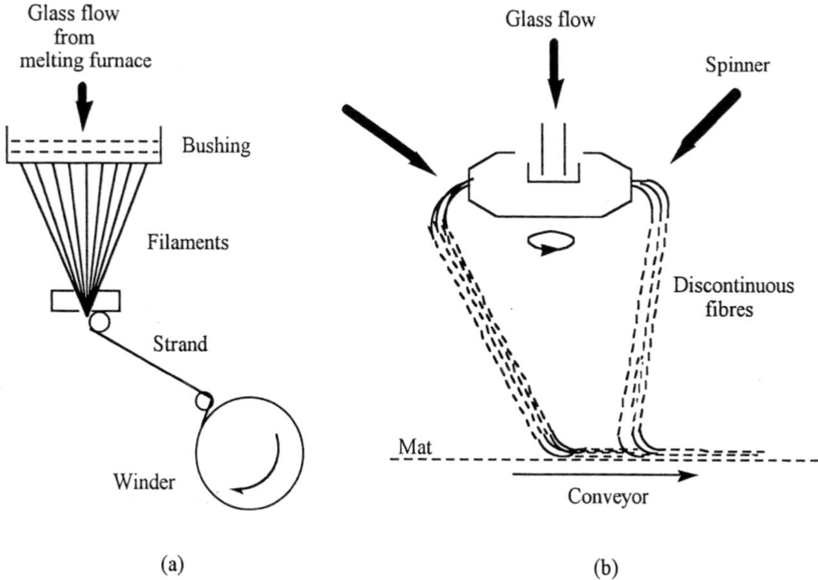

Figure 2.4 Glass fibre manufacture: (a) continuous filament; (b) glass wool

agent, the fibres are drawn by suction onto a horizontal moving conveyor. The mat of tangled fibres formed on the conveyor is carried through an oven which cures the bonding agent, then to trimmers and guillotines which cut the product to size. The mat may be further processed by rolling or pressing into rigid sections for pipe insulation. The mats are made into many products for heat and sound insulation in buildings, transport vehicles and domestic appliances.

2.5.5 **Optical fibre manufacture**

Optical fibre is a highly specialised form of glass fibre. Communications are increasingly based on electro-optic systems linked by fibre optic cables which carry information by laser light. Such optical fibres consist of two distinct glasses, a core of a highly refracting glass surrounded by a sheath of glass with a lower refractive index. If the beam of light strikes the end of the fibre at a certain angle, it is guided by total reflection at the core–sheath interface to the other end of the fibre. The fibre of glass must be so pure and free from defects that light inserted at one end will emerge at the other end a distance of more than 1 kilometre away. There are several manufacturing processes [1] being used to produce core fibre which include drawing the fibre from a double bushing and drawing from a doped rod preform made by chemical vapour deposition. All such processes require ultra-pure starting materials.

2.5.6 **Optical glass manufacture**

Glass for optical and ophthalmic applications must be totally homogeneous and possess an accurately-controlled refractive index. Heterogeneities in chemical composition must be eliminated in the melting and refining processes; those due to the thermal history of the glass must be eliminated by fine annealing, that is by very slow cooling. Melting takes place in small purpose-built tanks with a typical capacity of 5 tonnes. The melting, refining and stirring operations are usually carried out in separate sections to ensure the necessary optical quality. Platinum is used as the refractory glass-contact material to avoid contamination which may arise from the conventional refractories. The initial melting zone is often gas fired but all-electrical melting is used for the finishing stages. The glass composition may be changed to produce glasses of different refractive indices or may be coloured to produce tinted products. Glass emerging from the delivery orifice is sheared at a constant rate and gobs of the viscosity required for forming are produced. The gobs are pressed to produce standard curves and edge contours. Alternatively, the glass may be cast into blocks, extruded into bars or rolled into sheets for subsequent reforming. All shapes are subsequently cut, ground and polished to the closely-controlled shapes of the final products.

2.5.7 **Glass-tubing manufacture**

Glass tubing and rod is drawn continuously from molten glass [1, 2]. The glass flows from the furnace into a bowl in which a hollow vertical mandrel is mounted surrounded by an orifice ring. The glass is drawn through the annular space between the mandrel and the ring and then over a line of rollers to the drawing machine, which may be 120 metres away. Tubing is made by blowing air through the mandrel; rod is made without blowing. The dimensions of the tubing are controlled by the temperature of the glass, the rate of draw, the pressure of the blowing air and the relative dimensions of the bell and ring. In an earlier alternative method, the glass flows from the furnace in the form of a ribbon which wraps around a rotating hollow shaft or blowpipe to form a tube. Tubing can be made in diameters ranging from 1.5 to 65 mm and rods from 2.0 to 20 mm at drawing rates of up to 400 m per minute for the smaller sizes. A large amount of tubing is converted into technical products such as thermometers and components for, for example, the petrochemical industry.

2.5.8 Automatic glassware production

The forming machine copies the action of a handblower in gathering glass from the furnace and blowing the article in a wet cast-iron mould. It has been adapted for making drinking glasses, including stemmed ware, at up to 55,000 items a day. Spindles and blowpipes, together with their blowing air valves and moulds, rotate around a central column. The gathering equipment is carried on top of the column and cams control the sequence of operations. Glass is gathered by vacuum into a pair of blank moulds. The spindles are rotated and air is introduced to form each blank into a partly-formed product before blowing to the final shape in a wetted mould. The stem is added in a separate process. The articles then pass to the burn-off machine where oxygen-gas flames remove the waste glass and the finished piece is conveyed to the lehr for annealing.

2.5.9 Electric light bulb envelope manufacture

The ribbon machine was developed for the high-speed manufacture of bulbs for domestic lamps and vacuum flasks [1]. Production rates are in excess of 1000 items per minute. From the furnace, molten glass flows between two rotating water-cooled rollers and onto the ribbon machine. The ribbon is carried through the machine on a series of orifice plates which are linked into a continuous belt. A continuous chain of blowheads meets the ribbon from above, each blowhead coinciding with a hole in the belt. Air blows the glass through the hole to form a bulb inside a rotating mould which meets and closes around it from below. The shaped bulb is released from the mould, cooled by air jets and released onto a conveyor belt. This carries it through the annealing lehr to inspection and packing. The unused part of the ribbon passes directly to the cullet system for re-melting.

2.5.10 Pressed glassware

Pressing is used for objects with a simple basic shape and an opening wider than the base. A plunger forms the inner surface of the article by pushing the glass against the outer mould. Pressing can be hand operated or fully automatic; products range from simple bowls to the complex shapes of, for example, television face plates and optical components.

2.6 Secondary glass processing

When glass products have been formed, annealed and cooled to ambient temperatures, they may then be subject to a range of finishing processes. The following are illustrative.

2.6.1 Toughening or tempering

Glass has an extremely high strength in compression and when it breaks it does so because of tension at the surface. Glass can be thermally strengthened by inducing compressive stresses in the outer surfaces. In order to break toughened or tempered glass, the surface compression has to be exceeded and additional tension applied. Toughening is carried out by reheating the glass uniformly to a temperature just above that at which deformation could take place and then rapidly cooling the surfaces. The cooling chills the outer layers while the centre of the glass is still hot. As the centre of the glass cools it contracts, pulling the rigid surface layers into compression which is balanced by the tension in the inner layers. This method of strengthening can be applied to flat glass or simple shapes such as curved windscreens or tumblers. The thickness of the glass must be uniform and not too thin. The shape of the article must be such that all surfaces can be cooled uniformly. Bottles do not satisfy these conditions and they cannot be toughened in this way. Thermally toughened glass cannot be further processed since any damage to the surface will expose the central layer, which is in tension, and the glass will break. The shattering of a car windscreen is a good example of this phenomenon. It is possible to toughen certain articles chemically by ion exchange. The article is immersed in a molten potassium salt. The potassium ions replace sodium ions at the surface and, being larger, create a thin layer of compression. This method is limited to specialised products.

2.6.2 Coating

The coating of glass surfaces has been practised for centuries. Mirrors are a good example of this art. Coating processes have been developed in recent years for decoration, protection and strengthening, and for technical reasons, for example to control the transmission of light and other radiation through the glass. Most glass containers are coated to assist in the preservation of strength and to improve the handling of the product in manufacture and filling. Coatings are applied at two stages. Hot end coatings are applied immediately the containers have left the forming machine and before they enter the annealing lehr. Cold end coatings are applied just before the annealed containers leave the lehr. The coatings are applied either as a vapour or by spray. Examples of hot end coating materials are compounds of tin or titanium, while corresponding cold end coating compounds include organic waxes, polyethylene emulsions, polyethylene glycols and their fatty acid esters or fatty acids such as oleic acid. All cold end coatings must satisfy the health regulations of the countries where the containers are filled and distributed. Modern flat glass products are often coated. Patterned or textured glass carries a coating on the smooth side to protect the glass during handling and transportation. Advanced glazing products for solar control and for thermal insulation purposes are coated using both on-line and off-line processes. On-line pyrolytic coatings are applied during manufacture on the float line and consist usually of tin oxide doped with elements such as indium to obtain the selected transmission and reflection of wavelengths in the visible and infrared parts of the spectrum. Off-line coatings are applied by vacuum sputtering using a range of metallic and semiconductor materials. The coatings may be multilayer and many of the products are also tinted either in the body of the glass or in the coating.

2.6.3 Colouring and decolourising

Unless raw materials are very pure, glass made from silica sand and the other common raw materials will normally have a green tint, the colouration depending primarily on the amount of iron oxide and other impurities present. A sand containing as little as one-thousandth part of iron oxide will impart this effect. For some products manufacturers will decolourise the glass by adding small amounts of other colourants which produce a colour complementary to green so that the finished articles appear colourless. Thus selenium (which gives a pink colouration) and cobalt (which gives blue) can be added to soda-lime-silica glasses to offset the green or yellow due to the iron. Nickel, cobalt, erbium or neodynium are used similarly in the decolourising of lead crystal glass. The range of possible colours which can be produced is very wide. Colour may be used for purely aesthetic or for technical reasons in, for example, optical colour filters. Some of the most frequently used colourants and the colour they produce are listed in Table 2.1. The precise colour obtained with each colourant will depend upon the glass composition, the thermal treatment and the state of oxidation of the colourant and the glass.

2.6.4 Decorating

Formed and annealed glass may be further decorated by removing or adding glass to the surface. The former may be achieved by grinding or sand blasting of the surface of the

Table 2.1 Colourants used to colour glass

Colourant	Glass colours
Iron	Green, brown, blue
Manganese	Purple
Chromium	Green, yellow, pink
Vanadium	Green, blue, grey
Copper	Blue, green, red
Cobalt	Blue, green, pink
Nickel	Yellow, purple
Uranium	Yellow, brown, green
Titanium	Purple, brown
Neodymium	Purple
Praseodymium	Green
Cerium	Yellow
Carbon and sulphur	Amber, brown
Cadmium sulphide	Yellow
Antimony sulphide	Red
Selenium	Pink, red
Gold	Red

glass to produce a matt or obscured finish. Where a transparent surface is then required, the glass is polished on felt or wooden wheels or acid polished in a hydrofluoric acid solution. The complementary process makes use of vitreous enamels, which are glasses that melt at relatively low temperatures and can be applied to the surface of the glass. The article is reheated after the application so that the coating fuses permanently to the surface. Metal films can also be applied by spraying or by chemical or vapour deposition. Glass which has been formed and annealed may be reheated and formed into a new shape before being annealed again. Finally, surface textures or patterns can be imparted to the surface of glass by etching with hydrofluoric acid.

2.7 Principal glass types and applications

The following examples serve to illustrate the range of commercial glass compositions, the methods of manufacture and their applications.

2.7.1 Fused silica

Silica, SiO_2, can be melted and cooled to form a glass. Very high processing temperatures are needed but the products are characterised by good temperature resistance, low thermal expansion and hence good resistance to thermal shock and excellent chemical durability. They are widely used in scientific and industrial applications.

2.7.2 Soda lime silicate glasses

The principal constituents are silica (SiO_2), soda (sodium oxide, Na_2O) and lime (CaO). Other elements will be present to adjust the properties required for the manufacture and the end use of the products. There will be other elements added deliberately or adventitiously. The properties of these glasses make them suitable for a wide range of applications in container, flat and domestic glassware. The greatest tonnage of glass is for containers (bottles and jars). Container glass is made in three colours, white (also known as flint), green and amber. Typical compositions are listed in Table 2.2. Other trace elements may be present arising from colouring or decolourising or due to recycling of large quantities of containers, some of which may be of old manufacture. The total amount of such adventitious elements will be less than 100 ppm. Both flat glass and rolled glass are made to a similar composition, a typical example of which is given in Table 2.2. The recycling of post-consumer flat glass is limited and the presence of trace impurities is more easily controlled.

2.7.3 Borate silicate glasses

Borosilicate glasses, as the name implies, are composed mainly of silica and boric oxide, B_2O_3, together with smaller amounts of alkalis and aluminium oxide. They are characterised by relatively low alkali contents and have good chemical durability with low thermal expansion and hence good resistance to thermal shock. They have wide applications in industry, in the home, in kitchenware, and in laboratories and hospitals. The range of compositions is wide but a typical chemical analysis of a borosilicate glass is given in Table 2.2.

Table 2.2 Some glass compositions

Glass type*	SiO_2	Al_2O_3	Fe_2O_3	CaO	BaO	MgO	Na_2O	K_2O	Cr_2O_3	PbO	B_2O_3	As_2O_5
Container												
Flint	72.6	1.6	0.05	11.0	–	0.1	13.7	0.5	–	–	–	–
Amber	72.7	1.9	0.22	10.0	–	–	13.8	1.0	–	–	–	–
Green	72.0	1.1	0.96	8.4	–	2.1	15.1	–	0.19	–	–	–
Float and rolled	72.8	1.4	0.1	8.2	–	3.8	12.8	0.8	–	–	–	–
Borosilicate	80.2	2.6	0.07	0.1	–	–	4.5	0.3	–	–	12.3	–
Lead crystal	54.9	0.1	0.02	–	–	–	0.2	12.3	–	31.9	0.5	0.5
Optical glass	48.0	0.2	–	0.3	–	–	5.2	1.2	–	45.1	–	–
Insulating lead	63.0	0.6	–	0.3	–	0.2	7.6	6.0	–	21.0	0.2	–
Barium optical	36.2	3.5	–	0.2	44.6	–	–	0.2	–	–	7.7	–

*Values expressed as weight per cent.

2.7.4 Lead glasses

Lead oxide imparts excellent hand-working and optical characteristics and is widely used in the manufacture of crystal glass. An example of a composition for full lead crystal is provided in Table 2.2 but any glass containing at least 24 per cent PbO can legitimately be described as lead crystal under European directive 69/493 [6]. Lead glasses of somewhat different composition are used for electrical applications because of their good insulating and dielectric properties conferred by the lead oxide and by the combined use of potassium and sodium oxides. An example is listed in Table 2.2. Lead oxide is a common component of optical glasses as it imparts a high refractive index without colouring the glass and it can be used to control the dispersion of the glass (the dependence of the refractive index on wavelength). Optical glasses with a high dispersion are known as flint glasses and an example of their composition is given in Table 2.2. However, the composition of optical glasses varies very widely and reference should be made to specialised books [7] and manufacturers' catalogues. The substitution of lead oxide by other oxides allows for considerable variations in optical properties such as that exhibited for a barium optical glass in Table 2.2. A wide range of rare earth elements are common ingredients, while the transition elements including copper, chromium, manganese, iron, cobalt and nickel are used to produce strong absorption bands in different parts of the spectrum in the manufacture of optical colour filters. Silicate glasses with very high lead contents, up to 65 per cent by weight, are used in glasses for radiation shielding.

2.7.5 Aluminosilicate glasses

A small but important group of glasses, the aluminosilicates, contain some 20 per cent aluminium oxide (alumina, Al_2O_3), and often include oxides of calcium and magnesium oxides as well as boric oxide in relatively small amounts, and with only very small amounts of soda or potash. They tend to require higher melting temperatures than borosilicate glasses and are more difficult to work but have the merit of being able to withstand high temperatures with good resistance to thermal shock. Typical applications include combustion tubes, gauge glasses for high-pressure steam boilers, and the envelopes of halogen–tungsten lamps capable of operating at temperatures as high as 750°C.

2.7.6 Alkali barium silicate glasses

In normal operation the imaging system in cathode-ray tubes produces X-rays which must be absorbed by the glass envelope. This is afforded by glasses with heavy metal oxide components (lead, barium or strontium). Lead glasses are commonly used for the funnel and neck of the tube, while glasses containing barium are employed for the face plate or panel.

2.7.7 Borate glasses

There is a range of glasses, containing little or no silica, used for soldering glasses for metals or ceramics. When used to solder other glasses, the solder glass needs to be fluid at temperatures well below that at which the glass to be sealed will deform. Some solder glasses crystallise or devitrify during the soldering process and in this case the mating surfaces cannot be reset or separated. Lead borate glasses containing 60–90 per cent PbO, with relatively small amounts of silica and alumina to improve the chemical durability, do not crystallise in normal use. Glasses that are converted partly into crystalline materials when the soldering temperature is reached are characterised by the presence of up to about 25 per cent zinc oxide. Glasses of a slightly different composition (zinc-silicoborate glasses) may also be used for protecting silicon semiconductor components against chemical attack and mechanical damage. Such glasses must contain no alkalis (which can influence the semiconductor properties of the silicon) and must be compatible with silicon in terms of thermal expansion. These materials, known as passivation glasses, have assumed considerable importance in microelectronics technology.

2.7.8 Phosphate glasses

Most types of glass are good electrical insulators at room temperature, although those with a substantial alkali content may well be good ionic conductors in the molten state. The conductivity depends mainly on the ability of the alkali ions in the glass to migrate in an electric field. Some glasses that do not contain alkalis are electronic conductors. These are semiconducting oxide glasses and are used particularly in the construction of secondary electron multipliers. Typically they consist of mixtures of vanadium pentoxide (V_2O_5) and phosphorus pentoxide (P_2O_5).

2.7.9 Chalcogenide glasses

A highly specialised group of semiconductor glasses are non-oxide glasses. These may be composed of one or more elements of the sulphur group in the periodic table, combined with arsenic, antimony, germanium and/or the halides (fluorine, chlorine, bromine, iodine). Some of them are used as infrared transmitting materials and as switching devices in computer memories and as optical components for thermal imaging devices because of their transparency to long-wave infrared radiation.

A summary of all that has been described in this chapter can be found in Table 2.3.

Table 2.3 Summary of applications, compositions and properties of glass products

Product and applications	Specific qualities	Method of manufacture	Glass type – typical composition (all % by weight)
GLASS CONTAINER – Bottles – Jars for many food, drink and medicinal contents.	Relatively cheap. Inert in contact with contents. Can be sterilised. Can be recycled and reused. Three basic colours (white, green, amber).	Automatic blowing into moulds. Often coated.	Soda-lime-silica: SiO_2 74 Na_2O 14 CaO 11 Al_2O_3 1
FLAT GLASS Glazings for – Architecture – Transport including safety products.	Relatively cheap. Can be toughened. Durable and weather resistant. Scratch resistant.	Float process – for optical quality. Rolling – for patterned and wired products. Often coated or tinted.	Soda-lime-silica: SiO_2 71 Na_2O 16 CaO 9 MgO 3 Al_2O_3 1
DOMESTIC GLASSWARE – Domestic – Catering.	Good appearance. Durable for constant use. Inert in contact with contents.	Automatic pressing or blowing into moulds. Can be hand worked.	Soda-lime-silica: SiO_2 71 Na_2O 16 CaO 5 MgO 3 Al_2O_3 3 K_2O 1 B_2O_3 1
HEAT-RESISTANT OVEN TO TABLEWARE – Domestic – Catering.	Low thermal expansion. Resistant to thermal shock. Durable and easy to clean. Good appearance.	Automatic pressing or blowing into moulds.	Borosilicate: SiO_2 80 B_2O_3 12 Na_2O 4.5 Al_2O_3 4.5

Table 2.3 Continued

CRYSTAL GLASS – Domestic – Catering – Presentation pieces.	Long working range for hand working. Brilliant finish, attractive appearance. Easy to polish and engrave. Durable.	Hand forming by traditional methods or automatic blowing into moulds followed by cutting, grinding and polishing.	Lead glass – Full lead crystal: SiO_2 55 PbO 33 K_2O 11 – Lead crystal: SiO_2 6 PbO 25 K_2O 9 Na_2O 5
OPTICAL GLASS – Ophthalmic – Optical components.	Wide range of refractive indices. Wide range of dispersion coefficients. High transparency. Perfect homogeneity.	Pressing, extrusion, rolling following by grinding and polishing. May be tinted.	Silica based but wide range of compositions – see manufacturers' catalogues.
ELECTRICAL COMPONENTS Many applications – Cathode ray tubes – Insulators – Sealing glasses – Substrates.	High electrical resistance. Dielectric properties. High operating temperature. Impermeable to gases. Selected thermal expansion.	Automatic blowing and pressing. Drawing – rods, sheets. Pressing and sintering.	Silica based – lead alkali silicate – barium alkali silicate – borosilicate – aluminosilicate Wide range of compositions – see manufacturers' catalogues.
GLASS CERAMICS – Specialised electrical, thermal and chemical applications.	Wide range of thermal expansions, some zero expansion. Good electrical properties. High durability.	Melting of glass-making compositions followed by controlled crystallisation.	Range of compositions based on spodumene and eucryptite minerals – see manufacturers' catalogues.

LIGHTING GLASSWARE Many applications – Electric light bulbs – Tubing for fluorescent lighting – Sodium discharge lamps – Mercury vapour lamps.	High electrical resistance. Dielectric properties. High operating temperatures. Impermeable to gases. High chemical resistance. Selected colour or radiation transmission in visible, UV or IR.	High speed ribbon machine for electric light bulbs (also used for vacuum flasks). Automatic blowing, pressing or drawing.	Soda-lime-silica: – Electric light bulb SiO_2 72.5 Na_2O 15.9 CaO 6.5 MgO 3.0 Al_2O_3 1.3 K_2O 0.3 – Fluorescent light tubes: SiO_2 72.5 Na_2O 14.6 CaO 5.7 MgO 2.9 Al_2O_3 2.6 K_2O 1.2 B_2O_3 0.5
– Domestic shades – Industrial shades – Traffic, safety and signal lights.	High temperature and thermal shock resistance. Durable, weather resistant. Accurate and non-fading colour to BS and international specifications.	Hand or automatic pressing or blowing depending on quantity.	Wide range soda-lime-silica. Borosilicate.
LABORATORY GLASSWARE – Many applications.	High chemical durability. Low thermal expansion – good thermal shock resistance.	Hand or automatic pressing or blowing. Automatic drawing of rod and tube. Sintering.	Soda-lime-silica. Borosilicate or fused silica for high temperature capability and low thermal expansion.
SIGHT AND GAUGE GLASSES – Viewing windows in industrial plant.	High temperature and thermal shock resistance. Mechanical strength. Chemical resistance.	Cutting from flat plates. Pressing, followed by grinding and polishing, and sometimes toughening.	Soda-lime-silicate similar to flat glass. Borosilicate. High silica or fused silica.

Table 2.3 Continued

CHEMICAL INDUSTRY GLASSWARE Many applications – Tubing, heat exchangers, reaction vessels, pumps.	Transparency. High temperature and thermal shock resistance. Mechanical strength. Chemical resistance.	Drawing of tubing. Highly skilled hand working. Pressing and moulding.	Soda-lime-silicate glass similar to flat glass. Borosilicate. High silica or fused silica.
RADIATION SHIELDING Specialist applications in nuclear and other scientific fields.	Transparency. High density to block X-ray or γ-ray radiation. Transparent to visible wavelengths. High optical quality.	Extrusion or casting followed by cutting, grinding and polishing.	Soda-lime-silica with up to 80% by weight of lead oxide.
GLASS FIBRE – Continuous filament reinforcement of plastics, rubber and gypsum.	High strength. Relatively high elastic modulus. High durability. High electrical resistance in some applications.	Continuous filament drawing and coating. Can be woven into textiles or chopped and incorporated into composite materials.	Soda-lime-borosilicate: SiO_2 65 Na_2O 8 CaO 14 Al_2O_3 4 MgO 3 B_2O_3 5.5 K_2O 0.5 Low alkali-alumino-borosilicate: SiO_2 55 Al_2O_3 15 CaO 22 B_2O_3 7.5 Na_2O 0.5

– Glass wool; thermal and acoustic insulation.	High strength. Relatively high modulus. High durability.	Centrifugal extrusion and attenuation.	Soda-lime-borosilicate: SiO_2 59.0 CaO 16 Na_2O 11 MgO 5.5 Al_2O_3 4.5 B_2O_3 3.5 K_2O 0.5
BALLOTINI – Reflective glass spheres (1–60 μm) for road signs and markings.	High reflectivity with high durability.	Flame drawing.	Soda-lime-silica – similar to flat glass.

2.8 References

1. British Glass. (1992). *Making Glass*. Northumberland Road, Sheffield S10 2UA.
2. Doyle P. J. (1991). *Glass Making Today*. Portcullis Press, United Kingdom.
3. Rawson H. (1967). *Inorganic Glass-forming Systems*. Academic Press, London and New York.
4. Tooley F. V. (1984). *The Handbook of Glass Manufacture*, Vol. 1. Ashlee Publishing Co. Inc., New York.
5. Tooley F. V. (1984). *The Handbook of Glass Manufacture*, Vol. 2. Ashlee Publishing Co. Inc., New York.
6. EU Directive 69/493/EEC *on the approximation of the laws of the Member States relating to crystal glass*. Official Journal C108,19. 10. 65, p. 35.
7. Fanderlik I. (1983). Optical properties of glass. In *Glass Science and Technology*, Vol. 5. Elsevier, Amsterdam.

3

Microscopic techniques for glass examination

PAMELA S. HAMER

3.1 Introduction

Glass fragments can provide very significant forensic evidence. Glass as a mass-produced material is widely distributed and fragments from broken glass can provide very strong evidence to support a scenario in a specific set of case circumstances. In general, the more parameters identified for the glass fragments, the stronger the evidence provided by that glass. The prime methods for characterising glass fragments and comparing them with a source of glass broken in the commission of a crime are based on microscopic techniques. These provide quick and effective exclusion of unlike fragments and information on matching fragments which can be used to assess the significance of the match.

There is a vast amount of information in general and specialised texts on the manufacture of glass and its conversion into the variety of glass objects encountered in everyday life. Only a brief resume is presented here to develop the context in which the forensic scientist examines glass fragments. In composition, glass is invariably based on silica (silicon dioxide) obtained from sand. For most domestic use this is modified by a variety of other oxides. Soda (Na_2O) and potash (K_2O) are added to decrease the melting point of the silica, but this also makes the glass more unstable so stabilisers such as lime (CaO), magnesia (MgO) and alumina (Al_2O_3) are added to improve the chemical resistance. A variety of other materials are added to alter the properties of the glass and make it suitable for fabrication into objects which show acceptable optical properties. Flat glass is made in one of two ways. Flat, clear glass can be made by passing molten glass between two rollers. Unless the glass is polished flat the optical quality of this glass is not generally considered suitable for glazing modern domestic windows. A modification of rolled flat glass where a wire mesh is incorporated into the sheet is used as one type of safety glass as the mesh is very difficult to penetrate. Most modern flat glass is made by the float process in which a sheet of molten glass is floated on a bed of liquid tin to give a very flat, optically clear glass suitable for windows with no further preparation. Some of the tin is incorporated into the surface of the glass in contact with the tin bath and the fluorescent effect of this has been used by forensic scientists to provide rapid identification of small fragments of float glass. Another type of flat safety glass is laminated glass, usually formed from two sheets of flat float glass bonded together with a plastic film. Again this glass is very difficult to break and penetrate. A third type of flat safety glass is toughened glass. This is float glass treated by heating it and cooling it rapidly so the final solid glass sheet is stressed. When this glass breaks it shatters to form small cubes of glass which are less likely to cause cuts and damage to anyone or anything in contact with it. The stress in this glass alters its physical properties. Rolled flat glass can also be made with a pattern embossed in the surface for decorative properties or to make it translucent or opaque. Various elements can be added to the melt used for sheet glass to remove residual colour in the glass, or to form a specific colour or tint to the glass.

The other main sources of glass encountered by the forensic scientist are bottles, containers and domestic glassware. Different formulations of glass are used to provide different properties of the glass melt for easy fabrication of the object or to impart specific properties to the glass. Bottles are mass produced in an automatic plant blowing glass into moulds. Domestic glassware is produced by blowing or pressing into a mould and specialist items are often blown by mouth. Other manufactured glass items might also be encountered such as light bulbs, and lenses.

The different formulations used for glass items means that they have different chemical compositions. These govern the optical properties of the glass, including refractive index, the property usually investigated by forensic scientists. Glass made into different objects will be manufactured in different ways and this will give a shape to the final object. Small glass fragments broken from such objects might show the surface curvature of the original item or microscopic detail such as grooves or pits formed on the surface of moulded items.

Glass fragments recovered in casework are usually between 0.1 and 1 mm in length, although occasionally larger fragments are examined. The larger fragment might be seen on a garment recovered from a person suspected of acquiring glass from the scene of a crime, during the initial examination by the forensic scientist. With the aid of a simple stereo microscope mounted on a stand so it can be swung over the garment laid out on a bench, these fragments can be picked off with tweezers. Smaller fragments are usually recovered by shaking or brushing a garment over sheets of paper or a cone and by using a stereo microscope to examine the debris recovered The debris which falls off is then searched under a stereo microscope and the fragments are picked out with tweezers.

These glass fragments can come from a variety of sources. If a glass item is broken, fragments might scatter over people close to it. Hicks *et al.* [1] have recently studied glass fragments acquired on the clothing of a dummy placed at different positions in front of a breaking window. This repeats work done by Pounds and Smalldon [2]. Further work has been reported by Allan and Scrannage [3] and Underhill [4]. All these studies concentrate on glass transferred from breaking plain, sheet glass. A variety of other items might be broken during a crime, including patterned sheet glass, bottles or other containers, tableware or other domestic items like light bulbs. When these items break, the glass fragments produced show microscopic properties which relate to the glass used in the original item.

Many papers have been written about the acquisition of fragments from breaking window glass because that is the commonest type of glass broken in a crime. It has been suggested that the glass acquired accidentally by members of the public is not window glass. The survey of glass from the clothing of people unconnected with crime in Belfast [5] shows that most people have a few glass fragments on their clothing and when they do the glass tends to be of a random nature. The aim of the forensic scientist is to screen out glass fragments which will not provide any significant evidence in a case as quickly as possible and techniques based on microscopes provide very good methods for screening glass fragments. The microscopic characterisation of glass fragments that have not been excluded can be interpreted by the scientist to provide significant evidence in many cases.

The forensic scientist, therefore, uses microscopic techniques as follows:

- To find and recover small glass fragments.
- To examine the physical shape and colour of glass fragments to provide information on the source of the glass. To examine a surface on a fragment.
- To compare glass fragments with a sample of glass taken from the scene of a crime by comparing the refractive index values of the glass.

3.2 Recovery

The human eye–brain system sees detail and depth in an object because the light reflected from the same part of the object travels a slightly different light path to each eye. This gives a stereoscopic image. We are used to interpreting detail in objects and handling them using our stereo vision. We extend this facility to microscopic objects by using a stereo microscope. There are two types of stereo microscope available, the Greenhough type or the common main objective (CMO) type. Both can be made with a zoom magnification system. The optical system is designed so the light from one part of an object travels in different ways to the eye. There are, therefore, two light paths in these microscopes. In the Greenhough type the microscope body contains a pair of objective lenses and subsidiary lenses which pass the light to the two eyepieces. In the simplest systems the magnification is changed by sliding pairs of lenses into position in the microscope body, but more complex instruments are manufactured with lens systems which alter in position to provide zoom magnifications. In the CMO type of microscope there is a single large objective lens but the subsidiary lenses in the body of the microscope take the light paths from different parts of this lens to the two eyepieces. The subsidiary lenses can be arranged so the magnification changes due to alterations in their relative positions. This can be done smoothly so any magnification can be selected between two extremes. These are typically ×0.63 to ×4. The microscope eyepieces provide further magnification, usually ×10 but sometimes ×15. As stereo microscopes provide a stereoscopic image it is easy to see a small fragment and manoeuvre it using tweezers or a needle probe.

This type of microscope is used by forensic scientists to examine clothing, and the debris shaken or brushed from clothing, to recover glass fragments. The microscope is put on a bench stand or supported on a long arm so it can be swung over items (Figure 3.1). Illumination is provided by an external lamp. This might be a small tungsten or halogen lamp attached to a bracket on the side of the microscope pod, a ring illuminator pointing down from around the objective lens or a free-standing lamp or optical fibre guide. Each of these has advantages and disadvantages. The side-mounted lamp is the most adaptable giving a wide beam of illuminating light which can be adjusted to different angles and focused to a narrower beam. A great disadvantage is that these lamps tend to become very hot during prolonged use and this can make adjustments difficult. This can be overcome if a cold light source provided by an optical fibre from a remote lamp unit is used. These light sources give a very intense white light but the area of the beam is generally smaller than

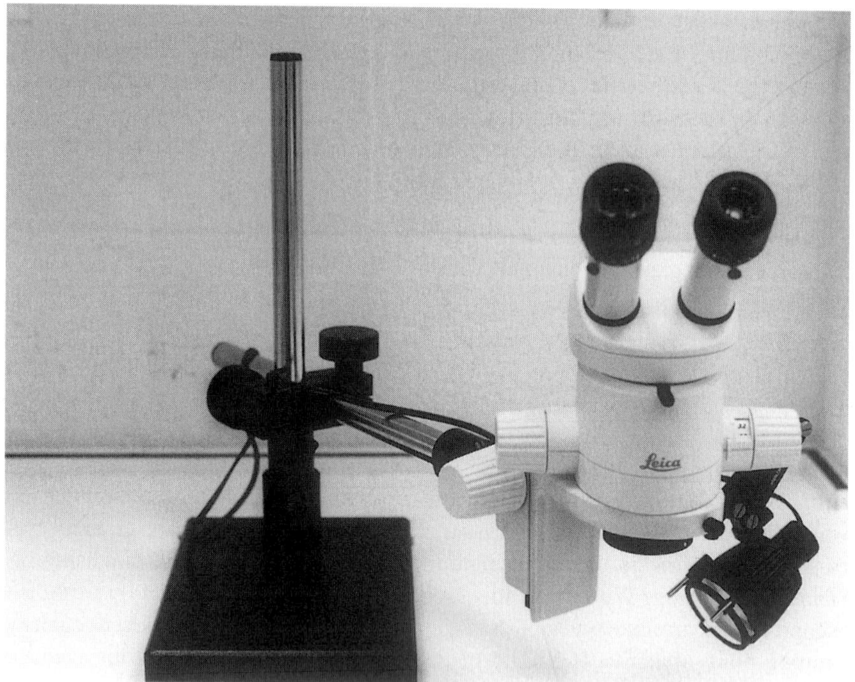

Figure 3.1 Stereo microscope on a long-arm stand

that with a bulb and it can be difficult to support the end of the fibre. Stiff fibres can be self-supporting, though still sufficiently flexible to allow the angle of the light to be varied. A ring illuminator might be a fluorescent light bulb or an optical fibre device. It is usually fitted around the base of the microscope pod, surrounding the lower lens of the objective. These sources give very good illumination coverage of the whole field of view of the microscope but the high illuminating angle gives a very flat light which cannot be adjusted. This type of illumination is not commonly used for glass examination, therefore, as there are many advantages in being able to adjust the illumination to highlight different parts of the object.

In the examination of debris, usually in a plastic Petri dish or other similar container, the scientist is continually shifting the debris with tweezers, a needle or probe to show all the faces of clear fragments. To assist in seeing the sharp edges or fracture surfaces of a glass fragment the dish is often put onto a dark surface. Many incident light stereo microscopes have a base containing a plate that is black on one side and white on the other. Most scientists search and examine their fragments against a dark background. A white background can, however, be used if fragments are coloured or tinted as it might be possible to see the faint colour against a white background, but not against a black one.

3.3 Surface examination

Much information on the source of glass fragments can be obtained by examining the shape of the fragments to identify those which have come from the surfaces of the broken glass object. A scientist refers to this as an 'original surface' on a glass fragment. It can be difficult to identify an original surface on a small fragment recovered from the debris swept from clothing viewed under a stereo microscope. The scientist usually inspects glass under a stereo microscope using magnifications from around ×6 to ×40 to detect fragments that have an original surface. Glass reflects about 4 per cent of the light that falls on it and marks and deposits on a surface can only be seen if the maximum amount of light is reflected from the surface into the microscope. This is generally described as specular reflection and is obtained by an angled light source that gives a variety of illumination positions to highlight different surfaces on a fragment. Freshly broken glass fragments will generally have clean fracture surfaces. Any surface which has a dirty or greasy deposit on it, might, therefore, be an 'original surface'. There might also be pits, scratches or scores on an original surface which have been formed by polishing or cleaning the surface. Some very distinctive parallel scratch marks can be made by windscreen wipers sliding across gritty deposits on the outside of car windscreens. An original surface can show other important features. For example, glass from the edge of a window which has recently been painted might show spots and smears of paint. Glass from the windows of a car that has been resprayed might also show tiny spots of paint. Many car windscreens and rear windows have a black coating at the edge that might be present on tiny fragments broken from the surface.

Freshly broken fracture surfaces of a fragment are clean and often show the typical fracture pattern caused by the concoidal fracture of the glass. It can be difficult to detect the original surface on small fragments broken from the surface of sheet glass since some areas of a fracture surface might appear flat. Locke and Elliott [6] suggested that surfaces less than 0.5 mm long might not be recognised as original since mirror fracture surfaces can give a flat surface over this sort of area. If a fragment larger than this is flat, it cannot be a fracture surface and, therefore, must have come from the surface of the broken flat object.

In recent years, a type of microscope described as a macroscope has become available. This has a different lens path which does not give a true stereo image but it has a large field of view and a long working distance. Macroscopes have better optical resolution than the simple stereo microscopes and magnifications of ×100 to ×120 are often used. These microscopes have internal incident illumination that can be directed down through the objective lens giving a bright focused beam of light onto the object. If a glass fragment is manipulated so that a suspected original surface is flat, facing towards the lens, this illumination gives strong specular reflection. Fine scratches, pits and tiny spots of paint can be easily seen.

For this examination the glass has to be supported at the correct angle and a method which works well is to mount the glass in a material described as 'Black-tac'. This is a substance sold in the UK as 'Blue-tac', with black fingerprint powder mixed with it to reduce the reflection. It can be put into the small aluminium caps used to seal vials to provide small, individual mounts for glass fragments. The substance is quite dry, it does not produce an oily deposit on the glass and fragments can be removed easily from it after the microscopic work has been completed.

3.4 Surface luminescence

Most modern sheet flat glass is made by the float process. In this the ribbon of molten glass is floated on a bed of molten tin. This gives the flat, defect-free surface required for sheet flat glass used for windows. Some of the tin is incorporated into the surface layer of the glass. Much work was done in the late 1970s by Lloyd [7] who characterised the luminescence produced by this float surface. The techniques used were not suitable for use on the small casework-sized fragments and luminescence is generally used now as a screening technique to identify fragments which might have come from the float surface of sheet glass, rather than as an analytical technique.

A typical microscope system consists of a quite basic stereo microscope fitted with a chamber containing a small lamp giving ultraviolet (UV) illumination at a wavelength of 254 nm. At this wavelength, the tin-rich surface of float glass luminesces, often with a greenish glow. The simplest chamber contains a small arc-discharge lamp fitted with a filter to pass only the 254 nm wavelength. The lamp is mounted in the chamber so that it points down onto the microscope stage where the glass fragment is mounted. The chamber must trap the scattered UV illumination so that it does not provide a hazard for the microscopist. In its simplest form the chamber is a simple box with a clear viewing window which can be placed on the baseplate of a stereo microscope. The glass fragment, mounted with the suspected float surface uppermost, is put onto the microscope stage and viewed in white light through the port in the chamber where it is brought into focus on the fragment. The lamp is then switched off, the chamber put over the glass fragment on the microscope stage and the UV illuminated. The luminescence might be weak so the microscope is usually used in a darkened room and the microscopist's eyes can take some time to accommodate to the low light level. The luminescence is usually quite distinctive but the intensity can be variable. It has been suggested that weathering of the glass surface reduces the intensity of the luminescence. If a control sample of glass broken at the scene of the crime is available, it should be mounted and the intensity checked.

Most of the microscopes used for identification of surface luminescence of small fragments are custom-made. Hamer [8] described a typical modification of a stereo microscope. In this instance, the chamber for this microscope contained four small arc lamps linked to a microswitch cut-off operated by a hinged door so the glass could be easily and rapidly positioned on the microscope, examined and removed without any hazard from scattered UV radiation. There are interesting recent developments in optical fibre technology which means that an optical fibre could be used to pipe UV illumination into a chamber from a remote lamp and this could be incorporated into a custom-made system.

The presence of luminescence from a flat glass surface provides good evidence that a glass fragment is from a float glass source. Using the microscope system described above, however, some glass appears to show weak luminescence from the bulk of the glass. When this is checked against a fragment of known float glass it is usually obvious that this is not surface fluorescence but care should be taken in interpreting weak luminescence and it is good practice to have authentic samples available. Other glass might show luminescent properties, for example the phosphor-coated glass used in television sets and the phosphor coating on fluorescent light bulbs. Some flat glasses also have bulk fluorescence but these are likely to be 'old' glasses and the presence of such glass will inevitably decrease. Surface treatment of other glass objects, such as bottles, can also provide a surface that fluoresces under these conditions.

3.5 Surface contour

An impression of the shape of the original surface on a small glass fragment is obtained when the fragment is viewed under a stereo microscope or a macroscope. Curvature can often be inferred by the shape of a shadow projected onto the glass and reflected into the microscope. To determine the contours of an original surface, particularly to confirm that a surface is flat, it is necessary to use an interference technique. A variety of systems are used by forensic scientists but they usually depend on interference between light reflected from the glass surface and light reflected from a reference surface. This is described as the Michaelson interferometric method. This original forensic method was described by Elliot *et al.* [9] although this has been modified in recent years. In a Michaelson interferometer, light enters an optical beamsplitter which directs part of the light to the glass surface and part to a mirror. Light is reflected from both these surfaces and recombines in the beamsplitter. If one of the surfaces, say the mirror, is flat but the glass surface tilted, the light which reflects from different parts of the glass has travelled a different distance from the light from the same point in the beamsplitter reflected from the mirror. These light rays interact giving constructive or destructive interference between the rays depending on the light path difference. Destructive interference gives dark areas and constructive interference gives bright areas. If the glass fragment is flat, an interference pattern is formed of straight dark bands. If the glass is curved, the dark bands are curved. More complex curvature on the glass gives curved and circular interference bands. Monochromatic light, usually green, is used to show the interference pattern clearly. The interference device described by Elliot was a Watson interference objective, but this has now been largely superseded by a modification introduced by Mirau (see Figure 3.2). This can be incorporated into a specially adapted incident light objective lens and fitted to an incident light

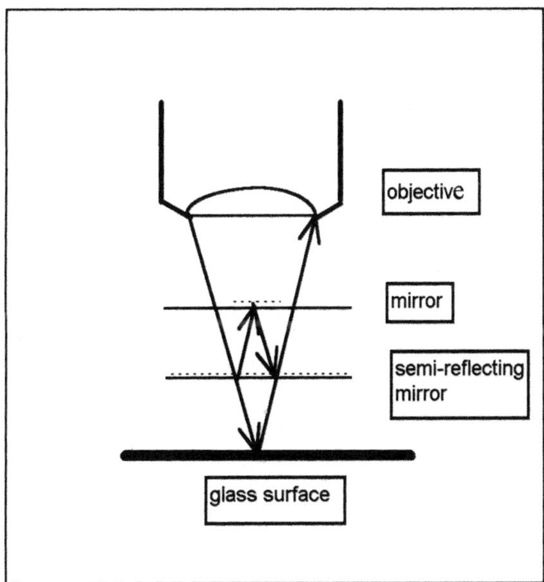

Figure 3.2 The optical system in a Mirau interferometer

compound microscope. A variety of designs are available from the major microscope manufacturers but they all have two semi-silvered mirrors acting as beamsplitters. The incident illuminating light beam passes to a mirror, smaller than the aperture of the light beam. Light passes round this mirror to a second semi-reflecting mirror. Part of the light is reflected from this to the mirror above. The rest of the light passes through and is reflected back from the surface of the glass fragment. This light interferes with the light reflected from the reference mirror to give an interference pattern apparently superimposed on the image of the glass surface. Again monochromatic light is used, usually by placing an interference filter in the light path of the microscope.

In practice, the interference pattern can be obtained by tilting the reference mirror or the glass surface. Most scientists prefer to tilt the glass surface on a small tilting stage. Elliot *et al.* [9] have described two versions. The more successful is a tilting stage from glassware ball-and-socket joints, cut in half with a diamond saw with the socket drilled so a small tilting lever can be fitted which protrudes through the hole in a standard microscope stage when the tilting device is mounted on a microscope stage. The glass fragment is supported on 'Black-tac' as described earlier.

The technique is very sensitive. Locke [10] has described a procedure for apparently flat glass where the glass fragment is tilted to set the interference lines to a set distance apart: 0.07–0.1 mm is usual. At this setting, sheet glass will generally give straight lines in the interference pattern. Small surface fragments of toughened glass can show slight curvature under these conditions as the surface distorts slightly because of the stresses induced by the toughening process. Very thin surface fragments also show slight curvature as they can distort slightly when mounted on the black-tac. Sheet glass which is not flat, such as the non-patterned side of patterned glass, has a very distinctive surface pattern (see Figure 3.3). It shows an interference pattern with gently curving lines and often includes small circular shapes. These are small dimples in the surface of the glass. A scientist can determine whether these dimples are pits or raised features by observing the direction of movement of the lines as the microscope stage is lowered to increase the distance between the objective lens and the glass surface. If they move into the centre of the dimple, they are raised features. Similar dimples are found on moulded items such as bottles. From observations of the surface it is therefore possible to suggest a possible source for small glass fragments. If a control glass sample is available, this technique enables the scientist to check that the surface contours are similar.

A further interference technique used to investigate the surface of glass fragments is incident differential interference contrast (DIC) microscopy. This technique is very sensitive to height differences on a surface which are shown as contrast differences (see Figure 3.4). The microscope is set up for incident light and a polar and analyser are used. A beamsplitter produces two slightly shifted beams of light reflecting from the surface of the glass. These interfere and height differences are shown by colour contrast following Newton's order of colours. Like the Michaelson, the beamsplitter is directional so the glass has to be rotated on the microscope stage to obtain the optimum contrast of all the features. A curved glass surface shows a colour pattern which follows the contours of the surface. Very fine scratches can be seen very clearly so it is a good technique for confirming surface abrasion.

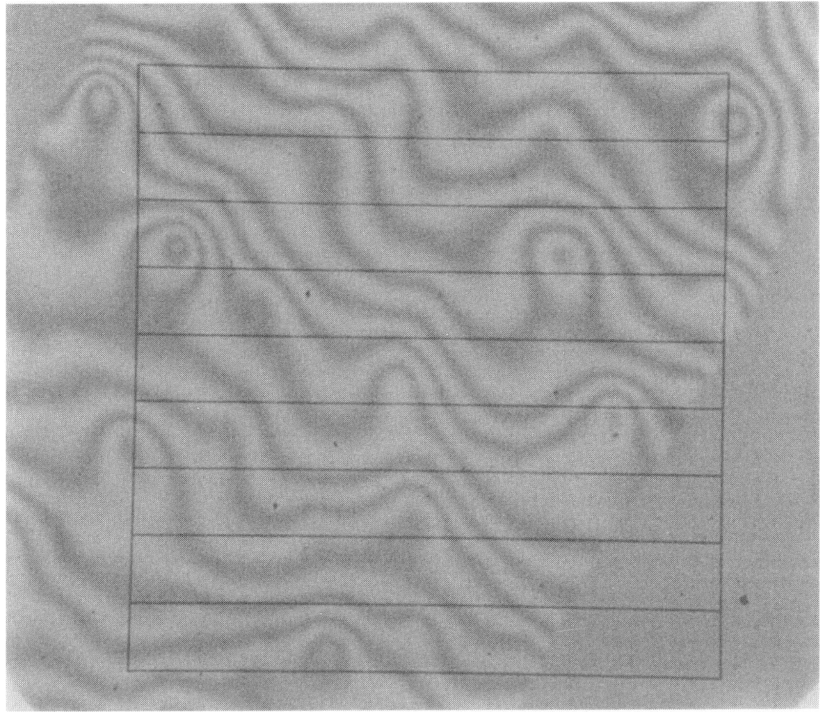

Figure 3.3 An interference pattern for a surface of patterned sheet glass; a grid graticule is superimposed

Figure 3.4 A differential interference contrast image from patterned sheet glass

3.6 Refractive index measurement

In the late 1960s a microscopic technique was developed to determine the refractive index of small glass fragments. The method was first published by Ojena and De Forest [11]. It relied on the temperature variation effect first described by Winchell and Emmonds [12]. The refractive index of a liquid changes as it is heated and cooled while the refractive index of solids is far less sensitive to temperature. If, therefore, a solid fragment is immersed in a suitable oil and heated, at some point the refractive index of the oil and the fragment will be the same. If the fragment is a clear transparent glass it cannot be seen when immersed in a liquid with the same refractive index. Therefore, the glass disappears. A wide variety of immersion liquids were used but clove oil was widely used in the UK to the mid-1960s. This had the disadvantage that the calibration shifted with time. Silicone oils avoided most of the difficulties associated with clove oil [13] and have been used routinely ever since. In the original method used by forensic scientists, the oil was heated on a small electric heater fitted to a microscope stage. Initially the temperature was monitored with a mercury thermometer but by around 1970 Mettler had introduced their FP1 hot stage, primarily designed for melting point determinations. It was, however, ideally suited to replace existing equipment. This was designed to heat a sample which could be viewed simultaneously down a microscope. The temperature of the hot stage was monitored by a thermocouple device. It now became possible to determine the temperature at which a glass fragment disappeared in an oil. The thermocouple measured the temperature of the heating panels in the hot stage and the temperature could be raised at 0.2, 1.0, 2.0, 3.0, 10.0°C per minute. The glass was mounted on a glass slide in an oil so the actual temperature of the glass was not the same as the temperature of the hot stage. As the hot stage was heated, the temperature of the oil lagged behind but this was not important for comparative purposes. Later models of the instrument also incorporated a cooling cycle so that the effect of this temperature lag could be eliminated by taking readings in both the heating and cooling modes and determining the mean. More accurate results were thus obtained for the disappearance temperature of the glass in the oil in this dynamic system. By devising a heating/cooling cycle, the match temperature was determined during the heating phase and then during a cooling phase and a mean match temperature could be established. It only remained to calibrate the oil and hot stage by immersing a set of glass samples with known refractive index in the oil and determining the mean temperature at which they disappeared. At the disappearance point the oil and glass had the same refractive index. It was then assumed that there was a linear relationship between the refractive index of the oil and the temperature so that a calibration line could be drawn. The refractive index (RI) of an unknown glass fragment could then be determined from this calibration provided its disappearance temperature was known. The calibration line was described as $RI = a + bT$ where a is the intercept of a graphical line at 0°C temperature and b is the slope of the graph. In practice the RI falls as the temperature (T) increases so the b is a negative quantity.

By the early 1970s the technique was becoming the standard method in the UK and a phase-contrast optical system was used on the microscope to improve the contrast of the glass near its match temperature. This allowed the microscopist, who was watching the glass disappear, to record the disappearance temperature as precisely as possible. In fact the human eye is not sensitive enough to determine a disappearance point: the glass disappears over a temperature range and the microscopist records when it disappears and then reappears and averages these values to get the match temperature. With a set of standard glass samples of known RI, such as those supplied by Schott, this is a very sensitive technique, recording the refractive index to the fourth decimal place. It is, however, very tiring for the microscopist staring at disappearing glass fragments and in the early 1980s the method was improved by the introduction of glass refractive index measuring (GRIM) equipment produced by Foster and Freeman (Foster & Freeman Ltd, 25 Swan Lane, Evesham, Worcs., UK) [14].

GRIM replaced the eyes of the microscopist with a video camera monitoring the contrast of the glass fragment in the oil. One of the early dedicated computer systems, GRIM monitored the video signal and the temperature of the hot-stage. The calibration data for the oil were stored in the computer memory so GRIM printed out the match temperatures during a cooling and heating cycle and converted the mean match temperature to refractive index. Over the years the original model GRIM1 has been largely superseded by GRIM2.

The GRIM2 instrument comprises a phase-contrast microscope, working with monochromatic light at 589 nm, fitted with a Mettler hot-stage. A video camera with CCD detector is attached to the camera port on the microscope. The hot-stage is controlled by the GRIM control unit which also connects to the camera to monitor the video signal. The video image is viewed on a monochrome monitor. The operator focuses the microscope to give an image of the glass fragments on the monitor taking care to select a temperature for the hot-stage which is less than the match temperature of the glass. This can usually be done by observing the image of the glass on the screen. If the glass is bright in comparison to the surroundings it is below the match temperature. If it is dark, the temperature is higher than the match temperature. If the hot-stage is set near the match temperature it is difficult to focus the microscope on a suitable glass fragment because the contrast of the fragments is very small. The operator then views the image on the monitor. In GRIM1 a small 'box'-shaped area was moved with a couple of control knobs on the front of the control unit to cover a vertical edge of a glass fragment. If there were no suitable vertical edges the video camera was rotated on its mount on the head of the microscope until an appropriate fragment was found. In GRIM2 the image of the glass is also displayed on a computer screen. The 'box' used to select the measuring area is moved using the computer mouse to cover an edge of glass. In this later version of GRIM the edge does not have to be vertical and the size of the 'box' can be altered. The operator then runs a 'search' program from a keyboard or using the mouse. GRIM ramps up the temperature of the hot-stage while monitoring the video signal from the 'box'. The contrast of the fragment changes and as the glass disappears GRIM recognises the point at minimum contrast of the edge, stops heating the hot-stage and allows it to stabilise at a temperature just above the mean match temperature of the glass. The operator then runs the 'Auto' program. The hot-stage heats and cools at a set rate, 2, 4 and 5°C per minute are used routinely, and the temperature equivalent to the minimum video signal in the 'box' from the edge of the selected fragment is recorded on screen. GRIM then heats the hot-stage and records a second match temperature. The mean of these is then calculated by GRIM. GRIM1 displays the data on the monitor, and prints them out on a small integral dot matrix printer. The operator then enters their name, the date, the case number and the oil and wavelength of light used. GRIM1 then automatically calculates the refractive index from calibration data for the oil stored in memory. GRIM2 is linked to a modern PC and is based on a Windows operating system. The match temperature results are displayed on the computer screen which also displays the image of the glass and the measurement box. Details of the oil and wavelength are entered into the computer at the start of the analysis and the refractive index is calculated and displayed immediately the match temperatures are determined. The whole sequence of measurement takes about 2–3 minutes.

The oils are calibrated for a particular wavelength of light following a calibration procedure program in GRIM. Three silicone oils are used to cover the refractive index range from 1.45 to 1.53. These oils are described by Locke (Locke Scientific, Pamber Heath, Newbury) who also gives details of their stability. A set of standard glasses are available for each oil. These are specialist glasses supplied with a certificate giving the refractive index of the glass at different temperatures. In the Calibration program, edges of glass are selected and the mean match temperature found from four or five separate heating and cooling cycles as in the Auto program. From the mean of these, the appropriate refractive index can be read off the table and entered into GRIM. The refractive index of glass alters with the wavelength of light used to measure it and with the temperature of the glass. In the forensic science community it is generally accepted that glass is measured at a nominal 589 nm wavelength and the RI is quoted at its match temperature. The RI values for the standard glass samples are, therefore, entered into GRIM from the tables supplied by reading the appropriate RI at 589 nm and the nearest match temperature. As the data are added, GRIM calculates the best straight line through the points relating RI to the temperature of the oil. The correlation coefficient is also calculated and a back-calculation displaying the refractive index of each of the control glasses calculated from the mean match temperature. Differences between this calculated figure and the given value are displayed as ΔRI values. Monitoring these values over a period of time on different GRIM instruments provides data on the accuracy of the certified data on the standard glass samples. Recent work by Collins in Australia [15] has suggested some discrepancies in the certified RI of some of the commonly used standards. These are given in the latest information provided by Locke with the glass standards. Locke describes his oils as A, B and C. The B oil covers the RI range most commonly encountered in forensic work. Twelve standard glass samples are provided for this oil although not all are used routinely by forensic laboratories. The lowest and highest RI standards are at the limits of temperature range of the hot-stage and might not be used, as glass with these RI values is rarely encountered in casework. Some of the standards have very similar RI values and some laboratories omit one of these from the calibration but use it for monitoring the calibration data over a period of time. With a well-chosen set of standard glass samples, a correlation coefficient of 0.999... is expected for the calibration of oil B. Oil A covers the high refractive index range and is usually calibrated with about five standard glass samples. Oil C covers the low RI range and little glass of this type is encountered by most laboratories. Only two glass standards are available for this oil so the GRIM will probably be re-calibrated every time a glass in this RI range is measured. The RI ranges of the Locke oils overlap. Silicone oils from other sources such as those designated DC 710 and MS 704 might not overlap in RI range. The intermediate range might be covered if these oils are mixed together in appropriate quantities but a separate calibration should be produced for the mixed oil.

The calibration of GRIM appears very stable. In casework use the calibration is usually checked by adding a standard glass to the glass samples examined. Several criteria are used to define whether the calibration is acceptable or not. One commonly used can be described as the ±0.2 degree level. In this, one of the standards used for calibration is run during the analysis and if the match temperature obtained differs by more than ±0.2° further checks are undertaken. A different edge of glass is chosen and the match temperature is found. If this still falls outside the 0.2° level a problem is indicated. The oil may be contaminated or there may be problems with the hot-stage or optical system as described later. The whole system should be checked and reset and a clean oil sample used. If the standard glass is still not within the 0.2° level, a different standard glass should be selected and the mean match temperature compared with that obtained during the original calibration. If this is also outside ±0.2° the instrument will require resetting and recalibration. Among GRIM users it is accepted that a GRIM machine usually needs recalibrating after 1 to 2 years' use.

Two methods have developed in mounting the glass fragments in a silicone oil. In one, the glass is put into the cavity of a dished cavity slide and immersed in a drop of the oil. This is placed directly onto the hot-stage. If necessary the glass fragment is cracked to give a fresh edge and a fragment is manipulated in the well. This is usually done under a stereo microscope fitted with a transmitted light base which gives oblique illumination. The microscopist manipulates the fragment with a needle to the centre of the well. This is the method used in many of the forensic laboratories in the UK. It is a rapid mounting method and the glass fragment can be recovered after RI measurement and examined by further microscopic tests or chemical analysis. The disadvantages include the degradation of the optical system by interposing the curved and ground glass surface into the optical system, the difficulty of putting and manipulating the wide, thick slides in the hot-stage and the presence of a pool of oil in the hot-stage which can contaminate the adjacent surfaces. The other method is to break or crack a tiny glass fragment on one of the special thin microscope slides provided by Mettler or an equivalent flat microscope slide. The glass fragments are immersed in a small drop of oil and a coverslip placed over. The advantage of this system is that the glass is mounted to give a good optical image in the phase-contrast microscope. This should provide optimum contrast near the match temperature. Since the glass is broken into smaller fragments it is easier to see heterogeneity within the glass and avoid areas which will not give a refractive index result which is representative of the glass object broken. The presence of these heterogeneities will be described in more detail below. If the glass has been broken into tiny pieces, however, they cannot be readily recovered from the oil and this mounting technique is essentially destructive of the glass. In practice, most glass fragments can be cracked on the slide when mounting and a portion retained. When using this method, the glass will have been examined prior to the RI determination for original surfaces and luminescence properties. The retained portion is used for further chemical analysis if required.

There are a number of factors which affect the RI of a glass fragment calculated with a GRIM system. GRIM records the minimum intensity response from a video or CCD camera as the glass is heated and cooled in the oil. The reproducibility of these results depends on the precise detection of this minimum. In the early GRIM1 machines the quality of the edge selected for measurement was shown by the edge contrast figure (ECF). This was a value calculated by the machine from 0 to 99. A good, high-quality edge gave an ECF of 99. A very poor edge where the rate of change of the edge signal was very low would give an ECF of 0. It was possible to connect a chart recorder to GRIM1 to follow the change of the signal from the edge. An ECF of 99 gave a response showing a very sharp minimum. An ECF of 0 gave a response showing a very ill-defined minimum. It was suggested that for maximum precision in the results the operator should aim for higher ECF figures. In the original documentation with GRIM1 it was suggested that further readings should be taken if the ECF was below 10. Over the years there has been little evidence that the low figures gave imprecise results and Locke suggested in some unpublished work

that lower figures were acceptable. The ECF does, however, give some indication of a problem with a result; for example, if the fragment of glass has moved in the oil such that the edge under examination falls outside the area of the box. Therefore, the ECF has been retained for GRIM2 with the addition of a display during the cooling/heating cycle of the change of contrast of the edge with time temperature. This demonstrates well when a poor quality edge has been chosen, since asymmetric peaks, distortions in the peaks and very indistinct minima are readily visible (see Figure 3.5).

Other instrumental problems which might affect the quality of the result are dislodging the interference filter and having oil on the heat shield or on the stage. The wavelength of illuminating light is selected using a narrow bandpass interference filter. Those usually supplied are nominally 589 nm though the maximum intensity of light might be ±2 nm from this. The filter usually used has a bandpass at half intensity of ±10 nm. The filter will only work to its specification if the light is passing normally through the filter. If the filter is tilted to the optic axis of the microscope it will pass light of a different wavelength. This then alters the match temperature of the glass in the oil. Locke investigated this [16]. A further problem is encountered if the alignment of the phase optics of the microscope is not checked for each sample preparation.

A further effect investigated by Collins [15] was a significant alteration in the match temperatures obtained if oil leaked round the edge of the microscope slide onto the heating pads of the hot-stage. This produced a significant temperature difference and was very difficult to remove. In most cases even after prolonged cleaning the GRIM had to be recalibrated. It has also been reported that the presence of a film of oil on the heat shield that protects the microscope objective can alter the match temperature of a glass in an oil.

The Mettler hot-stage is generally very reproducible in its cooling and heating rate. The stage is air cooled by forced air from the integral fan and it should be noted that the cooling rate will be slower if the ambient temperature is close to the temperature of the hot-stage. At faster cooling rates the fall-off from the recorded rate occurs at higher ambient temperatures.

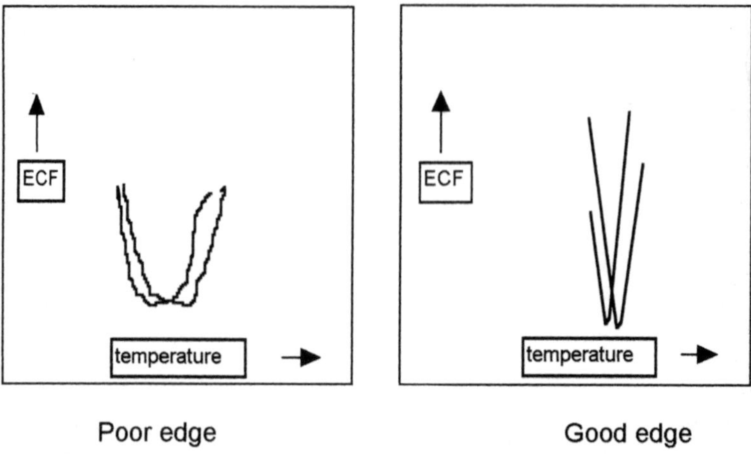

Figure 3.5 The video response of GRIM for a poor edge and a good edge

There are a variety of factors within a sample of glass that can also affect the refractive index result. One is the innate variation within a glass object. This can vary from about ±0.0001 RI units for a small modern float glass window to about 0.0004 for a thick toughened glass. Dabbs and Pearson provided the first data on the variation in the RI for a glass object [13]. Some glass shows distinct variation in RI. For example, patterned glass sometimes shows 'banding'. These are series of dark bands shown in a glass fragment when it is in an oil close to its match temperature. Welch *et al.* [17] investigated this phenomenon. A similar type of effect is shown by float glass. Underhill [18] described the tramlines sometimes seen in small crushed glass flakes and was able to measure the RI of both the float and non-float surfaces. This proved to be very difficult to do and is rarely attempted in casework. This RI anomaly at the surfaces of float glass can be a problem in measuring the RI of small glass fragments recovered from clothing of someone suspected of breaking a large float-glass window. Many of the fragments will be from the surface of the glass as the propagating cracks splinter along the glass surface. Measurement of the RI of a fragment from the surface will not give the RI of the bulk of the glass. This was investigated by Zoro *et al.* [19]. Care must be taken when examining recovered flakes of glass as the presence of part of a float or non-float surface in the edge covered by the GRIM measuring 'box' can give an anomalous result as the bulk RI of the glass is modified by the surface effect. It is usually possible to recognise glass fragments that are giving an anomalous result because of the presence of a float or non-float surface if the fragment is observed throughout the Auto sequence on the GRIM monitor. The edge does not disappear but gives a gradual change in light intensity as it passes through the match temperature for the bulk glass (see Figure 3.6).

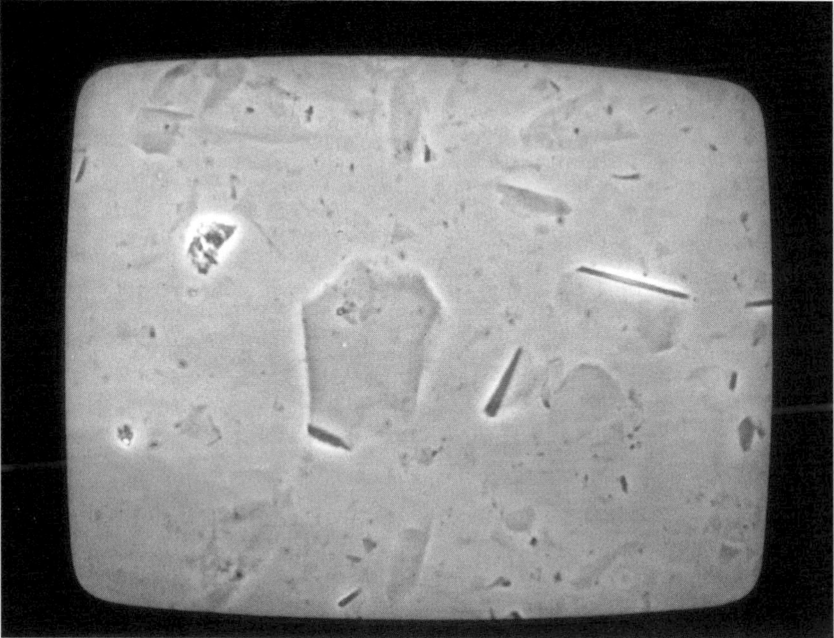

Figure 3.6 Surface effects on glass fragments seen on the GRIM monitor

Most forensic laboratories calibrate GRIM by correcting the refractive index of the calibration standards to their match temperature. The RI of recovered glass fragments are therefore correct at the match temperature. This would mean that a glass fragment that could be analysed in two oils, at a high temperature in one oil and a low temperature in the second oil, would have a different RI. This is not a problem in comparing glass samples within a case but may be a problem in retaining RI results and incorporating them into a database. With GRIM it appears that most laboratories are using very similar oils and the refractive index values obtained in inter-laboratory trials are generally very similar. This means there is the possibility of incorporating data from many different laboratories into a large database.

In this section it has been assumed that all RI results are obtained using an illuminating wavelength of nominally 589 nm. In the USA forensic scientists working with glass have used dispersion data to compare glass samples. The RI of the glass is determined with different illuminating wavelengths. In the UK Locke *et al.* [20] established that little additional discrimination was obtained by considering dispersion data.

3.7 Annealing

Annealing is the removal of internal stresses from the glass by the controlled application of heat. Basically it involves heating the glass above the transformation temperature and then cooling it slowly. In the forensic science context the control and suspect samples are heated in a small metal boat. The boat has a number of cavities drilled in the top and a suspect fragment and a sample of the control glass are placed in separate cavities. It is also prudent to add a known toughened glass standard to check that the correct annealing sequence has taken place.

In UK laboratories a small furnace is used that incorporates a horizontal tube. The boat is pushed into the centre of the furnace and the heating cycle started. The furnace cycle is fully automatic and a typical annealing cycle heats the boat up to 590°C, holds at this temperature for 12 minutes then cools at 4.5°C per minute until a temperature of 425°C is reached. The glass is maintained at this temperature for 1 minute before free cooling. Non-toughened glass typically shows an increase in refractive index after annealing of approximately 40×10^{-5} although higher values are sometimes recorded. Toughened glass typically shows an increase of 140×10^{-5} although this will depend on the thickness of the piece of glass from which the sample was taken. It should be noted that annealing will not remove the differences referred to above between refractive index readings on the surface and in the bulk of a sample of glass. Indeed, such differences can still be seen on annealed fragments and it is therefore still desirable to avoid surface fragments if at all possible [21].

3.8 Comparison of glass fragments

The forensic scientist is usually attempting to compare the physical and optical properties of a fragment of glass recovered from a person believed to have been involved in a crime with glass samples taken from broken objects at the scene of the crime. Very good discrimination is provided by the tests described above. Clearly if the surface shape of a fragment is different from that of the control item, the fragment could not have come from that item. If the fragment shows fluorescent properties typical of a flat float glass and the control glass has no fluorescent properties, the glass fragments could not have come from the same broken object.

Comparison of refractive index values is more difficult. Fragments of glass from a glass object will give a range of refractive index values. Conventionally, measurements of this type are assumed to follow a normal distribution but for some situations encountered by a forensic scientist this is an incorrect simplification. The scientist might not have a control sample that is fully representative of the broken object, the object might show heterogeneity or have a treatment such as a float surface which modifies refractive index values from some parts of the object. The work by Dabbs and Pearson providing data on the refractive index change across a sheet of glass was followed by a variety of other papers describing special cases in which variation occurred. Work by Underhill and by Zoro investigated the effect of float surfaces.

A variety of techniques are in use to compare the results of refractive index tests of recovered glass against control glass. Simple inspection of the data using a 'dotty' diagram to plot the actual refractive index values enables a scientist to view the scattering of values and provides visual information on clustering of data points. Statistical tests such as the *t*-test or modifications of the *t*-test can be used. In some situations a cluster analysis can provide a more incisive analysis of the data. There have been a number of papers published recently, such as that by Walsh *et al.* [22], which have raised interesting limitations of the *t*-test and suggested modifications based on a Bayesian approach to the evidence. This is explored further in a later chapter.

3.9 Acknowledgements

The author thanks Drs M. Dabbs and O. Facey for their contributions to this chapter.

3.10 References

1. Hicks T., Vanina R., Margot P. (1996). Transfer and persistence of glass fragments on garments. *Science and Justice* **36**:101–107.
2. Pounds C. A, Smalldon K. W. (1978). The distribution of glass fragments in front of a broken window and the transfer of fragments to individuals standing nearby. *Journal of the Forensic Science Society* **18**:197–203.
3. Allan T. J., Scrannage J. K. (1998). The transfer of glass – Part 1. Transfer of glass to individuals at different distances. *Forensic Science International* **39**:167–174.
4. Underhill M. (1992). The acquisition of broken glass. *Science and Justice* **37**:121–127.
5. McQuillan J., Edgar K. A. (1992). Survey of the distribution of glass on clothing. *Journal of the Forensic Science Society* **32**:333–348.
6. Locke J., Elliott B. R. (1984). The examination of glass particles using the interference objective. Part 2: a survey of flat and curved surfaces. *Forensic Science International* **26**:53–56.
7. Lloyd J. B. F. (1981). Fluorescence spectrometry in the identification and discrimination of float and other surfaces on window glass. *Journal of Forensic Sciences* **26**:325–342.
8. Hamer P. S. (1989). The use of microscopes in forensic science. *The Quekett Journal of Microscopy* **36**:277–287.
9. Elliot B. R., Goodwin D. G., Hamer P. S., Hayes P. M., Underhill M., Locke J. (1985). The microscopic examination of glass surfaces. *Journal of the Forensic Science Society* **25**:459–471.
10. Locke J. (1984). New developments in the forensic examination of glass. *The Microscope* **32**:1–10.

11. Ojena S. M., De Forest P. R. (1972). Precise refractive index determination by the immersion method, using phase contrast microscopy and the Mettler hot stage. *Journal of the Forensic Science Society* **12**:315–329.
12. Winchell A. N., Emmons R. C. (1926). Some methods for determining refractive indices. *American Mineral* **11**:115–119.
13. Dabbs M. D. G., Pearson E. F. (1970). The variation in refractive index and density across two sheets of window glass. *Journal of the Forensic Science Society* **10**:139–148.
14. Locke J., Underhill M. (1985). Automatic refractive index measurement of glass particles. *Forensic Science International* **27**:247–260.
15. Collins B. (1997). The investigation of the parameters affecting the determination of the refractive index of glass using GRIM. A report to the National Institute of Forensic Sciences, Australia.
16. Locke J. (1983). The use of interference filters in refractive index measurements. *Forensic Science International* **22**:57–63.
17. Welch A. E., Rickard R., Underhill M. (1989). The observation of banding in glass fragments and its forensic significance. *Journal of the Forensic Science Society* **29**:5–13.
18. Underhill M. (1980). Multiple refractive index in float glass. *Journal of the Forensic Science Society* **20**:169–176.
19. Zoro J. A., Locke J., Day R. S., Badmus O., Perryman A. C. (1988). An investigation of refractive index anomalies at the surfaces of glass objects and windows. *Forensic Science International* **39**:127–141.
20. Locke J., Underhill M., Russell P., Cox A., Perryman A. C. (1986). The evidential value of dispersion in the examination of glass. *Forensic Science International* **32**:219–227.
21. Locke J., Sanger D. G., Roopnarine G. (1982). The identification of toughened glass by annealing. *Forensic Science International* **20**:295–301.
22. Walsh K. A. J., Buckleton J. S., Triggs C. M. (1986). A practical example of the interpretation of glass evidence. *Science and Justice* **36**:213–218.

4

Elemental analysis of glass fragments

JOSÉ R. ALMIRALL

4.1 Introduction

Glass examiners in forensic laboratories have found it useful to incorporate elemental analysis as part of their routine examination and evaluation of glass evidence. The value of the measurement of major, minor and trace elemental composition of glass for its classification into glass types has been recognised for some time [1–4]. It is usually helpful to be able to classify the questioned glass into one of a number of possible categories, such as sheet (or 'float', the name of the process for the manufacture of most sheet glass), container, vehicle window (also made by the float process), vehicle headlamp (a borosilicate glass) or tableware (including leaded glass). One reason for classification is to facilitate the assessment of the association between the questioned and the known fragments by either classifying both as the same type of glass and then applying the appropriate comparison criteria or by eliminating straight away the questioned fragment from originating from the known source.

Technological advances in the manufacture of glass and the improved quality control in the glass industry has led to less variability in physical and optical properties between the manufacturers of products and also to less variability between the different plants from the same manufacturer. Through the use of the computer-controlled delivery of raw materials and the sophisticated on-line monitoring of the manufacturing lines, the glass manufacturing industry has aimed to control the differences in density, refractive index, thickness, colour and toughening (tempering) properties. The result of this improved quality in the manufacture of glass has been an industry-wide narrowing in the ranges of physical properties in glasses of the same type. Consequently, the *discrimination* potential, or the ability to distinguish between glass fragments, has been diminished and the sole reliance on measurements such as density and refractive index can lead to overstating the value of positive 'matches'. For example, recent studies show that for float glass sampled on three separate occasions over a period of 18 months from the same United States plant, the refractive index was found to be analytically indistinguishable [5] when using a precise method for the measurement of refractive index. Consequently, a number of publications have been devoted to the evaluation of different instrumental methods for the classification and discrimination of forensic glass fragments utilising chemical composition data. The purpose of this chapter is to summarise the work that has been done to date in this area and to describe briefly the theory, operation, advantages and disadvantages of each of the techniques used for the measurement of minor and trace element composition in glass. The instrumental methods of interest to forensic scientists are atomic spectroscopy, which includes atomic absorption (AA), and inductively coupled plasma emission (ICP-AES) techniques, X-ray methods, especially scanning electron microscopy coupled with energy and wavelength dispersive detectors (SEM/EDX/WDX) and the new techniques known as total reflection fluorescence (TXRF) and microbeam XRF. Radiochemical methods such as neutron activation analysis (NAA), although applicable, have for practical reasons been replaced with contemporary methods of inorganic mass spectrometry (ICP-MS).

4.2 Forensic and analytical implications of the composition of glass

The major raw materials employed for the manufacture of soda-lime-silica glasses are soda ash (Na_2CO_3), limestone (CaO) and sand (SiO_2). Table 4.1 lists the typical elemental composition of this class of glass as well as the average composition for borosilicate and leaded glasses, the three types of glass most often encountered in forensic casework. The components of glass are divided into primary formers, intermediate formers, modifiers, stabilisers, colourants and decolourants, accelerating and refining agents, opalisers and, finally, cullet. Oxides of silica and boron can form glass on their own and are therefore called primary formers. The primary source of silica is sand and, despite a huge supply of sand on the earth's surface, few deposits are suitable for making glass without some processing. Of particular concern is the iron and chromium contamination in some sand deposits that could lead to undesirable colouring in container, tableware and leaded glasses. Boric oxide (B_2O_3) is added to other oxides in the making of borosilicate glass, which is used in automobile headlamps, cookware and laboratory glassware. Its chemical stability and low expansion properties provide for a glass with good thermal shock and temperature-resistant characteristics. Aluminium oxide (Al_2O_3), usually referred to as alumina, is considered an intermediate former because it requires another oxide, although not necessarily a primary glass former, to make glass. Alumina is added to reduce devitrification (crystallisation) and enhance chemical durability. The modifier soda (Na_2O) is formed in the molten mix when sodium carbonate is used and combines with a primary glass former to lower the melting point and reduce the cost of the process. Oxides of calcium and magnesium (CaO and MgO) are added as modifiers in the form of dolomite (a mixed calcium/magnesium carbonate). Although strontium oxide (SrO) is also considered a modifier, it is generally found as a contaminant from calcium sources in glasses other than television screens, where it is added for its X-ray absorption characteristics. Cerium oxide, CeO_2, has replaced arsenic oxides in recent years as a decolourising agent and can be found in container, tableware and leaded glasses in trace amounts. The use of scrap glass or cullet is widespread in the manufacturing of glass as it lowers the melt temperature and the reuse of this waste glass due to breakage and line failure also increases the yield of the plant. Recycled glass is a source of cullet in some container plants operating in the United States and adds some measure of heterogeneity as trace contamination. This is a desired effect from the forensic scientists' viewpoint because it adds discrimination potential between batches originating from the same plant.

Table 4.1 Average values for the composition of some glasses (as % of element by weight)

Silica	Si	Na	Ca	Mg	K	Al	Fe	Ba	Pb	B
	99.5 (as SiO_2)									
Soda-lime (Float)										
(UK)[1]	34.08	9.50	5.86	2.29	0.66	0.74	0.07			
(US)[2]	33.90	10.24	6.41	2.29	0.11	0.18	0.12	0.01		
(Germany)[2]	33.58	10.14	6.60	2.40	0.13	0.28	0.16	0.12		
(Spain)[2]	33.46	10.24	6.62	2.37	0.21	0.37	0.07			
(Japan)[2]	33.54	9.66	5.80	2.36	0.83	0.90	0.14			
Soda-lime (Container)										
(UK)[1]	33.99	8.16	9.79	0.06	0.42	0.85	0.03			
Flint (US)[1]	33.90	10.21	7.10	0.64	0.43	1.04	0.04	0.03		
Green (US)[1]	33.75	10.29	7.40	0.39	0.46	1.01	0.11	0.04		
Amber (US)[1]	33.49	10.38	7.25	0.45	0.59	1.31	0.15	0.04		
Borosilicate[1]	37.55	3.34	0.07		0.25	1.11	0.05			3.82
Lead crystal[1]	27.39	0.96			10.88		0.01		23.39	0.47

Note: These concentrations reflect the average values for a selected number of plants during the period 1967–1987. The variation between all glass plants and manufacturers has not been well characterised in the literature because of proprietary concerns.
Sources: (1) Personal Communication, G. P. Warman and R. H. Keeley, 1983. (2) Personal Communication, R. R. Bell, 1989.

The measurement of magnesium concentration for the classification of glasses into float and non-float categories has been recognised by a number of workers [1–3]. It is usually possible to classify the float glasses by a magnesium level of about 2 per cent by weight or higher. Figure 4.1 illustrates the bimodal distribution for a population of glass representing a variety of containers and float glasses. It is worth noting the small overlap at the 1.0–1.9 per cent level attributed to four float and two container samples (out of a total of 255 classified glass samples). Magnesium oxide is added to the glass melt in concentrations of up to 4 per cent as a substitute for calcium oxide to act as a stabiliser and to improve the flow properties of the hot glass settling on the molten tin bed of the float process. The addition of magnesia has the added effect of considerably increasing the chemical durability of the glass, an important characteristic for window glass expected to withstand constant exposure to harsh weather conditions. Since even small amounts of iron produce a green colour in glass and it is desirable to make containers of food products and tableware as colourless as possible, great care is taken to eliminate the colour. Iron concentrations have been used [1–3] in classification schemes where coloured containers, both green and brown (or amber), show elevated iron concentrations (up to 0.3 per cent or higher) but colourless containers usually measure much less than 0.1 per cent of iron. A considerable amount of iron (of the order of 0.1 per cent by element weight or higher) is generally tolerated in float glass. Vehicle windows can be characterised as having even greater concentrations of iron than non-vehicle windows made by the float process (over 0.2 per cent) [6] but some coloured architectural glasses can also reach these levels of iron in float glass. For additional information on classification schemes showing differences between containers and

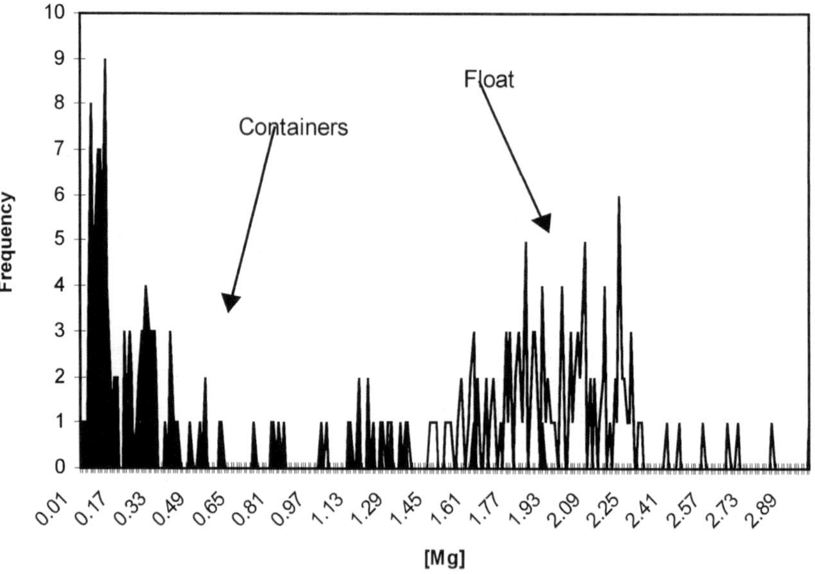

Figure 4.1 Magnesium concentration distribution in 255 container and float glass samples collected from Dade County, FL, USA

float glass, the reader is referred to Hickman's work [1, 3]. He used a combination of refractive index values and the variations in the concentrations of manganese, iron, magnesium, aluminium and barium obtained using ICP-AES. In 1986 Ryland [2] reported a scheme which compared Ca/Mg and Ca/Fe intensity ratios, all values being obtained by SEM/XRF analysis, while Koons *et al.* [4] used ICP-AES to observe the barium, strontium, magnesium, iron, manganese and aluminium concentrations to classify glasses.

Although the classification of glass fragments into categories does play an important role in the analytical scheme of forensic glass samples, the ability to discriminate between glass *of the same type* is more valuable to the forensic scientist. The objective of the glass examiner is first to determine if the glass from the questioned sample found on a suspect, for example, is indistinguishable from a known source of glass at the crime scene (as this would associate the suspect to the crime). Second, he is asked to express to the court an opinion as to the strength or meaning of that association. If, for example, the only measurement that could be performed on both the questioned and the known sample was the assessment of colour by simple visual methods and they were determined to be the same colour, he could state that the glass samples were indistinguishable. Given the fact that many glasses are not distinguished by colour alone, how much weight should be given to that association? If, on the other hand, he were able to measure quantities that he knew to be very discriminating, as is the elemental composition in glass, he could infer that it would be coincidental for a questioned and known sample to match in their composition if those samples did not originate from the same source. Additionally, given enough background data, he could estimate the likelihood for the event of a coincidental match for a set of composition data. This estimate would further assist the examiner to compose his opinion on the value or strength of the association of the questioned glass to the known source of the glass.

The discrimination potential of elemental composition analysis in glass has been documented as early as 1973 by Coleman and Goode [7] in their work using NAA for the quantitative analysis of 25 elements. These workers found that by evaluating the concentrations of 11 elements (Al, As, Ba, Ca, Hf, Mn, Na, Rb, Sb, Sc and Sr), they were able to distinguish all but two pairs of glass samples from a set of 539 samples. In 1976, Reeve *et al.* [8] used energy dispersive X-ray (EDX) spectroscopy for the analysis of 81 window glasses and found that they could distinguish all windows by using this technique, using their match criteria. In 1977, Andrasko and Maehly [9] reported that they were able to distinguish 38 out of 40 windows by EDX. Haney, in the same year, used spark source mass spectrometry (SSMS) to distinguish 25 of 28 windows in one set of samples and 182 of 190 windows in another [10]. Inductively coupled plasma atomic emission spectroscopy (ICP-AES) was reported by Koons *et al.* in 1991 [11] to be able to distinguish 78 out of 81 vehicle windows from a survey of cars found in a junk yard. These workers add that the three samples that were indistinguishable from each other originated from vehicles made by the same manufacturer, in the same year.

The reason for the use of so many different techniques for the analysis and comparison of glass samples is that forensic laboratories must often use the available instrumentation and adapt its use to as many types of evidence analyses as possible. A review of some of the methods used for the elemental analysis of forensic glass samples was reported by Buscaglia [12] who discussed the advantages and disadvantages of the individual analytical techniques.

4.3 Atomic spectroscopy

Atomic emission (AE) and atomic absorption (AA) are two types of atomic spectroscopy which have been applied to the analysis of forensic glass samples [13–16]. These techniques derive analytical information from the ultraviolet, visible and infrared regions of the electromagnetic spectrum. Atomic gases are formed when an aqueous mixture of dissolved inorganic species is aspirated into a hot flame. These techniques depend on the measurement of radiant power of a characteristic line which is emitted, in the case of AE, or absorbed, in the case of AA, by the sample in the flame. For AA, light energy is absorbed when a ground-state atom is elevated to an excited state. A hollow cathode tube generating the specific energy is the source of the radiation. Flame AA is very easy to use, relatively inexpensive, and provides exceptional sample throughput for the analysis of a small number of elements. AA is a well-documented technique [14–16, 22] generally considered best for the rapid analysis of single elements but which can be automated for multi-element analyses. The need for a different lamp for each element and the limited number of elements that can be analysed in a practical sense makes the technique somewhat inflexible and/or time-consuming for forensic work. Another major disadvantage of the technique is its relatively low sensitivity. For certain glass samples, some of the elements of interest are found close to the detection limits of the technique. Table 4.2 illustrates the comparison in detection limits for the atomic spectroscopy techniques and the ICP/MS method. Catterick

Table 4.2 Atomic spectroscopy detection limits for AA, GFAA, ICP/AES and ICP-MS techniques (micrograms/litre)

Element	AA	GFAA	ICP/AE	ICP-MS	Element	AA	GFAA	ICP/AE	ICP-MS
Ag	1.5	0.02	0.9	0.003	Mo	45	0.08	3	0.003
Al	45	0.1	3	0.006	Na	0.3	0.02	3	0.003
As	150	0.2	50	0.006	Nb	1500		10	0.0009
Au	9	0.15	8	0.001	Nd	1500		2	0.002
B	1000	20	0.8	0.09	Ni	6	0.3	5	0.005
Ba	15	0.35	0.09	0.002	Os	120		6	
Be	1.5	0.008	0.08	0.01	P	75000	130	30	0.3
Bi	30	0.25	30	0.0005	Pb	15	0.06	10	0.001
Br				0.2	Pd	30	0.8	3	0.003
C			75	150	Pr	7500		2	<0.005
Ca	1.5	0.01	0.02	0.05	Pt	60	2	10	0.002
Cd	0.8	0.008	1	0.003	Rb	3	0.03	30	0.003
Ce			5	0.0004	Re	750		5	0.0006
Cl				10	Rh	6		5	0.0008
Co	9	0.15	1	0.0009	Ru	100	1	6	0.002
Cr	3	0.03	2	0.02	S			30	70
Cs	15			0.0005	Sb	45	0.15	10	0.001
Cu	1.5	0.1	0.4	0.003	Sc	30		0.2	0.02
Dy	50		2	0.001	Se	100	0.3	50	0.06
Er	60		1	0.0008	Si	90	1	3	0.7
Eu	30		0.2	0.0007	Sm	3000		2	0.001
F				10000	Sn	150	0.2	60	0.002
Fe	5	0.1	2	0.005	Sr	3	0.025	0.03	0.0008
Ga	75		4	0.001	Ta	1500		10	0.0006
Gd	1800		0.9	0.002	Tb	900		2	<0.0005
Ge	300		20	0.003	Te	30	0.4	10	0.01
Hf	300		4	0.0006	Th				<0.0005
Hg	300	0.6	1	0.004	Ti	75	0.35	0.4	0.006
Ho	60		0.4	<0.0005	Tl	15	0.15	30	0.0005
I	0.008			0.008	Tm	15		0.6	<0.0005
In	30		9	<0.0005	U	15000		15	<0.0005
Ir	900	3	5	0.0006	V	60	0.1	0.5	0.002
K	3	0.008	20	0.015	W	1500		8	0.001
La	3000		1	0.0005	Y	75		0.3	0.0009
Li	0.8	0.06	0.3	0.0001	Yb	8		0.3	0.001
Lu	1000		0.2	<0.0005	Zn	1.5	0.1	1	0.003
Mg	0.15	0.004	0.07	0.007	Zr	450		0.7	0.004
Mn	1.5	0.035	0.4	0.002					

Source: A guide to techniques and applications of atomic spectroscopy, Perkin Elmer Inc.

and Wall [16] describe a rapid and accurate method for the analysis of glass samples (250–500 μg) for the measurement of the elements Mg, Mn and Fe using AA. Hughes *et al.* [14] also report an AA method for the determination of As, Fe, Cr, Mg, Mn, Na and Pb, and Tennent *et al.* [15] report the use of AA in the analysis of stained glass samples.

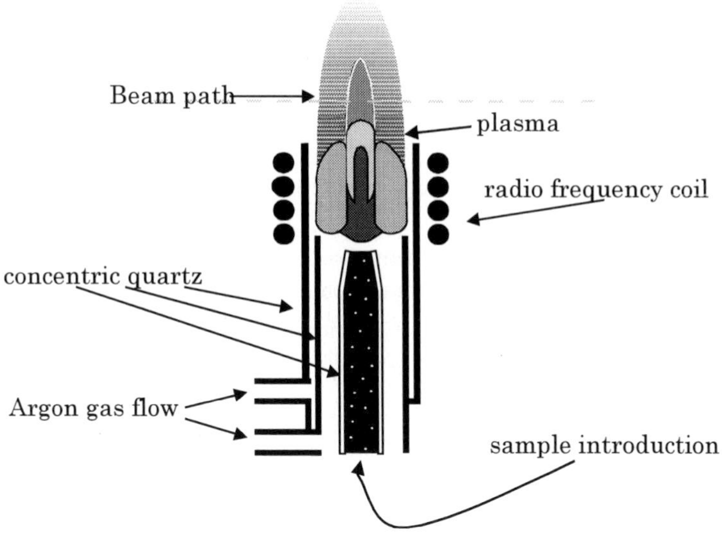

Figure 4.2 Inductively coupled plasma (ICP) torch assembly

Two versions of the emission technique have been reported for use in forensic glass analysis. The emission of light occurs when thermal energy (or electrical energy in the case of arc/spark spectroscopy) is available to excite a free atom or ion to an unstable energy state. The excited species returns to a more stable configuration, or the ground state, by emitting light at a specific wavelength. Simple flames generally do not provide sufficient thermal energy to act as an energy source; hence, arc/spark systems were developed. Other excitation sources used in AE have been direct current (DC) or alternating current (AC) arcs, voltage sparks, glow discharges, lasers, inductively coupled plasmas (ICPs), direct current plasmas (DCPs), capacitively coupled microwave plasmas (CMPs), microwave-induced plasmas (MIPs) and graphite furnaces. Blacklock *et al.* [13] report the use of spark source emission for measuring the concentrations of Al, Ba, Fe, Mg and Mn in glass samples weighing approximately 1 mg.

The development of ICP represented a significant improvement in the use of emission spectroscopy. The use of argon plasma maintained by a radio-frequency field and ionised argon gas produces sufficient energy to excite ions effectively. The plasma reaches temperatures of up to 10,000 K with the sample experiencing useful temperatures between 5000 and 8000 K. At these temperatures, complete atomisation of the elements occurs which minimises the chemical interference effects. The basic ICP torch consists of three concentric quartz tubes through which streams of argon gas flow (Figure 4.2). The sample is introduced into the innermost tube with argon acting as the carrier gas. The outer tube serves to thermally isolate the hot plasma from the outer components of the torch. Argon flows in this outer tube as a tangential stream and radio-frequency (RF) power is applied through a coil surrounding the outer tube, forming an oscillating magnetic field. The plasma is made conductive by injecting electrons from a Tesla coil into an argon mixture, and the resulting ions and their associated electrons interact with the fluctuating magnetic field produced by the induction coil. The ionic species flow in closed annular paths and the resistance caused by this movement results in ohmic heating, producing very high temperatures. For an excellent source of in-depth information about the theory, operation and application of ICP/AES, the reader is referred to Bouman's books [17, 18] on the topic.

The major advantages of an argon plasma over a simple flame as an excitation source are:

- complete atomisation which produces an environment where atomic and ionic species dominate
- spectral lines can be found for elements having widely differing ionisation potentials where they can be detected and measured simultaneously
- the large dynamic range of detection provides for the calibration of the method for samples containing elements present at widely differing concentrations.

Recent advances in sample introduction systems are of particular interest to forensic glass examiners. One notable example is Cetac Corporation's development of a microconcentric nebuliser. This new sample introduction system offers an advantage over pneumatic nebulisers. The self-aspirating, hydrofluoric acid (HF)-resistant nebuliser uses flow rates of 30 μL/min to yield the same detection limits as a pneumatic nebuliser which must operate at flow rates of up to 1000 μL/min. This introduction system reduces the required sample sizes to less than 1 ml. Sample volume is always an important consideration for the preparation of the glass samples normally encountered in casework.

Hickman [1] first reported a scheme for classifying glass samples as sheet, container, tableware or headlamp using ICP/AES. He analysed the quantitative data for the five elements manganese, iron, magnesium, aluminium and barium for 349 glass samples and, by also incorporating refractive index measurements, he was able to correctly classify, using defined criteria, 91 per cent of the glass samples. Koons *et al.* [4] reported an analytical procedure using ICP/AES, applied to the classification of 182 sheet and container glasses. Except for two container glasses from the same manufacturing plant which were classified as sheet glass and two sheet glasses classified as container glasses, the samples were all correctly classified. Wolnik *et al.* [19] reported an ICP/AES procedure for the analysis of container glass. This work, undertaken at the Food and Drug Administration Laboratories, arose from the need for analytical data to interpret the results of glass fragments found in baby-food jars. A procedure was developed to characterise the glass used in the manufacture of the baby-food containers and for the characterisation of the 'adulterant' glass found inside the jars to determine if the 'adulterant' glass had the same source of origin as the containers themselves. Results for the analyses of 89 baby-food containers from five different manufacturers in the USA did not show substantial variation in the concentrations of the major components. However, the concentrations of the trace elements Al, Ba, Ca, Fe, Mg, Na, Sr, Ti and Mn can discriminate between the manufacturers. Table 4.3 provides a summary of the composition data for the five manufacturers. The researchers report on a case where four glass jars containing different products were purchased from the same store but manufactured by two different manufacturers and processed at two different locations. The compositions of the fragments allegedly found by the consumer in all four products were found to be analytically indistinguishable, indicating the same source of origin. In addition, the found fragments were distinguishable from the actual container compositions. There were no other complaints concerning these products and the combined information indicated that the complaint was fraudulent. Koons *et al.* [11] commented on

Table 4.3 Elemental composition of baby-food containers (means, SDs and range in mg/g) (from Wolnik *et al.* [19])

Manufacturer (Location)	Parameter	Al	Ba	Ca	Fe	Mg	Na	Sr	Ti	Mn
Brockway (Jacksonville, FL, USA)	Mean	9.03	1.67	76.3	0.335	0.7	105	0.163	0.293	<DL*
($n = 31$)	SD	0.49	0.07	3.0	0.077	0.159	3	0.04	0.042	–
	Min.	8.38	1.48	68.5	0.231	0.47	97	0.106	0.209	–
	Max.	11.4	1.79	82.4	0.522	1.09	111	0.242	0.388	–
Owens-Illinois (Toledo, OH, USA)	Mean	9.92	0.132	76.3	0.384	2.57	97.7	0.203	0.137	0.053
($n = 24$)	SD	0.9	0.065	3.3	0.147	0.81	4.9	0.051	0.097	0.021
	Min.	6.05	0.035	67.6	0.234	0.76	86.4	0.14	0.053	<DL
	Max.	11.1	0.233	84.2	0.667	4.17	111	0.326	0.304	0.085
Ball (Muncie, IN, USA)	Mean	9.36	4.34	74.5	0.481	0.792	106	0.1	0.044	0.054
($n = 14$)	SD	0.92	0.16	3.5	0.062	0.064	4	0.013	0.008	0.009
	Min.	8.05	3.99	66.9	0.392	0.614	93	0.078	0.033	<DL
	Max.	12.3	4.62	81.4	0.711	0.87	114	0.138	0.065	0.079
Kerr (Los Angeles, CA, USA)	Mean	9.14	0.024	76.6	0.356	2.45	105	0.073	0.169	0.167
($n = 12$)	SD	0.32	0.009	2.8	0.042	0.26	3	0.007	0.011	0.018
	Min.	8.45	<DL	67.7	0.311	2.03	99	0.062	0.151	0.133
	Max.	9.67	0.056	80.6	0.477	2.94	112	0.086	0.193	0.194
Anchor Hocking (Lancaster, OH, USA)	Mean	6.97	0.087	81.8	0.309	1.51	105	0.131	0.284	0.056
($n = 8$)	SD	0.3	0.007	3.7	0.056	0.44	3	0.011	0.044	0.006
	Min.	6.3	0.078	71.2	0.239	1.17	99	0.115	0.225	0.04
	Max.	7.52	0.096	86.5	0.426	2.54	109	0.156	0.353	0.062
Total ($n = 89$)	Min.	6.05	<DL	66.9	0.231	0.614	93	0.062	0.033	<DL
	Max.	11.4	4.62	86.5	0.711	4.17	114	0.362	0.388	0.194

* Less than detection limit (this level varies depending on the dilution factor).

Table 4.4 Frequency of indistinguishability of 81 sheet glass samples taken in pairs (3240 comparisons) (from Koons *et al.* [11])

Comparison parameter and criteria	No. of indistinguishable pairs	Frequency
(1) nD ±0.0002	648	1:5.0
(2) nD ±0.0001	418	1:7.8
(3) (1) and nC ±0.0004 and nF ±0.0004	487	1:6.7
(4) (2) and nC ±0.0002 and nF ±0.0002	178	1:18.2
(5) EDXRF	305	1:10.6
(6) (5) and (3)	81	1:40
(7) (5) and (4)	33	1:98
(8) ICP-AES	3	1:1080
(9) (8) and (3)	3	1:1080
(10) (8) and (4)	2	1:1620

the discriminating capacity of ICP-AES for differentiating 81 vehicle windows. This previously mentioned report compared analytical data for the elements Al, Ba, Ca, Fe, Mg, Mn, Na, Sr and Ti. The same paper compared the discrimination capacity of refractive index alone, and in combination with EDXRF and also with ICP-AES. This work is summarised in Table 4.4.

Two articles by Ducreux-Zappa and Mermet [20, 21] describe the application of the relatively new laser ablation technique coupled to ICP-AES. Laser ablation introduction systems are particularly suited to solid samples such as glass because there is little sample preparation, the analysis is fast and it may be considered a non-destructive technique because of the extremely small amount of sample that is consumed in the analysis. In practice, a laser atomises microscopic segments of the glass surface for introduction into the plasma. Further details of this sample introduction technique can be found in the section on inorganic mass spectroscopy (Section 4.6).

4.4 X-ray methods

X-ray radiation is produced when the inner electrons of an atom are displaced from the atom by a beam of high energy such as an electron or X-ray beam. This causes electrons at higher energy states to relax to the energy state of the originally displaced electron. The differences in the energy between the two electronic states is given off as X-rays. Because the relaxed electron has left vacant another electron 'site', other electrons in higher electronic states relax into these lower energy 'sites'. This is a kind of domino effect which leads to one element emitting X-rays of different energy values at different wavelengths. This technique of X-ray fluorescence is widely used for the qualitative and quantitative analysis of elements having atomic numbers greater than oxygen (>8). Instrumentation to detect the emission can be divided into two types, wavelength dispersive and energy dispersive; the first discriminates the various parts of the spectrum using the wavelength of the emission and the second translates the emission to energy quantities with electronic components which are used as a method of discriminating the energy.

Wavelength dispersive instruments use collimated beams of X-rays as a source of

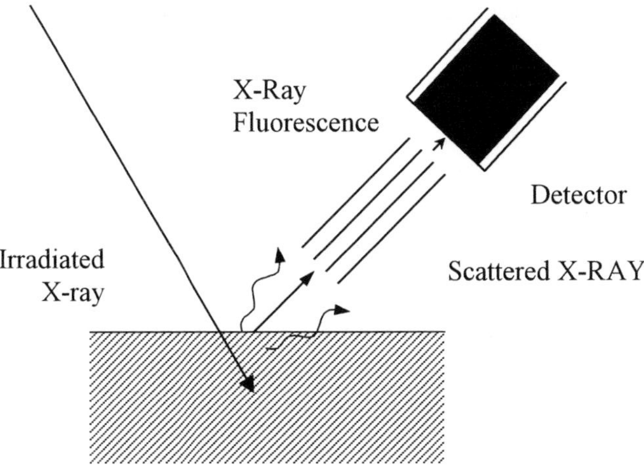

Figure 4.3 Typical XRF detector geometry

energy that is then dispersed into its component wavelengths. Discriminating these wavelengths with a monochromator requires more sophisticated (and expensive) configurations than the energy dispersive instruments (see Figure 4.3 for a diagram of the geometry for the detector components of a simple energy dispersive instrument). The advantages of X-ray fluorescence are that it is non-destructive, spectra are relatively simple and there is little spectral line interference. In addition, small samples can be analysed and the multi-element analysis capability make it a speedy and convenient technique for forensic samples. The main disadvantages are that these methods are not as sensitive as the optical ones, where the detection limits commonly range around the low fraction of a per cent (~0.01 per cent), at best, for most elements. Owing to competing processes such as Auger emission, the detection and measurement of elements with atomic numbers below 23 produce reduced fluorescent intensities. In practice, quantitative analysis of forensic glass samples is best achieved by an evaluation of the ratios of elements rather than by the measurement of absolute concentrations. Very small and irregularly shaped samples (such as are commonly found in forensic casework) are not amenable to this type of analysis. Samples with flat surfaces and known working angles are necessary for quantitative determinations. For a detailed explanation of the theory of both these methods, refer to Skoog and West [22]. Sample preparation involves embedding the glass fragments in a plastic resin and then polishing the surface until it is flat using grinding methods. The surface is usually coated with a carbon layer and the fragment is sampled at different locations. Quantitative analysis with and without standards can be used for the analysis of elements such as Na, Mg, Al, Si, K and Ca assuming the rest of the sample matrix is oxygen.

The broad use of energy dispersive instruments in forensic applications has made them a popular instrument in forensic laboratories. Reeve *et al.* [8] first reported the use of SEM/EDX as a means to further discriminate glass samples that were indistinguishable by refractive index and density only. They reported that out of 81 glass samples analysed, only two were indistinguishable when calcium concentration and the elemental concentration ratios to calcium (Ti/Ca, Mn/Ca, Fe/Ca, Cu/Ca, Zn/Ca, As/Ca, Rb/Ca, Sr/Ca and Zr/Ca) were recorded but when physical properties were also considered, all the glass samples were distinguishable. Andrasko and Maehly [9] later reported on the incorporation of SEM/EDX in an analytical scheme that was able to distinguish all but two of 40 window-glass samples. The scheme included the analysis of refractive index, density, Na/Mg, Na/Al, Mg/Al, Ca/Na and Ca/K concentration ratios. Howden *et al.* [23] reported the ability to standardise the peak area ratios from different XRF instruments in the Forensic Science Service by using a set of 12 reference standard glasses. They reported the use of correction factors to standardise the peak area ratios of Mg/Ca, Si/Ca, K/Ca, Fe/Ca and As/Ca so that the data collected by the different instruments could be related and compared. Ryland [2] reported a classification scheme for sheet vs. container glass samples using the SEM microprobe determination of Ca/Mg intensity ratios and the XRF Ca/Fe concentration ratio.

Total reflection X-ray fluorescence (TXRF) is a relatively new technique designed for surface analysis in the semiconductor industry and has found some application in forensic science. In XRF, the source X-rays irradiate the analytes at high incident angles ($\sim 45°$), and penetrate deep (several hundred micrometres) into the sample. As a result, fluorescence X-rays and scattered X-rays are emitted from the sample. The TXRF technique uses a primary beam with a very low glancing angle ($< 0.1°$) such that the X-rays reflect from the surface and penetrate only the top surface (several tens of Angstroms). This reduces the scattering and improves the signal-to-noise ratio. The basic components of the instrument are very similar to XRF (see Figure 4.4 for a diagram of the geometry of the TXRF detector components) but the sample preparation, data analysis and detector performance are more similar to optical methods such as atomic spectroscopy techniques. A detailed description of the TXRF technique can be found in a paper by KlockenKamper *et al.* [24]. Kubic and Buscaglia [25, 26] reported the application of TXRF and a related technique, elemental X-ray total reflection analysis (EXTRA), to the analysis of small glass samples. They reported the simultaneous analysis of elements ranging from phosphorus to uranium with instrument detection levels less than or equal to 10 pg for 50 of these elements. Little sample preparation is needed for semiquantitative screening while dissolved samples give

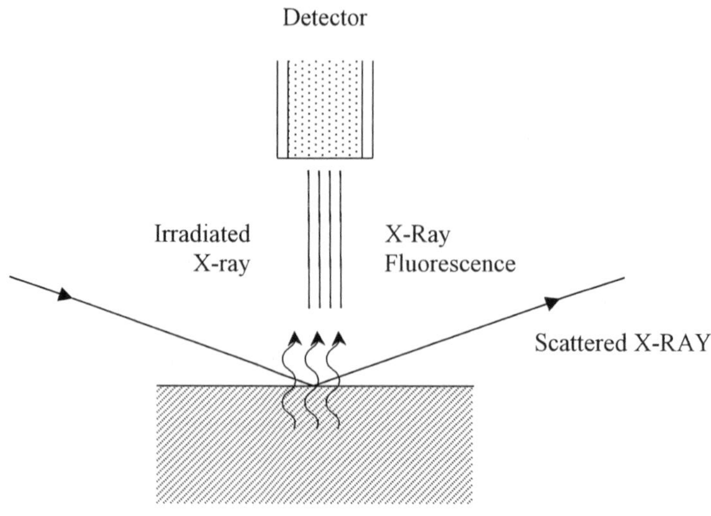

Figure 4.4 Total reflection X-ray fluorescence (TXRF) detector geometry

results comparable to other quantitative analysis techniques and a high dynamic range without the need for multiple dilution.

A promising X-ray technique for the analysis of small, irregularly shaped glass fragments is the μ-beam XRF technique developed in Sweden and reported by Rindby and Nilsson [27]. This technique incorporates a very narrow beam of X-rays focused by conical capillaries (see references 28 and 29 for a description of the capillary optics system). The advent of X-ray capillary optics (XCO) has enabled the development of inexpensive table-top microbeam instruments that have a sensitivity comparable with synchroton instruments using conventional optics. Engström *et al.* [30] report a minimum detectable amount (MDL) of 40 fg of calcium in a paper sample using the microbeam instrument. The capillary configuration allows for maximum transmission at the energy corresponding to the characteristic lines. In addition, the National Forensic Science Laboratory in Sweden has incorporated the linearisation of conventional correlation models to permit the quantitative analysis of irregular topology and surfaces. These workers conclude that the precision, in terms of standard deviation/range, is much higher for the microbeam XRF compared to the conventional SEM measurements. This translates to a lower probability of determining that a randomly selected pair of samples match by the XRF measurement compared to the SEM measurement. A comparison of SEM and microbeam XRF results reveals that in more than 30 per cent of all cases the XRF measurements produce sub-groups within the main groups identified by the SEM measurement [27].

4.5 Radiochemical methods

Neutron activation analysis (NAA) is a very sensitive technique for elemental composition. Concentration determinations in the parts-per-billion range are common. Before the development of the more sensitive optical and X-ray methods, NAA was used to explore the discrimination potential of elemental composition determination of the minor and trace elements in forensic glass samples. The basis for activation analysis is the measurement of the radioactivity induced in a sample as a result of irradiation by nuclear particles, such as neutrons from a nuclear reactor. The lack of accessibility to a nuclear reactor makes this technique out of reach to most forensic laboratories. For a detailed explanation of the principles of NAA, the reader is referred to Kruger's book [31].

Schmitt and Smith [32] were the first to report the application of NAA to a forensic case, while Atalla and Lima [33] used NAA for the analysis and comparison of an automobile windshield in a hit-and-run homicide case. Coleman and Goode's [7] early work (1973) with NAA reports the generation of analytical data from concentration measurements of Al, As, Ba, Ca, Hf, Mn, Na, Rb, Sb, Sc and Sr. These elements were selected because all measurements were shown to be independent of each other, that is, the concentration of one element did not correlate with the concentration of any of the others. For a varied population of glasses these workers found a high degree of discrimination with only two of 539 glass samples being indistinguishable.

4.6 Inorganic mass spectrometry

Inductively coupled plasma-mass spectrometry (ICP-MS) is a combination of the ICP technique with a quadrapole mass spectrometer. This technique takes advantage of the efficient ionising environment of the ICP to simultaneously atomise all the analyte species and the identification power of a mass spectrometer detector where the species can be analysed quantitatively. The mass spectrometer replaces the monochromator in the AA and AES techniques, allowing for the speciation of elements. Rather than separating light according to its wavelength, the mass spectrometer separates the ions introduced from the ICP according to their mass-to-charge ratio. Detectors can quantify the number of ions present with a high degree of sensitivity. ICP-MS combines the multielement capability and the broad dynamic range of ICP emission with the enhanced sensitivity and ability to perform quantitative analyses of the elemental isotopic concentrations and ratios. A diagram (Figure 4.5) illustrates how the basic components of an ICP are coupled to the mass detector. A relatively new introduction system known as laser ablation (LA) is particularly well suited for forensic glass samples because of its non-destructive nature and little sample preparation requirement. This introduction system generates particle aerosols from a solid material by pulsing a laser beam onto the target surface. Ablated material is swept into the ICP-MS by the nebuliser argon gas flow. The source is usually an ultraviolet laser with a high-frequency pulse-to-pulse repetition rate. Increasing the laser repetition rate can maximise the signal intensity. Sample viewing and simultaneous laser focusing are facilitated and controlled through an automated x-y-z translation stage and a high-resolution camera. A diagram (Figure 4.6) illustrates the components of the LA introduction system into the ICP. The advantages of this technique are its ability to analyse small spot sizes ($<10\,\mu m$), the provision for an interface to an ICP-MS (or ICP-AES) which results in all the advantages of an ICP-MS (speed of analysis, multielement capability, high sensitivity, broad dynamic range and isotopic discrimination, as summarised in Table 4.6), minimum sample handling because of its ability to analyse solids directly and to avoid solution-based interference found in quadrupole ICP-MS and fast analysis times.

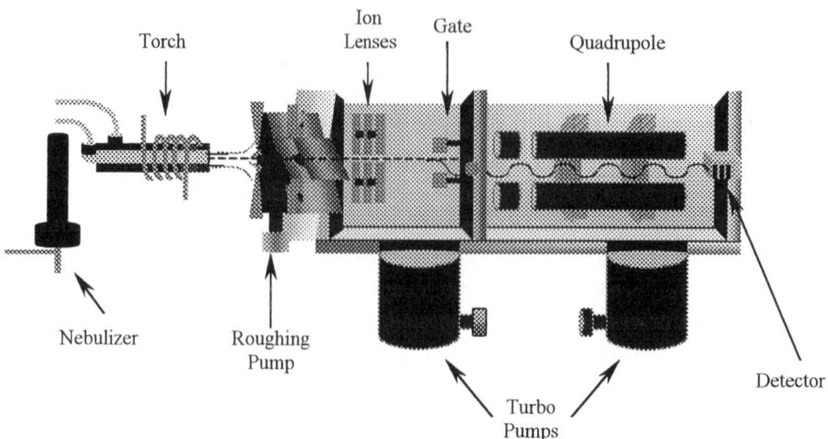

Figure 4.5 Diagram for the components of an inductively coupled plasma mass spectrometer (ICP-MS)

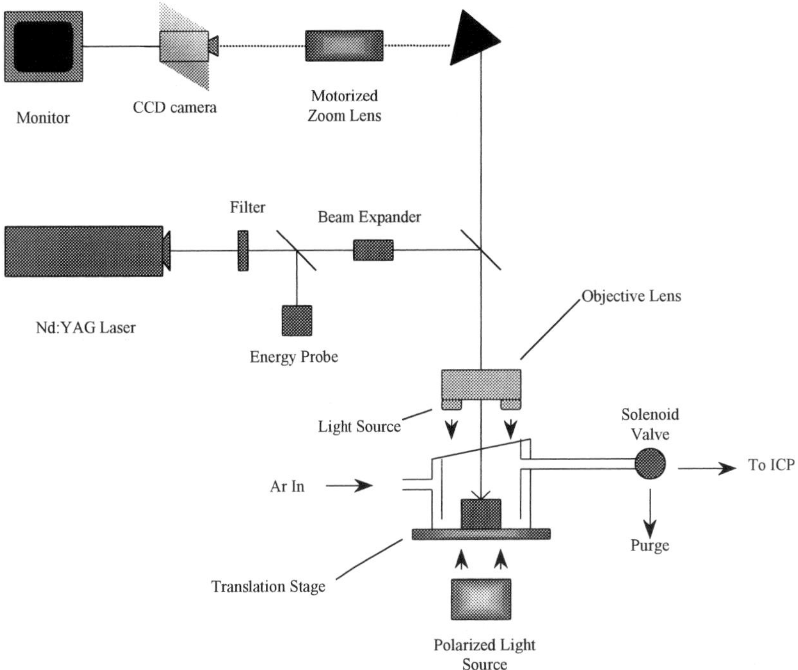

Figure 4.6 Laser ablation schematic diagram for the Cetac LSX-100 system

Parouchais *et al.* [34] reported the application of ICP-MS to glass analysis using solution nebulisation to monitor 62 elements (Table 4.5) in the atomic range from 7 (lithium) to 238 (uranium). These workers used a three-acid digestion procedure ($HF:HNO_3:HCl$, 2:1:1) to minimise the physical, spectral and chemical interferences arising from the high concentrations of dissolved solids and other factors. They reported using elemental ratios to eliminate the requirement for weighing and the ability of this technique to distinguish glass samples that were indistinguishable by refractive index (RI). Suzuki *et al.* [35] reported the discrimination of 22 windshield glass samples by pairwise comparison of the elements Co, Rb, Sr, Zr, Ba, La and Ce. All 231 pairs could be discriminated by composition and RI comparisons while 15 pairs were indistinguishable by RI comparison only. These workers also report the successful discrimination of all 138 pairs of 17 samples of headlight glasses by comparison of the Zr, Ba, Sr, Sb, Hf, As, Mo and Pb concentrations and RI, while 19 pairs could not be discriminated by RI comparison only. These workers used a relatively fast (2 hour), $HF/HClO_4$ microwave digestion procedure for the dissolution of the glass samples. Recent studies by Stoecklein *et al.* [36, 37] and Becker [38] of 45 glass samples from float plants representing the United States, Europe and Asia glass productions produced 990 possible pairwise comparisons. These workers report 100 per cent discrimination by measuring the composition of 30 elements. They used a range overlap criteria (comparing the mean ± 2 standard deviations) to determine discrimination. If the ranges for one or more elements are separate, then the pair is considered distinguishable and if the ranges overlap for every element, the pair is reported as indistinguishable. The mean is calculated from triplicate runs and the standard deviations are those calculated

Table 4.5 Elements monitored for discrimination between glass samples in Parouchais ICP-MS study [34]

Elements in working standards			Additional elements
Li	Y	Eu	Be
Mg	Zr	Gd	B
Al	Pd	Tb	Ca
Ti	Ag	Dy	Sc
V	Cd	Ho	Zn
Cr	In	Er	Ga
Mn	Sn	Tm	Nb
Fe	Sb	Yb	Mo
Co	Te	Hf	Lu
Ni	Cs	Re	Ta
Cu	Ba	Au	W
Ge	La	Tl	Os
As	Ce	Pb	Ir
Se	Pr	Bi	Pt
Rb	Nd	Th	
Sr	Sm	U	

from long-term experiments. They report the most useful elements for discrimination to be Al, K, Ti, Mn, Fe, Rb, Sr, Zr, Sn, Bi, Sb, Eu, Ho and Pb. Other studies in the US by Almirall *et al.* [39] and Duckworth *et al.* [40] involving inter-laboratory validation of the ICP-MS method for glass analysis have also been reported.

An Australian group [41] has also reported the application of LA-ICP-MS to the analysis of forensic glass samples. Preliminary work suggests great promise for the fast and accurate multielement comparison of small samples in a non-destructive manner. Outridge [42] has written on the potential application of LA-ICP-MS to other types of forensic samples, including biological materials and geochemistry. This technique may very well be the discriminating tool for the small samples examined in the forensic laboratory, as the author suggests.

4.7 Conclusions

Table 4.6 summarises the characteristics of the instrumental techniques described in this chapter. Each of the techniques has its advantages and disadvantages and all of the techniques have found an application in forensic science laboratories. The recent developments in ICP-MS and its sample introduction systems, such as laser ablation, show great promise for the application of this technique to forensic glass analysis. In spite of its relatively high cost, ICP-MS should be given careful consideration in a laboratory where the routine elemental analysis of glass fragments is being contemplated.

Following the retrieval of good elemental composition data, there is the task of interpreting the data. Recent attempts at evaluating the multivariate data using traditional statistical tools by Bayne [43] such as principle component analysis and clustering techniques and by Koons *et al.* [44], as well as the use of a Bayesian approach (a continuous approach to the

Table 4.6 Characteristics of the instrumental methods for the elemental analysis of glass fragments

Characteristics	AA	EDX/XRF	ICP-AES	ICP-MS
Detection limit	1 (μg/l)	varies	0.1–1 (μg/l)	<0.01 (μg/l)
Multielement	No	Yes	Yes	Yes
Destructive	Yes	No	Yes	Yes (No w/LA)
Analytical range (order of magnitude)	3.5	6	5.5	8
Sample required	mg range	mg range	μg–mg range	μg range
Sample preparation	Involved	Low	Involved	Involved (Low with LA)
Sample throughput	Low	High	High	High
Cost	Low	Moderate	Moderate	High (Very high w/LA)
Ease of use	Easy	Intermediate	Intermediate	Difficult

multivariate problem) by Curran *et al.* [45, 46], has the potential for quantifying the discrimination power of elemental analysis. It is to be expected that this evaluation will develop to the point where forensic scientists can better communicate to the courts the value of the associations between a questioned glass and a known source.

4.8 References

1. Hickman D. A. (1981). A classification scheme for glass. *Forensic Science International* **17**:265–281.
2. Ryland S. (1986). Sheet or container? Forensic glass comparisons with an emphasis on source classification. *Journal of Forensic Sciences* **31**:1314–1329.
3. Hickman D. A. (1987). Glass types identified by chemical analysis. *Forensic Science International* **33**:23–46.
4. Koons R., Fiedler C., Rawalt R. (1988). Classification and discrimination of sheet and container glasses by inductively coupled plasma-atomic emission spectrometry and pattern recognition. *Journal of Forensic Sciences* **33**:49–67.
5. Almirall J., Cole M., Furton K., Gettinby G. (1996). Characterization of glass evidence by the statistical analysis of their inductively coupled plasma/atomic emission spectroscopy and refractive index data. *Proceedings of The American Academy of Forensic Sciences Meeting, Nashville, TN, February 19–24*, **2**:50.
6. Almirall J., Cole M., Carrasquilla E., Furton K. (1994). The use of GRIM2 and ICP/AES for the analysis of glass evidence. *Presentation to the Southern Association of Forensic Scientists, Orlando, FL.*
7. Coleman R., Goode G. (1973). Comparison of glass fragments by neutron activation analysis. *Journal of Radioanalytical Chemistry* **15**:367–388.
8. Reeve B., Mathiesen J., Fong W. (1976). Elemental analysis by energy dispersive X-ray: a significant factor in the forensic analysis of glass. *Journal of Forensic Sciences* **21**:291–306.
9. Andrasko J., Maehly A. (1978). The discrimination between samples of window glass by combining physical and chemical techniques. *Journal of Forensic Science* **22**:250–262.
10. Haney M. (1977). Comparison of window glasses by isotope dilution spark source mass spectrometry. *Journal of Forensic Sciences* **22**:534–544.
11. Koons R., Peters C., Rebbert P. (1991). Comparison of refractive index, energy dispersive X-ray fluorescence and inductively coupled plasma atomic emission spectrometry for forensic characterization of sheet glass fragments. *Journal of Analytical Atomic Spectrometry* **6**:451–456.

12. Buscaglia J. (1994). Elemental analysis of small glass fragments in forensic science. *Analytica Chimica Acta* **288**:17–24.
13. Blacklock E., Rogers A., Wheals B. (1976). The quantitative analysis of glass by emission spectrography: a six element survey. *Forensic Science* **7**:121–130.
14. Hughes J., Catterick T., Southeard G. (1976). The quantitative analysis of glass by atomic absorption spectroscopy. *Journal of Forensic Science* **8**:217–227.
15. Tennent N., McKenna P., Lo K., Ottaway J. (1984). Major, minor and trace element analysis of medieval stained glass by flame atomic absorption spectrometry. In *Archaeological Chemistry* (ed. Lambert J.), Chapter 8, pp. 133–150.
16. Catterick T., Wall C. (1978). Rapid analysis of small glass fragments by atomic-absorption spectroscopy. *Talanta* **25**:573–577.
17. Boumans P. (1987). *Inductively Coupled Plasma Emission Spectroscopy. Part I. Methodology, Instrumentation and Performance.* John Wiley and Sons, New York.
18. Boumans P. (1987). *Inductively Coupled Plasma Emission Spectroscopy. Part II. Applications and Fundamentals.* John Wiley and Sons, New York.
19. Wolnik K., Gaston C., Fricke F. (1989). Analysis of glass in product tampering investigations by inductively coupled plasma atomic emission spectrometry with a hydrofluoric acid resistant torch. *Journal of Analytical Atomic Spectrometry* **4**:27–31.
20. Ducreux-Zappa M., Mermet J. (1996). Analysis of glass by UV laser ablation inductively coupled plasma atomic emission spectrometry. Part 1. Effects of the laser parameters on the amount of ablated material and the temporal behavior of the signal for different types of laser. *Spectrochimica Acta. Part B: Atomic Spectroscopy* **51**(3):321–332.
21. Ducreux-Zappa M., Mermet J. (1996). Analysis of glass by UV laser ablation inductively coupled plasma atomic emission spectrometry. Part 2. Analytical figures of merit. *Spectrochimica Acta. Part B: Atomic Spectroscopy* **51**(3):333–341.
22. Skoog D., West D. (1980). Principles of Instrumental Analysis (2nd edn). Saunders College/Holt, Rinehart and Winston, Philadelphia, PA.
23. Howden C., Dudley R., Smalldon K. (1977). Standardization of the peak area ratios obtained from the analysis of small glass fragments using energy-dispersive X-ray fluorescence spectrometers. *X-Ray Spectrometry* **10**:98–102.
24. KlockenKamper R., Knoth J., Prange A., Schwenke H. (1992). Total-reflection X-ray fluorescence spectroscopy. *Analytical Chemistry* **64**:1115A–1121A.
25. Kubic T., Buscaglia J. (1996). Total reflection X-ray fluorescence spectrometry (TXRF)–Further application to elemental profiles on trace evidence. *Proceedings of The American Academy of Forensic Sciences Meeting, Nashville, TN, February 19–24*, **2**:50.
26. Kubic T., Buscaglia J. (1996). The application of high pressure scanning electron microscopy (ECO-SEM) to the analysis of forensic samples. *Proceedings of The American Academy of Forensic Sciences Meeting, Nashville, TN, February 19–24*, **2**:50.
27. Rindby A., Nilsson G. (1996). Report on the XRF μ-beam spectrometer at the Swedish National Forensic Science Laboratory, with special emphasis on the analysis of glass fragments. Personal communication.
28. Engström P., Larsson S., Rindby A., Stocklassa B. (1998). *Nuclear Instrumentation and Methodology* **B36**:222.
29. Rindby A., Engström P., Larsson S., Stocklassa B. (1989). *X-ray Spectrometry* **18**:109.
30. Engström P., Larsson S., Rindby A., Stocklassa B. (1989). *NUCL. INSTRUM. and METH.* **B36**:222.
31. Kruger P. (1971). *Principles of Activation Analysis.* Wiley-InterScience, New York.
32. Schmitt R., Smith V. (1970). Identification of origin of glass by neutron activation analysis in a forensic case. *Journal of Forensic Sciences* **15**:252–260.
33. Atalla L., Lima F. (1970). An actual case of glass debris identification by NAA in an automobile accident. *Radiochemical and Radioanalytical Letters* **3**:13–22.
34. Parouchais T., Warner I., Palmer L., Kobus H. (1996). The analysis of small glass fragments using inductively coupled plasma mass spectrometry. *Journal of Forensic Sciences* **41**:351–360.
35. Suzuki Y., Sugita R., Suzuki S., Kishi T. (1997). Application of ICP-MS for the forensic discrimination of glass fragments. *Proceedings of The American Academy of Forensic Sciences Meeting, New York, NY, February 17–22*. **3**:43.

36. Stoecklein W., Kubassek E., Fischer R., Chadzelek A. (1996). The forensic analysis of float-glass characterization of glasses from international sources. Forensic Science Institute, BKA, Trace Evidence Meeting – San Antonio (July)..
37. Stoecklein W., Fischer R., Becker S., Chadzelek A. (1998). In *International Workshop on the Forensic Examination of Trace Evidence*, pp. 71–79 (Tokyo).
38. Becker S., Chadzelek A., Stoecklein W. (1999). Classification of float glasses with respect to their origin by chemometric analysis of elemental concentrations and the use of LA-ICP-MS as a tool in forensic science *Proceedings of European Winter Conference on Plasma Spectrochemistry*, Pau, France, pp. 113–114.
39. Almirall J., Duckworth D., Bouza A., Furton K., Morton S., Bayne C., Koons R. (1999). Discrimination power of ICP-MS in the forensic analysis of glass fragments. *Proceedings of the American Society of Mass Spectrometry Meeting* (June).
40. Duckworth D. C., Bayne C. K., Morton S. J., Almirall J. R. (2000). Analysis of variance in forensic glass analysis by ICP-MS: Variance within the method. *Journal of Analytical and Atomic Spectrometry* **15**(7):821–828.
41. Personal Communication, Alister Ross (1996). National Institute of Forensic Science, March 4.
42. Outridge P. (1996). Potential applications of laser ablation ICP-MS in forensic biology and exploration geochemistry. *Spectroscopy* **11**:21–26.
43. Personal Communication, Charles Baynes (1996). Oak Ridge National Laboratory, April 1.
44. Koons R., Buscaglia J. (1999). The forensic significance of glass composition and refractive index measurements **44**(3):496–503.
45. Curran J., Triggs C., Almirall J., Buckleton., Walsh K. (1997). The interpretation of elemental composition measurements from forensic glass. *Science and Justice, Journal of the Forensic Science Society* **37**(4):241–245.
46. Curran J., Triggs C., Almirall J., Buckleton J., Walsh K. (1997). The interpretation of elemental composition measurements from forensic glass II. *Science and Justice, Journal of the Forensic Science Society* **37**(4):245–249.

5

Statistical interpretation of glass evidence

NIAMH NIC DAÉID

5.1 Introduction

Different methods of data analysis are used in many aspects of forensic science. Mathematical techniques of dealing with experimental data and specifically statistical and probabilistic analysis of data are frequently used in the interpretation of scientific results produced in a forensic context, including those resulting from glass evidence. This may be to measure associations between two variables, to determine standard deviations, calculate confidence intervals, generate accurate calibration curves, or calculate relative risks and probabilities. In applications to forensic science, usually a mixture of both the classical statistical approach and a probabilistic or Bayesian approach involving likelihood ratios is taken. Other types of data interpretation are also used, including techniques such as factor and cluster analysis. The type of analysis applied depends upon various factors such as:

- the type of data generated (for example, refractive index, elemental concentration)
- the questions being asked (for example, how many fragments should be analysed; do two fragments have a common origin?)
- other relevant background information which may have an influence on the evaluation of the scientific results (for example, persistence and distribution of fragments on clothing).

There are a number of questions which should be considered before choosing the type of mathematical approach to apply to a particular situation:

1. If there are several fragments found on a suspect's clothing, how many should be tested in order to be representative?
2. Do all the glass samples recovered come from the same source and is this source the same as that of the control samples?
3. What are the chances of finding several different glass fragments on the clothing of any given person either associated or not associated with criminal activity?
4. What is the probability that an equivalent number of samples of the same glass type (e.g. window) selected from anywhere in the world will have the same refractive indices?

The answers to these questions involve complex statistical analysis. Generally, the sample group (or a statistically correct number of samples from the overall sample group) is compared with a randomly selected control group. Statistical tests have been used to determine if all the samples can be considered as coming from one or more sources. Further statistical tests can be used to determine if the samples are consistent with coming from the same source as the control samples.

5.2 Occurrence of glass on clothing and choosing sample sizes

Research [1–3] has indicated that it is not uncommon to recover glass fragments from clothing of individuals both associated and not associated with criminal activity. Furthermore, it is not uncommon to recover multiple fragments and to recover fragments from multiple sources. This suggests it may be necessary to select a representative subgroup of recovered glass fragments for analytical measurement. Because of this, it may be helpful to have a method for determining the probability of recovering glass fragments which match those of a control glass within the total number of recovered fragments given the analytical results of a subgroup of the total. One type of mathematical expression which may help to accomplish this task is the hypergeometric distribution [4] and its use in determining matching probabilities in respect of glass evidence is illustrated below.

$$P(m/n, M, N) = \frac{({}^{M}C_{m})({}^{N-M}C_{n-m})}{{}^{N}C_{m}} \quad (1)$$

where
N is the total number of recovered fragments,
n is the total number of recovered fragments which are analysed,
M is the total number of matching fragments in N, and
m is the total number of matching fragments in n
and

$${}^{a}C_{b} = \frac{a!}{(a-b)!(b!)} \quad (2)$$

Example 1
Thirteen glass fragments were recovered from the clothing of an individual accused of breaking a specific window. Five fragments were chosen at random and subjected to refractive index (RI) measurement. Of these, three were found to have RI values which indicated that they could have come from the control window.

1. What is the probability that these three fragments represent all the matching fragments within the recovered fragments (i.e. all 13 fragments)?

   $N = 13$
   $M = 3$
   $n = 5$
   $m = 3$

   giving $p = 0.0349$ (3.49 per cent).
2. What is the probability that these three fragments represent 75 per cent of the matching fragments within the recovered fragments?

$N = 13$
$M = 4$
$n = 5$
$m = 3$

giving $p = 0.119$ (11.9 per cent).

3. What is the probability that these three fragments represent 50 per cent of the matching fragments within the recovered fragments?

$N = 13$
$M = 6$
$n = 5$
$m = 3$

giving $p = 0.3264$ (32.64 per cent).

This type of approach provides some crude measure as to the value of taking a specific number of fragments for analysis but one major flaw is that it does not take into account any chance matching between control and recovered fragments. Various modifications to this approach have been suggested to take into account the likelihood of chance matches with the control fragment and any other fragment [5].

In reality it may be more practical to choose the fragments for analysis on the basis of their distribution on clothing given the alleged event or activity. In an approach such as this it is important to choose fragments randomly from all the various locations where fragments are found on the submitted items.

5.3 Grouping glass fragments

Assuming that RI measurements have been made on a set of recovered fragments, the next task is to attempt to group these in terms of their refractive index values. There are various ways in which this can be achieved: comparing the values and grouping them visually, (either by eye or by some graphical method) or through the application of statistical tests. This is illustrated below in Example 2.

Example 2
Given glass RI values can these fragments be considered as one statistically similar group of fragments or are there more than one grouping of fragments present?:

RI
1.52548
1.52552
1.52555
1.52546
1.52544
1.52555
1.52556
1.52555
1.52556
1.52554

First the data are ranked in ascending order and the median (middle) value determined. In this case the median value is found to be 1.525545.

Next a test criteria is devised. One proposed by Evett and Lambert [6–10] suggests that, if the standard deviation of the glass multiplied by some pre-decided critical limit value is greater than the range of refractive index values of two test pieces of glass, then the fragment with the RI furthest from the median can be excluded and the remaining group tested again. The value of 4×10^{-5} is used as an approximation of the standard deviation regardless of the number of fragments recovered.

From normal distribution tables [11] the critical value at a 95 per cent level of confidence is 3.92. The test criteria for inclusion of fragments in one group at a 95 per cent confidence level is therefore:

$$4 \times 10^{-5} \times 3.92 > \text{range of RI}$$

Starting with the values closest to the median the RI range is:

$$1.568 \times 10^{-4} > 1 \times 10^{-5}\ (=1.52555 - 1.52554)$$

This means that these two fragments can be grouped together. In fact all of these fragments form one group using this method. If this were not the case then the fragment with the value furthest from the median is removed from the group and the grouping repeated.

Once the analysed glass fragments have been grouped using whatever chosen method, comparative statistical analysis between the control glass and the suspect glass can be carried out.

5.4 Classical statistical approach

For glass refractive index measurements the test statistic which is proposed as most appropriate is a weighted *t*-test [12]:

$$t = \frac{\sqrt{\dfrac{n \times m}{n} + m}\,(\bar{x} - \bar{y})}{\sqrt{\dfrac{(n-1)s_x^{\,2} + (m-1)s_y^{\,2}}{(n+m-2)}}} \qquad (3)$$

where
n and m are the number of glass fragments in each group respectively,
s_x and s_y are their respective standard deviations, and
$\bar{x}$ and $\bar{y}$ are their respective mean values.

If the calculated test statistic is greater than or equal to the tabulated test statistic then the null hypothesis is rejected and there is a significant difference between the data.

$$(t_{calc} \geq t_{tab} = \text{significant difference})$$

Example 3
Given the following data of RI measurements taken from fragments of glass from two control windows and those recovered from the clothing of an accused person, is it possible to state that the recovered samples came from the same source as either of the control groups?

	Control 1 = m_1	Control 2 = m_2	Recovered = n
	1.51813	1.51822	1.51818
	1.51811	1.51825	1.51819
	1.51810	1.51815	1.51821
	1.51812	1.51817	1.51817
	1.51813	1.51814	1.51818
Mean	1.518118	1.518186	1.518188
s.d.	1.304×10^{-5}	4.722×10^{-5}	1.483×10^{-5}

Using the weighted t-test statistic:

For recovered sample and control 1 $t_{calc} = 7.926$
For recovered sample and control 2 $t_{calc} = 0.09$

At the 95 per cent level the tabulated value of t with $(n + m - 2)$ degrees of freedom is 2.30. This means that at the 95 per cent level there is no significant difference between the recovered sample and control group 2 ($t_{calc} < t_{tab}$) but there is a significant difference between the recovered sample and control group 1 ($t_{calc} \geq t_{tab}$).

This suggests that the recovered sample and control 2 could have a common origin. This does not mean that they definitively originated from the same source but that they could have come from the same source or any other similar source.

5.5 Coincidence probabilities

The question of coincidence probabilities is more difficult to resolve and requires the accumulation of large amounts of data on refractive index values of numerous glass fragments of different types (float glass, container glass, vehicle glass, etc). These values can be used and compared with the samples recovered to establish a coincidence probability.

5.5.1 Databases

In order to attempt to assess the range of RI values and frequencies with which they occur in any given population it is necessary to use some sort of database. There are various databases which exist worldwide, two of which are held by the Federal Bureau of Investigation (FBI) in the United States and the Forensic Science Service (FSS) in the United Kingdom. The FBI database contains RI measurements of samples from 2029 cases between 1978 and 1990 and the FSS database contains approximately 8950 RI measurements from case samples between 1977 and 1983. The differences in RI values between the two has been attributed to the wider and older range of glass sources available in the UK over the past 50 years as well as differences in manufacturing processes and raw materials. Despite this, in both cases the majority of casework samples fall into the refractive index range of 1.5153 to 1.5193.

5.5.2 Probability

Probability theory allows the calculation of the 'chance' of a result occurring and is the proportion of occurrences of that sample point in repeated experiments, denoted $P(x)$. This value lies between 0 and 1, where 1 means that the sample point always occurs and 0 means that the sample point never occurs. In determining the coincidence probability of a match between two sets of glass fragments, in other words the chances of a random person having the same number of glass fragments on their clothing which statistically match the control, probabilistic theories can be applied. The Bayesian approach is best suited to forensic evidence and a full explanation is given in reference [13].

5.6 Bayesian approach

5.6.1 Baye's theorem

For any two events E_1 and E_2, the probability of event 1 happening given that event 2 happens is expressed as:

$$P(E_1|E_2) = P(E_1)\frac{P(E_2|E_1)}{P(E_2)} \tag{4}$$

This links the probability of event 1 happening with the probability that event 2 happens given event 1 happens. Baye's theorem is sometimes written in an odds form where:

$$\text{Odds ratio} = \frac{P(\text{event happens})}{P(\text{event doesn't happen})} = \frac{P(\text{event})}{P(\text{not event})}$$

Baye's theorem becomes

$$\frac{P(E_1|E_2)}{P(\text{not } E_1|E_2)} = \frac{P(E_1)}{P(\text{not } E_1)} \times \frac{P(E_2|E_1)}{P(E_2|\text{not } E_1)} \tag{5}$$

odds for event 1 occurring = prior odds × likelihood ratio = posterior odds

For our discussions, the likelihood ratio is given by the probability that glass fragments occur on the suspect's clothing given that the suspect broke the window divided by the probability that glass fragments occur on the suspect's clothing given that someone else broke the window. We need to determine a numerator and denominator of this expression for various different cases.

5.6.2 Single recovered fragment [14–16]

Denominator

Let b_i represent the probability that a random person will have glass fragments on their clothing at any given time such that $i = 1$ represents one fragment, $i = 2$ represents two fragments and so on.

Then b_1 is the probability that a person has one fragment of glass on their clothing at ant given time and has a probability density function of p^1. So for a fragment of refractive index, y, the denominator of the likelihood ratio can be expressed as:

$$p(E_2/\text{not } E_1) = b_1 p^1(y) \tag{6}$$

Where E_1 is the control window being broken and E_2 represents that there is one fragment of glass on the clothing. Equation (6) then represents the probability of finding glass of refractive index y on the suspect's clothing given that the control window was not broken.

Numerator
Here there are two options: either the fragment was on the clothing before the event occurred and no fragments were transferred; or one glass fragment was transferred from the window and none was present on the clothing beforehand.

Let t_i be the probability that i fragments would have been transferred from the control source where $i = 0, 1, 2, \ldots$ (note: this does not take any account of fragment persistence or ability to recover that fragment).

Let $p(y/\mathrm{Cx})$ be the probability of the refractive index measurement being y given the measurements made from the control sample.

Then the numerator can be expressed as

$$P(\mathrm{E}_2/\mathrm{E}_1) = t_0 b_1 p^1(y) + t_1 b_0 p(y/\mathrm{Cx}) \tag{7}$$

which represents the probability of the glass fragment being on the clothing given that the control window was broken.

Thus, for one fragment on the clothing of a suspect the likelihood ratio is expressed as

$$\mathrm{LR} = \frac{t_0 b_1 p^1(y) + t_1 b_0 p(y/\mathrm{Cx})}{b_1 p^1(y)} \tag{8}$$

5.6.3 Two recovered fragments

Denominator
The denominator is essentially the same as in the previous case with $i = 2$, the refractive indices of both fragments being represented collectively by $y = (y_1, y_2)$. Then the denominator of the likelihood ratio can be expressed as

$$p(\mathrm{E}_2/\text{not } \mathrm{E}_1) = b_2 p^2(y) \tag{9}$$

Numerator
Here there are four options: no fragments were transferred and there were two already present, fragment with $\mathrm{RI} = y_1$ was transferred while fragment with $\mathrm{RI} = y_2$ was already present, fragment with $\mathrm{RI} = y_2$ was transferred while fragment with $\mathrm{RI} = y_1$ was already present or both fragments were transferred and none was already present.

This gives the expression:

$$P(\mathrm{E}_2/\mathrm{E}_1) = t_0 b_2 p^2(y) + t_1 b_1 p(y_1/\mathrm{Cx}) p^1(y_2) + t_1 b_1 p(y_2/\mathrm{Cx}) p^1(y_1) + t_2 b_0 p(y/\mathrm{Cx}) \tag{10}$$

and the likelihood ratio becomes:

$$\mathrm{LR} = \frac{t_0 + t_1 b_1 [p(y_1/Cx) p^1(y_2) + p(y_2/Cx) p^1(y_1)] + t_2 b_0 p(y/Cx)}{b_2 p^2(y)} \tag{11}$$

The calculation of the probabilities and parameters involved in these equations is possible through the use of the available databases; however, various assumptions must also be made which restrict the use of this approach for casework.

A practical approach to the use of Bayesian statistics has been described by Evett and Buckleton [17]. This uses data relating to the presence of glass on non-crime clothing [2] and includes the following assumptions:

1. Some criteria have been used to group the recovered fragments.

2. There is a criterion for deciding that two groups of fragments match.
3. There is some means for testing the frequency of occurrence of groups of glass of differing RI values in the population of non-crime clothing.

The probability of a suspect having groups of fragments on their clothing is p_i ($i = 1, 2, 3, \ldots$), the probability that these groups contain more than three fragments (i.e. a large group) is s_L, the frequency of occurrence of this RI on clothing (from the database) is f_i ($I = 1, 2, \ldots$), the probability of no transfer is T_0 and of large group transfer is T_L. All can be calculated from existing data [1, 2] and used to determine the likelihood ratios.

Example 4

A suspect is observed breaking a window, arrested 30 minutes after the event, their garments examined and four glass fragments found. The RI values show the fragments to be of one group which match the control window and which occur in 3 per cent of the database. What is the likelihood that the recovered fragments came from the control window?

The denominator will be the probability that a person selected at random would have a large group of fragments on his clothing with the desired RI value:

$$P(\mathrm{E}_2/\text{not }\mathrm{E}_1) = p_1 s_L f_1 \tag{12}$$

The numerator takes into account that the suspect could have had either no glass on his clothing before the incident and that the group was transferred from the control window or that the glass was present on the clothing prior to the incident:

$$P(\mathrm{E}_2/\mathrm{E}_1) = p_0 T_L + p_1 s_L f_1 T_0 \tag{13}$$

And the likelihood ratio is given by:

$$\mathrm{LR} = p_0 T_L / p_1 s_L f_1 + T_0 \tag{14}$$

Example 5

The same as above but there are two groups of four fragments each found on the suspect. The denominator will be the probability that a person selected at random would have two large group of fragments on his clothing with the desired RI value

$$P(\mathrm{E}_2/\text{not }\mathrm{E}_1) = 2 p_2 s_L f_1 \tag{15}$$

The numerator takes into account the fact that the suspect could have had four non-matching fragments and four matching fragments present and none transferred or four non-matching fragments present and four matching fragments transferred:

$$P(\mathrm{E}_2/\mathrm{E}_1) = p_1 T_L + p_2 s_L f_1 T_0 \tag{16}$$

And the likelihood ratio is given by

$$\mathrm{LR} = p_1 T_L / 2 p_2 s_L f_1 + \tfrac{1}{2} T_0 \tag{17}$$

Similar equations can be derived using this approach for cases where two windows are broken and controls from one window matches the group of four fragments:

$$\mathrm{LR} = p_0 T_0 T_L / p_1 s_L f_1 + T_0^2 \tag{18}$$

and for cases where two windows are broken and controls from both windows each match one group of four fragments from the suspect:

$$\mathrm{LR} = p_0 T_L^2 / 2 p_2 s_L^2 f_1 f_2 + p_1 T_0 T_L^2 / 2 p_1 s_L f_1 + p_2 T_0 \mathrm{T}_L^2 / 2 p_2 s_L f_2 + T_0^2 \tag{19}$$

The final terms in these expressions can be ignored without any significant loss of accuracy giving expressions for the likelihood ratios for each example as:

Scenario 1 (one recovered group, one window)	$p_0T_L/p_1s_Lf_1$
Scenario 2 (two recovered groups, one window)	$p_1T_L/2p_2s_Lf_1$
Scenario 3 (one recovered group, two windows)	$p_0T_0T_L/p_1s_Lf_1$
Scenario 4 (two recovered groups, two windows)	$p_0T_L^2/2p_2s_L^2f_1f_2$

Using existing databases numerical values may be given to these variables. The various parameters quoted by Evett and Buckleton [17] are:

P_0 = 0.636 the probability of a person having no glass on their clothing.
P_1 = 0.238 the probability of a person having one group of fragments on their clothing.
P_2 = 0.087 the probability of a person having two groups of fragments on their clothing.
s_L = 0.029 the probability that a group of fragments found on a person's clothing is large.
f_1 = 0.03 the frequency of occurrence of glass with the RI of group 1 on clothing.
f_2 = 0.03 the frequency of occurrence of glass with the RI of group 2 on clothing.
T_0 = 0.2 the probability of no glass transferred, retained and found.
T_L = 0.6 the probability of a large group of glass transferred, retained and found.

When values for the likelihood ratios are calculated the following is obtained:

	1 group	2 groups
1 window	1843 (Scenario 1)	943 (Scenario 2)
2 windows	368 (Scenario 3)	1,738,000 (Scenario 4)

This means that if eight fragments of glass are recovered from a suspect and they are found to be two groups of four fragments each where one group matches one control window and the second group matches the second control window, then the likelihood ratio for this to occur is 1,783,000.

5.7 Elemental analysis

Since narrow limits are set for the main components used in the manufacturing process of float glass, the ability to discriminate between such glass samples by refractive index alone may be considerably reduced. In recent years the improvement in the quality of glass production has permitted control of the physical and optical properties of glass to a high degree. This has caused a decrease in variation of refractive indices for the population of float glass. It has become necessary to develop a reliable chemical analysis for its discrimination.

In order to aid the discrimination between microtraces of glass with the same or similar refractive index values, elemental analysis can be carried out. This is most often achieved using techniques such as ICP-MS, energy dispersive X-ray fluorescence (XRF), scanning electron microscopy (SEM) or neutron activation analysis (NAA). Many of these techniques are capable of multielement analysis of the samples. Once the elemental concentration is determined, the analytical results are subjected to chemometric analysis using procedures such as factor and cluster analysis [18, 19].

5.7.1 Factor analysis, principal component analysis and cluster analysis

These are all techniques designed to summarise data in terms of functions or combinations of variables which reveal patterns within the data. Principal component analysis is designed to reduce a large number of components to a smaller number of principal components which are ordered to the extent to which they occur within the data set with the theory that these will be responsible for most of the variation within those data. Factor analysis is similar in that it attempts to group correlated variables together into 'factors'. Cluster analysis is used to summarise the information contained within large sets of data by attempting to fit together items of a data set which best fit a specifically defined cluster. Such data analyses, because of their complexity, are carried out by computer programs such as Minitab or SAS. This type of approach to elemental glass evidence is proving useful in grouping and clustering fragments together.

5.8 Conclusion

Glass, because of its ubiquitous nature, is a common feature in forensic casework and as a consequence much has appeared in the literature regarding glass as evidence and also the interpretation of that evidence. Like many areas of forensic science, glass evidence is being subjected to the application of the Bayesian approach to evidence interpretation and this is proving useful in providing answers with regards to coincidence probabilities of glass. Over the years the production and manufacture of glass has improved and as a result the discriminating power of refractive index measurements may be affected. Currently, and in the future, developments in factor analysis and chemometric interpretation of the elemental results of glass analysis may become more and more common.

5.9 References

1. Lambert J. A., Satterwhite M. J., Harrison P. H. (1995). A survey of glass fragments recovered from clothing of persons suspected of involvement in crime. *Science and Justice* **35**:273–281.
2. McQuillan J., Edgar K. (1992). A survey of the distribution of glass on clothing. *Journal of Forensic Sciences* **32**:333–348.
3. Hicks T., Vanina R., Margot P. (1996). Transfer and persistence of glass fragments on garments. *Science and Justice* **36**:101–107.
4. Bates J. W., Lambert J. A. (1991). The use of the hypergeometric distribution for sampling in forensic glass comparison. *Journal of Forensic Sciences* **31**:449–455.
5. Curran J. M., Triggs C. M., Buckleton J. (1998). Sampling in forensic comparison problems. *Science and Justice* **38**:101–107.
6. Evett I. W. (1977). The interpretation of refractive index measurements. *Forensic Science International* **9**:209–217.
7. Evett I. W. (1978). The interpretation of refractive index measurements II. *Forensic Science International* **12**:37–47.
8. Evett I. W., Lambert J. A. (1982). The interpretation of refractive index measurements III. *Forensic Science International* **20**:237–245.
9. Evett I. W., Lambert J. A. (1984). The interpretation of refractive index measurements IV. *Forensic Science International* **24**:149–163.
10. Evett I. W., Lambert J. A. (1985). The interpretation of refractive index measurements V. *Forensic Science International* **27**:97–110.

11. Miller J. C., Miller J. N. (1994). *Statistics for Analytical Chemistry*. Ellis Horwood Ltd, Chichester, UK.
12. Evett I. W., Lambert J. A. (1983). The interpretation of refractive index measurements. Estimation of coincidence probabilities. CRE Report 477, Home Office.
13. Aitken C. G. G., Stoney D. A. (1991). *The Use of Statistics in Forensic Science*. Ellis Horwood Ltd, Chichester, UK.
14. Evett I. W., Buckleton J. S. (1989). The use of the Bayesian approach to evidence evaluation. *Journal of Forensic Sciences* **29**:317–324.
15. Evett I. W., Lambert J. A., Buckleton J. S. (1995). Further observations on glass evidence interpretation. *Science and Justice* **35**:283–289.
16. Buckleton J. S., Evett I. W., Weir B. S. (1998). Setting bounds for the likelihood ratio when multiple hypothesis are postulated. *Science and Justice* **38**:23–26.
17. Evett I. W., Buckleton J. S. (1990). *Journal of Forensic Sciences* **30**:215–223.
18. Curran J. M., Triggs C. M., Almirall J. R., Buckleton J. S., Walsh K. A. J. (1997). The interpretation of elemental composition measurements from forensic glass evidence. *Science and Justice* **37**:241–244.
19. Almirall J. R., Cole M. D., Gettinby G., Furton K. G. (1998). Discrimination of glass sources using elemental composition and refractive index: development of predictive models. *Science and Justice* **38**:93–100.

6

Interpretation of physical aspects of glass evidence

JOHN I. THORNTON

6.1 Glass as physical evidence

Glass occurs with moderate frequency as physical evidence. The ubiquity of glass in our everyday environment – in architectural situations, in automobile windows, in beverage bottles and other liquid containers, and in incandescent light bulbs – results in many situations in which evidence and exemplar samples of glass are subjected to forensic examination. From a forensic standpoint, the presiding property of glass is its susceptibility to breakage. The glass may be broken purposefully, as in the case of a forcible entry into a building, or it may be broken inadvertently incidental to a violent struggle. In either event, glass found with a suspect may assist in associating the suspect with a particular scene in which glass was broken. In less common situations, glass at a crime scene may provide other types of information, such as the sequence of gunshots, or whether a petroleum accelerant was used in the commission of arson.

Most forensic issues involving glass evidence are addressed through the application of not just one approach, but rather an ensemble of methods. Optical methods may involve the determination of refractive index and dispersion. Chemical methods may determine the elemental signature of a specimen of glass. Physical methods may address the following considerations:

- An evidence specimen of glass may be determined as having come from a particular type of object, e.g. a beverage bottle, a flat window pane, or an eyeglass lens.
- A fragment or shard of glass may be established as having come from a particular glass object that has been broken, or from a particular region of a glass object.
- A fracture may be determined as being consistent with impact, or with thermal stress.
- The origin of a fracture may be established, along with the direction of force and relative amount of energy responsible for the fracture. In some instances, the order of multiple fractures may be determined.

6.2 Recovery of evidence glass fragments

The recovery of glass fragments as evidence is usually straightforward and poses no particular problems. Conspicuous particles may be picked off with forceps, either with the unaided eye or with help of a stereoscopic binocular microscope. If a rigorous search for glass in clothing is to be made, however, a visual search will not be sufficient; vacuuming of the clothing will need to be conducted. This is a simple procedure involving a trap in a vacuum line that will capture fragments of less than a millimetre. The procedure will of course capture glass as well other trace evidence, e.g. hair, fibres, paint and soil. The glass may then be separated from other detritus recovered by the vacuum trap by picking it out under the stereo microscope. At the same time, the glass may be given a preliminary sorting on the basis of colour and thickness. In some instances, the fluorescence that is generally associated with glass produced by the tin float method may be of assistance in separating glasses of different origins. Glass is isotropic, and will appear dark in the field of a polarising microscope with crossed polars; this provides a quick and easy test to discriminate between glass and most other translucent materials that might be confused with glass, whether coloured or not.

6.3 Physical matches of fractured glass

In many instances, it is possible to establish that two pieces of glass were originally joined. This is accomplished usually by means of a 'physical match', i.e. a palpable match of irregular surfaces, although there are other, less frequently employed methods as well. These other methods include the correspondence of conchoidal or ribmarks or of hackle marks seen under the microscope, the continuity of ream or cord as shown by shadowgraphs, and irregularities in surface topography as shown by laser interferometry.

6.3.1 Significance of physical matches

The most conclusive evidence of a commonality of source in any evidence material is a 'physical match', that is, an exact fit of the broken edges of two pieces of evidence. In fact, a physical match is generally considered to be the zenith of all forensic identifications.

Glass is an amorphous solid, with no definite structure and with no favoured cleavage as determined by a crystalline lattice. A fracture is a rupture of atomic bonds. In glass the atoms themselves are arranged in no particular order, and the fracture is therefore between atoms that are positioned uniquely in the glass. In addition, in glass, as with any other brittle solid, electrons in the outermost orbitals of a particular atom are held loosely. With the close atomic packing of a solid, some electrons may be associated now with one atom, but an instant later with another nucleus, and still later with another nucleus. There is therefore an electron cloud that meanders through the glass, with areas of relative strength and areas of relative weakness that are changing continually. These phenomena together make it essentially impossible for fractures in different pieces of glass and at different times to be replicated exactly. No realistic probability model exists to associate a physical match with a probability of a chance replication, but there is universal acceptance of the uniqueness of a match of fracture surfaces, even over a short distance.

6.3.1.1 Demonstrating a physical match

If the broken edges of two pieces of glass can be fitted together exactly, the conditions of a physical match have been achieved. Since almost any glass likely to be encountered probably has a thickness of at least a few millimetres, this physical match is in three dimensions. To illustrate the match of fracture margins for purposes of court presentation, typically only the two-dimensional match of irregular fracture margins is depicted. In many instances where the fracture margins are irregular, the demonstration of the match is trivial. But in other instances it is difficult to convey the match convincingly by means of photography, even though the 'feel' of the 'fit' leaves no doubt in the mind of the examiner that a physical match exists. In some instances this is because the fracture is along clear straight lines and consequently possesses little apparent character. In other instances involving small flakes of glass that have been separated from a larger piece, the surface that matches is often convex on the flake and concave on the parent piece of glass. This renders photography of the match problematic. O'Hara and Osterburg [1] suggested photography by oblique lighting and the superimposition of fracture surfaces by means of transparencies. Radley [2] suggested a similar approach, but used photographic prints positioned adjacent to one another instead of transparencies. Stapleton [3] prepared a replica of the parent glass and compared the replica with the small paint flake. Nelson [4] advocated the photographic comparison, not of the fracture margins of the glass, but of the hackle lines present in the parent and daughter samples.

6.3.2 Demonstration of continuity of ream or cord

Cord is defined by ASTM Standard C162-52 as 'an attenuated glassy inclusion possessing optical and other properties differing from the parent glass'. When it occurs in sheet glass, it is more often referred to as ream; in containers, it is generally termed cord. Whether cord is present in glass to the extent that it may be visualised is largely dependent upon the care with which the glass is manufactured, as cord is attributed to minute differences in chemical composition. Rarely will cord be readily apparent to the eye; special techniques will be required to demonstrate it; although it is nearly always present in container glass, in sheet glass it may be significantly less pronounced. The subject of cord is discussed extensively in a five-part article by Knight [5] in a review that, while dated, is still the governing work on the subject.

In drawn sheet glass, the ream manifests itself in the form of linear striae. Apparently no background information is available from the glass manufacturing industry as to the changes to be expected over time, but indisputably the striae are persistent over a short production interval and certainly within a particular window pane.

The ream of sheet glass and cord of containers was used by von Bremen [6] to show a commonality of source, even in instances in which non-contiguous fragments were compared. An intense point source (0.8 mm) of light was suspended over high-contrast black-and-white film at a distance of 2 to 4 m. The sample of glass in question was suspended above the film plane at a distance of 8 to 90 cm. The optimum distance of glass and light source must be determined in each case by experimentation. When the film is developed after exposure of up to 20 seconds, the cord or ream is apparent on the photographic negative. von Bremen has used this technique to show continuity in ream and cord in sheet glass, in bottles, and in automotive light bulbs. This is a very powerful means of matching glass. Unfortunately, however, the technique is underutilised in most forensic laboratories as the cord or ream is visualised only after the film is developed, and considerable experimentation is required to establish the optimum distance of the source, the distance of the glass above the film plane, and the exposure time. When undramatic results are obtained, it is not clear whether the experimental conditions are less than optimum or whether the glass is inherently devoid of significant cord.

Welch *et al.* [7] successfully demonstrated what appears to be cord by means of phase contrast microscopy. The banding seen by Welch and co-workers is at right angles to the cord described by von Bremen. While this apparent contradiction has not been resolved, it would appear that the phase contrast method offers a benefit over the von Bremen method in terms of ease of analysis.

6.3.3 Demonstration of surface topography

Thornton and Cashman [8] have shown that contiguous and in some instances non-contiguous samples of glass can be matched on the basis of surface topography by means of laser beam interferometry. When a beam of monochromatic light is divided into two separate beams that travel different paths and are then recombined, interference may occur and interference fringes observed. These fringes, generally referred to as Fizeau fringes, when applied to problems of glass comparison may demonstrate either a difference in refractive index, or a difference in the thickness of the glass, or both. The continuity of the Fizeau fringes will establish a commonality of source between two or more fragments.

A 5 watt argon ion laser delivering monochromatic energy at 514.5 nm is passed through a negative lens to spread the beam and enlarge the area to be examined. The sample of glass in question is placed at a distance of 75 cm from the diverging lens, and at an angle of 15° from the axis of the laser beam. Any interference figure will then be reflected backwards, and will be seen on a projection screen placed at a distance of 275 cm from the glass specimen.

The appeal of this technique is that the Fizeau fringes can be observed immediately and manipulated when the laser is powered down to 1 or 2 watts; it is not necessary to photograph the fringes and wait until the film is developed to determine whether the experimental design is optimised or if there are fringes to be seen. By using the interferometry in a reflectance rather than in a transmittance mode, the fringes are not washed out by the intensity of the illumination beam superimposed on the Fizeau fringes.

This technique is particularly suitable for the comparison of pieces of glass that have been cut by means of a glass cutter. Even a few angstroms of difference in thickness, together with any microheterogeneity with respect to refractive index differences within the glass, will show interruption and continuity of Fizeau fringes which will establish that the pieces of glass had originally been joined. The method is not suitable for 'crazed' tempered glass.

6.4 Fractures in flat glass

The examination of fracture surfaces, a process that is frequently seen in the engineering literature and is referred to as 'fractography', has been used widely in the study of brittle solids [9, 10], including glass, ceramics, metals and polymers. It is often possible to tell from the appearance of a fractured surface where the fracture originated specifically, in which direction it has travelled, whether the fracture was due to thermal stress or impact, and in the latter case, the relative energy and direction from which the force was applied.

6.4.1 History of forensic interest in flat glass fracture

The earliest discussion of the nature of glass fracture as an aid in the resolution of forensic issues is the discussion of high-velocity projectile fractures in the 1906 edition of the text by Gross [11]. The forensic implications of the manner in which flat glass breaks was described by Matwejeff in 1931 [12], followed shortly by the FBI laboratory [13] and Tryhorn [14, 15]. The subject has been reviewed from a forensic standpoint by McJunkins and Thornton [16] and by Thornton and Cashman [17]. From an engineering standpoint, the fracturing of glass has been reviewed by Shinkai [18], Orr [19], Phillips [20], Mecholsky [21], Mencik [22], and Kepple and Wasylyk [23].

6.4.2 Failure upon tension

This discussion will first pertain to flat glass, and then will be extended to other types of glass. Glass breaks under tension, not compression. In terms that suffer somewhat from imprecision (but which nevertheless may be more meaningful to a lay jury), when you push on a pane of glass it does not begin to break on the front side to which the force has been applied, but rather from the back side that is stretched. It breaks under tension because the tensile strength of glass is always less than the compressive strength. As stress is applied to the glass, the tensile strength will invariably be reached before the compressive strength, and at that point the glass will fracture. Glass undoubtedly could break under compression, but before it has an opportunity to break under compression, it has already broken under tension.

The classical explanation of glass fractures, illustrated in Figure 6.1, has proceeded along the lines of the following:

- As a force is applied to the near surface of the glass, the glass bends elastically with the near surface under compression and the far surface under tension.
- A fracture will initiate at a point defect on the far surface. The fracture will proceed toward the near surface and at the same time will radiate outwards from the central point of origin. The radiating fractures are termed radial fractures.
- The fracture may be complete at this point. But in many other instances, there is still an accumulation of stress that has yet to be relieved. If this is the case, the continued stress applied to the segments between the radial fractures places tension on the near surface of the glass.
- The segments between the radiating fractures break, with the fractures starting on the near side of the glass and extending between the arms of the radial fractures. These new fractures tend to form the boundary of a circle concentric about the fracture origin, and are consequently termed concentric fractures.

At the fracture surface, that is, the *edge* of the glass that has been broken, features may be

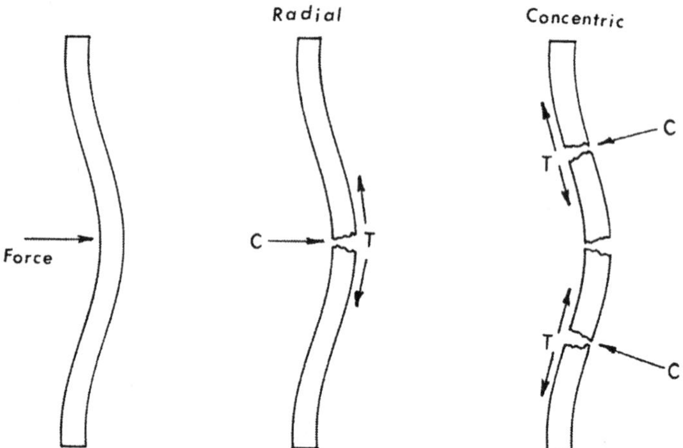

Figure 6.1 Processes involved in the breaking of flat glass. The classical explanation for the fracturing of glass depends upon the glass failing upon tension. A force applied to a plane piece of glass places the glass under compression on the side facing the force, but the compression on the forward side of the glass causes tension to be placed on the far side. The glass then fails under tension and radial fractures are created. If the energy cannot be relieved by radial fractures alone, subsequent compression is then placed on the far side of the glass, with consequent tension on the forward side. The glass then fails again under tension to create concentric fractures, with the initiation of the fractures on the forward side of the glass

observed that enable the analyst to determine the direction of the force responsible for the fracture. As discussed below, a relationship exists between fracture behaviour and certain topographic features of the crack. The Griffith theory of fracture propagation [24] demands that a flaw or defect be present before a fracture can progress. This defect may be so small as to be below the power of even an electron microscope to resolve, but it must be present. While the evidence of Griffith flaws or microcracks is largely indirect, they are generally conceded to exist [20–23].

Although the Griffith theory has gained the greatest acceptance within the glass engineering discipline, it is not the only theory. Poncelet [25] advanced an atomic approach to fracture propagation. Unlike the Griffith theory that requires a point defect, the Poncelet theory requires only the application of stress for a critical period of time. In this theory, there is a normal equilibrium rate of atomic bond rupture and re-formation, and this rate is influenced by stress. When the rate of bond rupture exceeds the rate of bond formation, a fracture will be initiated. There has not yet been an entirely adequate melding of these two competing theories but both appear to have merit. Although it is desirable that the forensic scientist be aware of the two theories, it is unlikely that it will make any difference in the practical interpretation of glass fractures in actual case situations.

In both theories, stress is applied to an area of the glass in a quantity sufficient for the strain-release energy to exceed the surface free energy of the glass, that is when the tensile strength of the surface is exceeded, a crack will spread very rapidly. As a crack progresses from its origin, there is a progression of different features that may be seen on the fracture surface, that is, the broken edge. The fractures are not chaotic but have characteristic features that are capable of being interpreted. The different features are generally referred to as mirror, mist, hackle (also called striations), and Wallner lines (also called conchoidal lines or ripple). There is no universally accepted nomenclature, and unfortunately there is some confusion in the literature. This confusion is very much apparent in comparing the works of Preston [26], Murgatroyd [27], Andrews [28], Rice [29], Poncelet [25] and Shinkai [18]. An effort at standardisation is seen, however, in the applicable ASTM standard [30].

6.4.2.1 Mirror

Near the fracture origin, the surface will be flat and featureless. This represents an area of relatively slow fracture propagation. It is also consistent with low stress fractures, such as those seen in thermal fractures. Since the surface is flat, reflection of light is a salient aspect of this area, and the surface is consequently termed 'mirror'. In impact fractures, the mirror may exist over only a few millimetres; in thermal or other low-stress fractures, it may continue for many centimetres.

6.4.2.2 Mist

As the fracture continues and builds up speed, the crack tip cannot dissipate the accumulated stress fast enough. Consequently, the surface area affected is increased in the effort to reduce the surface free energy. This is accomplished by a multitude of exceedingly minute cracks formed at the crack tip. These surface disruptions are so small that even under low magnification they appear as a 'frosted' or misty area. These areas are termed 'mist'.

6.4.2.3 Hackle

It is not altogether clear whether 'hackle' is formed as a result of a further extension of the phenomenon of the reduction of surface free energy by means of an increase in surface area, or whether it is formed on a fracture surface as a result of a localised realignment in an effort to remain perpendicular to the tensile fracture stress [31]. In forensic science, no other term has been used to describe these marks. In some engineering discussions of glass fracture, e.g. Kepple and Wasylyk [23], the term 'striations' is preferred. Hackle consists of rather coarse marks more or less parallel to one another. This is illustrated in Figure 6.2.

In some instances the hackle marks may be so coarse as to be visible with the unaided eye. Branching may be seen in some instances, but hackle often occurs without any obvious branching. Hackle is observed perpendicular to the curved conchoidal or Wallner lines. Hackle marks lie parallel to the direction of fracture propagation. In some instances [32] it may be possible to demonstrate a physical match of hackle marks between two pieces of glass that at one time had been joined.

6.4.2.4 Conchoidal marks

The most conspicuous of all of the glass edge markings are the curved shell-like or conchoidal marks. In fracture analysis discussions in the engineering literature they are

Figure 6.2 An illustration of hackle marks. The parallel lines on the fracture edge are hackle

generally referred to as 'Wallner lines' and less frequently as 'ripple'. In some forensic literature they are referred to as 'rib marks'. The marks, illustrated in Figure 6.3, are generally visible with the unaided eye. In some instances [32] it may be possible to demonstrate a correspondence of conchoidal marks between two pieces of glass that at one time had been joined.

It is axiomatic in all glass fracture discussion that a fracture will propagate only at right angles to the principal tensile strength. As a consequence of this universal and obligatory requirement, fractures tend to orient themselves to continually remain perpendicular to the principal tensile stress. This is certainly true of the Wallner line conchoidal marks and this phenomenon may be exploited to determine the direction of force resulting in the fracture.

The 'direction of force resulting in the fracture' is not necessarily synonymous with 'the side of the window to which the force is applied'. When examining a radial fracture, the conchoidal lines meet the side opposite the force at very nearly a right angle and curve

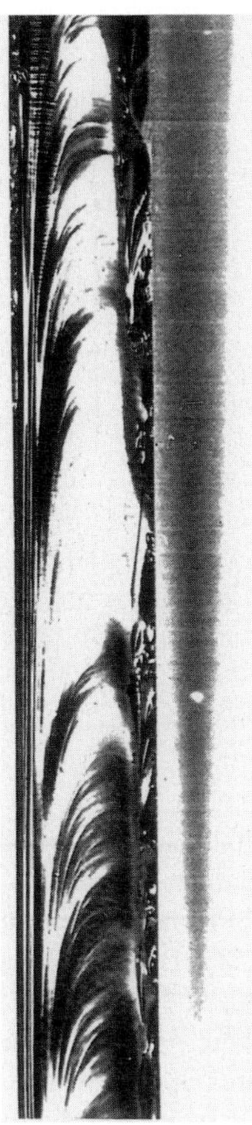

Figure 6.3 An illustration of conchoidal marks. The Wallner lines, also commonly referred to as conchoidal lines or rib marks, allow for a determination to be made of the direction of force applied to the fractured glass

back asymptotically toward the side from which the force was applied, but with concentric fractures, the tension is on the surface facing the force. Since the glass fails on tension and the fracture initiates now on the side facing the force, the curved Wallner lines now meet that side, that is the side facing the force, at nearly right angles. In Figure 6.4, the direction of the force is from left to right if we are looking at 'radial' fracture, but from right to left if we are looking at a 'concentric' fracture.

The significance of this is patent. It is exceedingly important to know the proper orientation of the fracture that is being examined. Confusion as to the type of fracture that is being examined will certainly lead to an erroneous interpretation as to the direction of force. It may be possible to select an unambiguous piece of glass by simple inspection, but if there is any doubt, the entire window (or at least those portions close to the point of impact) should be collected in order to fit the pieces back together to establish which fractures are the radial fractures and which are the concentric fractures. A photograph of the window

Figure 6.4 The breaking of flat glass. The direction of force applied to a plane piece of glass may be interpreted on the basis of the Wallner or conchoidal lines. In this figure, the direction of force is from left to right if the fracture is a radial fracture. The direction of force is from right to left if the fracture is a concentric fracture. The necessity of being able to orient a fracture as being a radial or a concentric fracture cannot be overemphasised

before it is collected may be invaluable, since some shards may have fallen to the ground and fractured as a result of this translational impact.

In the determination of direction of force, only those pieces that are close to the point of origin are suitable for interpretation. There is at least a suggestion that upon blunt impact, vibrational nodes are set up in a pane of glass, with the conchoidal marks on the more distant side of a vibrational node being reversed from what would typically be expected. Care should be taken that a shard broken off by hand in the collection process be clearly labelled as such and oriented properly by means of a diagram. It is inadvisable to select for examination a piece that is very close to a sash or frame, as the rigidity with which such a piece was held may influence the nature of the fracture features [18]. In the case of multiple impact fractures where one fracture will not cross over another, care must be taken that a radial fracture from a previous impact is not mistaken for a concentric fracture of a subsequent impact.

6.4.3 Failure by Hertzian fracture

Failure upon tension is not the only mode of fracture of glass. If it was, it would be impossible to break glass that was shored from behind, since the glass would be unable to bend and create tensile stress. The second mechanism by means of which glass may be broken is by the formation of Hertzian fractures [18]. In the case of rapidly applied high-energy impact, such as is commonly encountered in firearms projectile impact, a compressive stress wave will pass through the glass from the point of impact to the opposite side. This is the situation that in engineering parlance is referred to as 'dynamic loading' and takes place over an interval of a few microseconds. 'Quasi-static loading', on the other hand, the phenomenon that results in the breaking of glass from a handheld or thrown object, operates over an interval of a few hundredths of a second.

6.4.4 Propagation of mechanical waves

When a projectile strikes a pane of glass, longitudinal compression waves are generated, beginning at the point of impact and radiating outwards in a series of spherical wavefronts. The waves generated in water by an impacting pebble are a felicitous analogy. These wavefronts travel through the glass at the speed of sound in glass, approximately 5000 m/s (16,400 fps) to 6000 m/s (19,680 fps) The projectile is travelling at a much lower velocity. With a stress wave of 5500 m/s (18,040 fps), the wave is travelling approximately 20 times as fast as a bullet fired from a .45 automatic handgun, and six times as fast as a bullet fired from a .308 Winchester. These waves are known under several different names – mechanical waves, stress waves, compressive waves, shock waves, sound waves, and orsonic waves.

When the stress wave arrives at the opposite side of the glass, it is reflected, but upon reflection it becomes a tension wave [33, 34]. Upon striking the impact side, it will be reflected again as a compression wave and again launched toward the opposite side. This interactive process may result in the interference of tension waves when they are coincidentally in phase. The amplitude of the waves will then be additive and the accumulation of tensile strength results in the failure of the glass. A flake or 'scab' of glass is ejected from the opposite side. The largest flakes are at the edge boundary and successive flakes, closer to the point of application of stress, are smaller. A characteristic of glass broken in this fashion is a crater, opening away from the direction of the origin of the force. This is illustrated in Figure 6.5.

The fractures resulting in the formation of the flakes or scabs, which in turn result in the crater, are referred to as Hertzian fractures [35] or Hopkinson fractures [36]. The overall form of the crater, that is, the appearance and dimensions of the 'skirts' of the crater and the bevelling of the sides of the crater, are manifestations of the velocity of fracture propagation. The crater and the fractures associated with the crater margins do not require the actual perforation of the glass, although in most instances involving a high-speed projectile this will certainly occur. It is common in the case of low-velocity impact by pebbles or pellet gun projectiles to observe a crater on the opposite side of the glass but with either no hole on the impact side or a hole with a smaller diameter than the projectile.

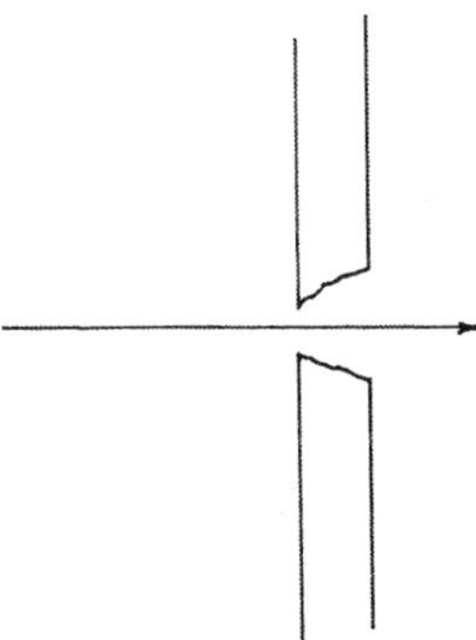

Figure 6.5 An illustration of a crater generated on the breaking of flat glass. A Hertzian crater is caused by high-energy impact. The cone invariably opens away from the point of application of force

6.4.5 Failure of tempered or disannealed glass

Glass may be strengthened by placing its surface in a state of compression by thermal means. This type of glass cannot be manufactured in large sheets and cut to fit a particular application. Instead, each piece is fabricated individually, with jets of air played over the surface while the glass is suspended. The surface of the glass then cools at a more rapid rate and when the interior cools, the compressive stress of the outside is approximately twice that of the centre.

When the glass fails, it fails catastrophically, with the glass being 'crazed' to the maximum extent of the pane.The energy required for producing fractures is stored elastically in the glass, and when a fracture reaches the centre tensile area, the entire item fails. The mechanism by which this catastrophic failure occurs, as demonstrated by high-speed photography at a million frames per second, has been advanced by Chaudhri and Liangyi [37]. Upon impact by a 'point-source', which may be a projectile or other object of small presentation area, a stress fringe is propagated. The form of this stress is dissimilar to that observed with ordinary annealed glass. Within two or three microseconds, this stress wave is travelling toward the centre of the glass at a velocity of about 1500 m/s (4950 fps). Upon reaching the interior tensile zone, a circular fracture front then travels along the tensile zone at a more or less uniform rate of 1500 m/s, although the fracture velocity back toward the surface of impact is much less (no more than 200 or 300 m/s and possibly approaching zero as the fracture nears the surface).

Hertzian fractures do not dominate the fracture processes, as they do with high-velocity impact of annealed glass. This is apparently because the skirts of the Hertzian crater tend to turn and run parallel to the surface of impact, and consequently are not propagated into the centre zone of tension.

6.5 Fractures in other items

6.5.1 Fractures in container glass

Glass fabricated into beverage bottles and other containers fracture in accordance with the fracture phenomena discussed previously in connection with flat glass. While the principles remain the same, the presentation of fractures may differ somewhat because of the particular geometry of these items. The classic work on the subject is that of Preston [38] and the subject has been reviewed by Kepple and Wasylyk [23].

The failure of glass containers under various conditions of stress may result in a product liability issue that may have a forensic component. In addition, glass containers may be broken in connection with criminal matters. An obvious example of the latter would be the reconstruction of a bottle suspected of having been used as a Molotov cocktail, that is a bottle filled with a flammable liquid that would shatter upon impact and ignite the liquid.

To an inexperienced observer, a broken bottle may appear to be a confused jumble of meaningless fragments. These fragments, however, can generally identify the fracture origin and provide some insight into the manner in which the bottle was broken, for example whether the bottle was broken by overpressure from within or by impact from without, or whether thermal stresses were involved in the fracture process.

As with flat glass, there is a definite sequence to the fracture process. With every failure of a glass container, there is a single location at which the fracture begins. This is termed the fracture origin. With very rare exceptions, the fracture will traverse the thickness of the glass wall to the opposite surface as well as being propagated away from the fracture origin. As the fracture proceeds, various characteristic markings are created. These markings are identical to those described for flat glass fracture, and the patterns of these markings may be interpreted in terms of the cause of the failure. These are described in the following subsections.

6.5.1.1 Internal pressure

If a glass container fails because of excessive internal pressure, the typical fracture appearance is a straight, vertical split in the sidewall of the container. This may result from excessive carbon dioxide in the case of carbonated beverages, from thermal expansion of liquids subjected to excessive heat, or from the generation of gas through spoilage or decomposition of the contents. Pressure fractures rarely originate in the neck of a bottle, as the neck is generally stronger than the sidewall portion. They may, however, originate at the heel or bottom, where the stresses are not usually greater than in the sidewall.

In sidewall pressure fractures, the fracture surfaces are likely to show a mirror surface which may or may not be associated with hackle. This is in keeping with other forms of relatively low stress, such as thermal fractures.

After the initial fracture along the long axis of the container, the fracture often forks at both ends, producing a profusion of fractures that are symmetrical about the initial longitudinal fracture in the sidewall. This is illustrated in Figure 6.6. As the fan-shaped pattern of fractures diverge, they travel around the circumference of the bottle and frequently cause the bottom and the neck to be separated from the remainder.

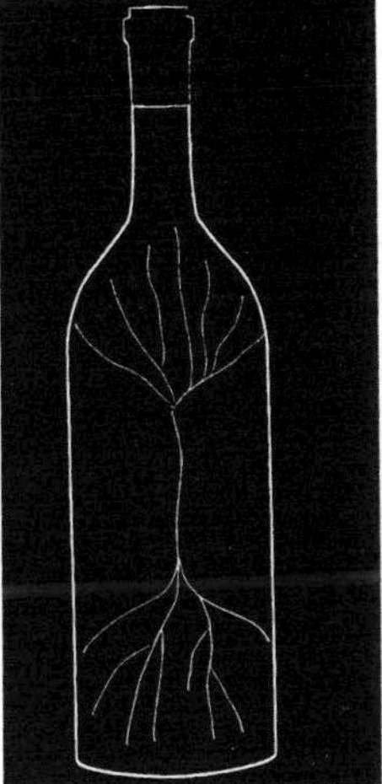

Figure 6.6 An illustration of a sidewall fracture of a glass container. A typical pressure-induced fracture of a beverage bottle, with a profusion of fractures symmetrical about the initial longitudinal fracture

6.5.1.2 Thermal fracture

Glass containers may fail if very hot liquids are introduced when the glass is cold, or if a hot container is placed into a cool liquid. Most thermal shock fractures occur in the heel or bottom of the container, and show few if any surface markings. The principal stresses in the sidewall of a container are oriented around the circumference of the container, resulting, in the case of pressure fracture, in a fracture along the long axis of the container. Since thermal fractures most often originate at the heel or bottom of the container, the fractures are often around the circumference of the base with no evidence of fan-shaped branching.

6.5.1.3 Impact fracture

Impact fractures generate the most complex stresses of any type of fracture. Mould [39] has developed the concept that there are three distinct phases of impact fracture, namely, contact stress, flexure stress and hinge stress. The actual fracture appearance will be determined by which one of these three stress types dominates the fracture.

Contact stresses are generated at or very close to the point of application of the impact force. These stresses result in a Hertzian fracture, the type of fracture that is responsible for cratering in flat glass. They are less likely to be encountered in the sidewall of a typical container than in the heel or in thick-walled bottles. They may also result in chips along the mouth of the container.

Flexure stresses arise from a localised defect on the inside of the glass. In these instances, only a very slight impact is necessary to cause the container to fail. This type of fracture results in a stellate or star-like appearance on the inside of the container. This is illustrated in Figure 6.7. At the same time, flexure stresses on the inside of the container

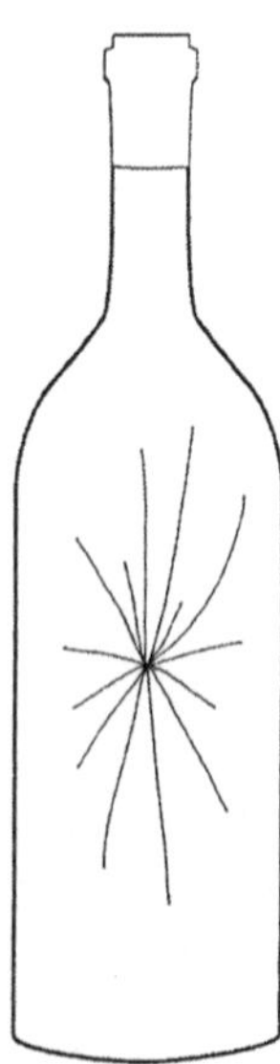

Figure 6.7 An illustration of a star-like fracture in a container vessel. A typical impact fracture of a beverage bottle, with radial fractures emanating from the initial point of impact

cause a bending of the glass wall on the outside. The stress created by this outward bending is the hinge stress. This phenomenon has been described from a forensic standpoint by Thornton [40] and is a manifestation of the same principle that gives rise to concentric fractures in flat glass. With the impact on the outside of the glass, the flexure fractures then originate from the inside but the hinge fractures always originate on the outside surface of the container. If only an isolated piece of glass is collected for examination, the danger exists that the fracture will be interpreted erroneously as having been caused by a force applied to the inside of the glass.

Hinge stress fractures generally occur at the same height as the initial flexure fracture and at about 45° on either side. Since the hinge stresses are associated with a bending of the glass, any hackle marks take on the wedge-shaped appearance typical of bending stresses.

6.5.2 Fractures in glassy polymers

The increase in the use of glassy polymers such as acrylics and polycarbonates as glass substitutes in architectural situations has resulted in instances where it is necessary to interpret high-velocity projectile fractures in these materials as well.

There are significant areas in which fracture appearances differ between glass and glassy polymers. One conspicuous difference is the occurrence of very coarse, curved hackle marks in polymers, that could be mistaken for conchoidal rib marks. These marks are rather treacherous in their interpretation. If these marks are interpreted as conchoidal marks, an incorrect attribution of the direction of force may result. A second difference is in the appearance of the bevelled edges bordering the cratering on the side opposite the origin of the force. With polymers, the crater opens to the side away from the force just as with glass, but the edges of the crater are convex rather than concave. A third difference is that with polymers, there is a greater tendency toward an inverse relationship of muzzle velocity and energy to the length of radial fractures and the degree of curving along crater boundaries. Finally, a fourth difference is that with polymers, concentric fractures are rarely encountered.

All of these phenomena can be explained on the basis of the increased flexibility and flow of the polymers relative to brittle glass. The subject has been considered from a forensic standpoint by Rhodes and Thornton [41] and from an engineering standpoint by Andrews [42].

6.5.3 Fractures in automobile glass

6.5.3.1 Laminated safety glass

In many parts of the world, automobile front windshields are of the laminated safety glass type. A thin layer of plastic is sandwiched between two pieces of annealed glass. There has been a gradual reduction in the thickness of these windows. The symmetric type is now typically of two panes of glass of 2.1 mm thickness, with a plastic interlayer of 0.76 mm. The asymmetric type is now 2.1/0.76/1.8-mm thick. In the USA little variation is noted with the front windshield being of the laminated safety type, and the side and rear windows of the tempered or disannealed type. A discussion of the fracturing of glass ply laminates is given by Shinkai [18] and a discussion of the strength of automotive window glass will be found in Manfré [43].

6.5.3.2 Tempered safety glass

Differentially tempered front windshields are nominally 4 mm in thickness, with an expected range of 3.96 to 4.17 mm, while side and rear windows are typically 3 mm in thickness with an expected range of 2.92 to 3.41 mm. Within a given window, the thickness range of any tempered glass automobile window is unlikely to be greater than ±0.075 mm. Tempered front windshields will not normally be encountered in the USA, but are common in much of the world. Tempered glass windows have a strength in the range 100–120 MPa, compared to a strength of 10 MPa for annealed laminated glass. The edges of tempered or disannealed glass fragments typically will show Wallner or conchoidal lines on both sides with an intermediate zone of hackle. Unlike annealed glass, the conchoidal lines will not indicate the direction of application of force.

6.6 Glass fractures produced by firearms

The fractures produced in annealed glass by firearm projectiles are different from those produced by blunt, low-energy impact. While projectiles may result in the catastrophic failure of an entire window, the typical manifestation of a high-velocity projectile impact into annealed glass is the formation of a crater by means of Hertzian fractures.

The crater resulting from the impact of a projectile may provide some information concerning the projectile and its velocity. The subject is not trivial, however. The subject was considered in depth in the early work of Tryhorn [14, 15] and for laminated automobile safety glass by Turfitt [44].

The crater invariably opens out on the emergent side. In the case of low-velocity projectiles such as that fired from a pellet gun, or a small pebble, a crater may be formed even though the projectile does not pass through the glass. Indeed, it is possible that the crater may not extend all the way from the back side of the glass to the front.

If the approach of the projectile to the glass is very nearly perpendicular, the crater will be symmetrical. If the approach of the projectile is at an angle, however, there is likely to be bevelling of the edges of the defect, with a greater amount of flaking seen on the emergent side of the glass, but in the direction opposite from which the projectile approached.

There does not seem to be a clear correlation of the calibre of the firearm with the size and appearance of the crater. This reflects the fact that the form of the crater is not a function of the velocity of the projectile nor of its diameter, but instead is a function of the velocity of crack propagation that results in the Hertzian fractures. Some generalisations are reasonable, however. Typically, a high velocity projectile, such as a rifle bullet, will yield a small hole with minimal flaking around the circumference. A slow-moving bullet, however, will deliver a more irregular hole with substantially more flaking. The size of the hole in the glass may be greater than the diameter of the bullet, since glass in the immediate environs of the bullet may be ground almost into dust and lost to the crater. At very low velocity where great distance or passage through an intermediate target has reduced the velocity of the projectile to under 200 fps, the glass may show evidence of tension-propagated impact fracture with no crater formation at all. In many instances, the bullet form may exert a determinative effect upon the appearance of the crater. To obtain the maximum amount of information from evidence of this type, experimentation with the same thickness of glass and various types of ammunition is essential in case situations.

With disannealed or tempered glass, the glass will break into small cubes. The location of the bullet hole through the window may be readily apparent, but the 'crazed' fractures extend throughout the entire window and the window is then very susceptible to subsequent damage which will obliterate the precise point of impact of the projectile. In instances where the precise point of impact cannot be determined, the flow of fracture markings in the remaining portion of the window may permit some opinion to be formed concerning the approximate point of impact. These markings may be mediated by curvature in the window, as is often the case with automobile rear windows, and in such instances an opinion concerning the point of impact and direction of approach of the projectile should be made only with the utmost diffidence.

6.6.1 Sequence of gunshots

Fractures in glass will not cross one another since if a fracture intersects an existing fracture, it will be arrested in its forward travel. This may facilitate the determination of the sequence of gunshots. For example, in Figure 6.8, the gunshot toward the bottom of the figure is the second gunshot, since the radial fracture of this shot that extends from the impact origin in a 2 o'clock direction intersected an existing fracture and came to an immediate halt. While one fracture will not cross another, there may be instances where spalling is noted at the intersection of the two fractures. This may occur when there is sufficient energy associated with the second fracture to create an asymmetrical, fan-shaped segment of a Hertzian fracture of the type seen in the cratering resulting from a projectile impact. The spalling will have the fustrum of the fan-shaped segment at the point of intersection of the two fractures, and the divergence of the fan will point in the direction of approach of the second fracture.

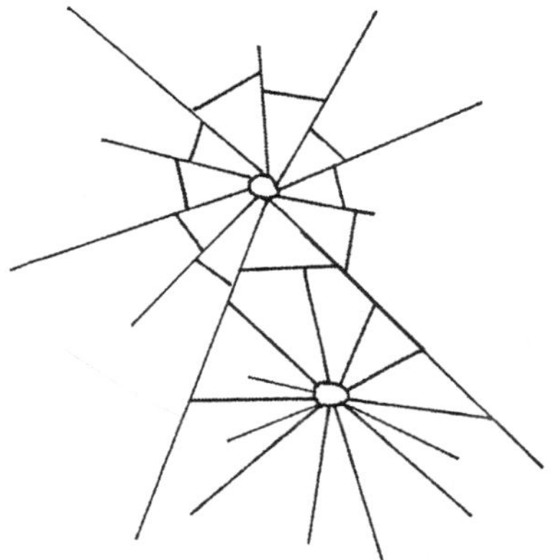

Figure 6.8 An illustration of the distribution of gunshot fractures which allow the sequence of shots to be determined. In the present example, the gunshot toward the bottom of the figure is the second of the two shots, since the radial fractures at 2 o'clock, 3 o'clock, 9 o'clock and 11 o'clock intercept radial fractures from the first shot and are arrested immediately

The spalling is almost entirely a surface effect, and is likely to be confined to a depth of only a millimetre or a fraction of a millimetre. It is associated with very high energy fractures typically generated by gunshot, and is unlikely to be encountered in blunt impact fractures.

6.6.2 Effect of glass on projectile trajectories

When a bullet passes through a pane of glass of even moderate thickness, such as ordinary window glass, the bullet typically is destabilised. For short distances on the emergent side of the glass any deflection from its original trajectory is likely to be small, but the projectile will have lost its true poise. As a consequence, over long distances following the penetration, the bullet may display significantly erratic behaviour.

6.7 Backward propagation of glass fragments

When a pane of glass is broken by impact, the majority of the broken fragments will fall in the direction of the force, that is on the side of the window opposite the application of the force. Some small fragments, however, are propelled backwards, toward the direction of the applied force. This phenomenon was described by Kirk [45] and amplified upon by Nelson and Revell [46] and Luce *et al.* [47]. The significance of this from a forensic standpoint is patently obvious. If a person breaks a glass pane by standing close and striking the glass with an object, fragments may subsequently be recovered from the clothing of that person. And since some fragments may be propelled as far as 3 metres, the possibility exists that some fragments may be recovered from the clothing of a person standing away from the glass and breaking the glass by means of some object hurled at the pane. Zoro [48] observed that a high proportion of the glass propelled backwards shows the flat surface facing the person breaking the window, and that the flat side is likely to be fluorescent if the glass was manufactured by means of the float method. Underhill [49] also noted that much of the backward fragmentation glass shows a flat side, but concluded that apart from this observation, the shape of the fragment does not indicate the method of acquisition.

An additional consideration may apply as well. After the glass is broken, there may be a profusion of small particles on the ground extending away from the broken window. These particles may be picked up on the soles of the shoes of anyone walking over that area.

6.8 Glass in fires

The interpretation of glass fractures associated with fires may involve several different phenomena, including thermal stress, impact fracture, and overpressure stress. An understanding of these processes, which may act independently but which may also act in concert, will enable the analyst to address such issues as whether the glass broke as a result of thermal stress or because of impact. Did the glass break before the fire started or did it break during the fire? Did the glass break as a result of a slow burning or did it break as a result of a sudden increase in temperature? Was an accelerant, that is a flammable solvent, used to set the fire? Glass is non-combustible, and consequently will be found in some form following the fire. In this sense, fire investigators have often stated that glass is a 'spectator' to a fire and may assist in establishing the factual circumstances at the time of the fire. The subject of the behaviour of glass in fires has been considered by De Haan [50], Cooke and Ide [51], and Schudel [52].

A thermal fracture of glass is, like an impact fracture, due to the failure of the glass upon tension but the process is different and the appearance of the fracture is different as well. When glass is heated, it expands. The temperature differential on one side of the glass may cause sufficient expansion, relative to the opposite side of the glass, that the resulting tension causes the glass to fail. A temperature differential of 60 to 70°C between the centre of a pane of glass and the surface will cause the glass to break. If the heating is very sudden and the temperature differential is exceedingly great, as when a flammable solvent is ignited, the tension and resulting fracture may act over many small domains. In such instances, the glass will fracture in a 'crazed' pattern, with irregular pieces measuring less than 10 mm on each edge. (Parenthetically, crazed glass is generally observed to be relatively free of soot, since the high heat discourages particulate aerosols of carbon from being deposited on the very hot surface. This is in contrast to glass surfaces in accidental fires where the fire smoulders for a time. In these instances the glass, whether fractured or not, tends to show a thick layer of soot.)

In a fire where no accelerant is involved, for example the typical accidental fire in which the temperature increases more gradually, the thermal stress to the glass acts over a greater area. The stress gradually accumulates to the point where the glass fails, but does not exceed the stress necessary to cause a fracture. Relative to impact fractures, the stress is low. The fracture originates at a weak spot or flaw. This is typically at the edge of the glass pane, but can be at a flaw at some other location. The fracture extends across the glass until the stress is relieved or until the fracture intercepts the edge of the pane. Initially the fracture travels in a straight line, but soon begins to meander. The most salient aspect of a thermal fracture is the simple meandering nature of the fracture. Many such fractures exhibit a sinuosity that is clearly distinguishable from impact fracture. This is illustrated in Figure 6.9.

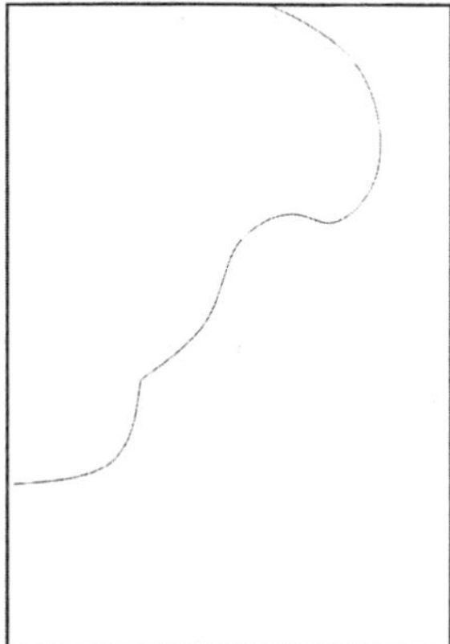

Figure 6.9 An illustration of the meandering nature of a thermal fracture. Unlike impact fractures, thermal fractures are sweeping, meandering fractures that frequently show abrupt changes in direction

The fracture surfaces are also different in character from those observed with impact fractures. In low-stress thermal fractures, there are few or possibly none of the markings associated with high-stress impact fractures – rib marks, hackle, or Wallner lines. These markings may be conspicuous, however, in those fractures resulting from the higher stresses associated with fires started with a flammable solvent.

Fractures resulting from thermal differential may also result from the sudden cooling of hot glass. These fractures have the appearance of small dish-like pits in the glass, and do not extend completely through the thickness of the glass. These defects in the surface of the glass may be the result of firefighting activities in which water is sprayed onto a hot glass surface. They occur on the side of the glass exposed to the water, and may assist in reconstructing the firefighting activities. In some instances water cooling of a hot glass surface may create a 'crazed' appearance, with the fracture extending completely through the thickness of the glass.

6.9 Retention and persistence of glass fragments in clothing

A number of experiments have been conducted to ascertain precisely how prevalent glass is on clothing. Different conclusions can be drawn from these experiments, primarily as a reflection of the different experimental modalities and the different types of fabric used.

In 1971, Pearson *et al.* [53] examined 100 suits submitted to a dry-cleaning establishment, and found glass fragments in 63 of them. A total of 551 fragments was recovered from the 100 suits. The greatest number of fragments (291) were found in trouser pockets, with the second greatest number (182) found in the jacket pockets. Seventy-eight fragments were found in the trouser cuffs (turn-ups), but only 70 per cent of the suits had cuffs on the trousers. Eighteen fragments were longer than 1 mm and 128 were longer than 0.5 mm. The mode of the size distribution was about 0.3 mm. The distribution of particles within the population of 100 suits is of considerable interest, however. Two suits contained 46 per cent of all of the glass recovered, and 50 per cent of the suits contained from one to five fragments.

Although this study by Pearson *et al.* was well structured and well executed, some diffidence must be exercised in applying it to many case situations at the present time. The study is now somewhat dated, and involves only suits. The age of the suits was not determined in this study, nor the occupation of the wearers. It is not established what type of recovery might be encountered with other types of clothing. One may posit that glass is more likely to be encountered with more casual work clothing than with suits, but specific information on this point is lacking.

Pounds and Smalldon [54] conducted an experiment, previously discussed in connection with backward scattering of glass, in which they counted the number of fragments of glass trapped in the clothing and hair of test subjects. In 1985, Brewster *et al.* [55] extended the work of Ponds and Smalldon by addressing the issue of the persistence of particles of glass once they are captured by clothing. The premise of this work was twofold, first that the type of cloth would influence the initial retention of glass particles, and second that the size of the particle would influence the retention. Both premises were supported by experimental work. Wool captured more particles than denim, and small particles (0.25 mm and less) were retained over larger-sized particles.

In 1992, McQuillan and Edgar [56] conducted a two-year survey of 432 items of clothing. None of the items was from a person suspected of having committed a crime. The items consisted of 216 pairs of trousers, 191 jackets, and 25 pullovers. Seventy-one per cent of the trousers, 58 per cent of the jackets, and 56 per cent of the pullovers had no glass present. Apart from one garment belonging to a unique occupational caste, the largest number of fragments found on a single garment was 57, representing 27 different sources. Considering all of the different types of garments together, when glass was found the mode would have been one or two fragments.

In 1995, Lambert *et al.* [57] reported on glass fragments recovered from the clothing of persons suspected of involvement in crime. They examined the clothing of 589 criminal suspects. Twelve per cent had no glass in their clothing, 45 per cent had 1 to 10 fragments, 20 per cent had 11 to 20 fragments, and 9 per cent had more than 50 fragments. In descending order, the prevalence of glass in garments followed the order of shoes > trousers > jacket > knitwear > shirt. The next in this sequence was the hair of the suspect. While this study reported a much higher incidence of glass in clothing, it should be recognised that the study was directed at persons who were suspected of harbouring evidence glass, and in fact 59 per cent of the 589 had glass that matched exemplar samples. The population studied clearly influenced the reported distribution of glass.

Hicks *et al.* [58] conducted experiments in which flat glass was broken in order to study the transfer and persistence of glass fragments in clothing. The results of this experimentation confirmed in general the research reported previously by other workers, and confirmed the common-sense considerations that would appear to be applicable. Several different fracture modalities were employed, and it was determined that although the number of fragments lodging in clothing and persisting over time was variable from one experiment to another, several general trends were confirmed. Among these is that the number of fragments recovered from clothing, or indeed whether any fragments will be recovered at all, is determined by:

- the number of strikes
- the distance between the glass and the person breaking the glass
- the size of the fragments
- the composition and weave of the fabric
- the time interval between the breaking of the glass and the subsequent recovery of glass from the clothing.

The number of fragments appeared to be influenced by the number of strikes, but not particularly determined by the instrument used to break the glass. A clear relationship was seen between the number of particles recovered and the distance between the glass and the person breaking the glass or an 'accomplice' standing nearby. Most of the glass was lost from the clothing in the first half-hour after the breaking. After 8 hours, the glass remaining consisted of small (0.2–0.5 mm) fragments, whereas immediately after breaking, many particles larger than 0.5 mm were observed. Glass was retained better in coarse woven cloth than in tight woven material. The authors of this work stressed that the number of fragments recovered is of little significance because the number reflects the effort exerted to recover them and because in actual case situations, one would never know the number that were present at time zero.

In 1997, Lau *et al.* [59] reported on the frequency of occurrence of paint and glass on the clothing of high-school students. In a study of 213 students, glass was found in 2 per cent of the outer clothing and 5 per cent of footwear. These data are significantly lower than the incidence of glass reported in other studies, and are particularly at variance with the study by Davis and De Haan [60]. The Davis and De Haan study examined 1300 shoes, representing 650 matched pairs, donated to a charitable organisation for sale at a charity shop. The study reported the presence of colourless glass in 20 per cent of the 1300 shoes examined.

A Bayesian statistical approach to the retention and persistence of glass fragments in clothing has been described recently by Curran *et al.* [61, 62]. The authors enunciate the factors that influence the final number of observed fragments of glass:

- The degree of fragmentation.
- The type and thickness of the glass.
- How many times the window was struck.
- The position of the person relative to the window that was struck.
- The size of the window.
- The type of clothing worn by the person breaking the window.
- The activities of the person between the time of the commission and the time of apprehension.
- The time between apprehension and confiscation of the clothing.
- The efficiency of the laboratory searching process.
- Whether or not the person gained entry to the premises.
- The mode of clothing confiscation, i.e. was force required or not.
- The weather at the time of the incident.

Curran *et al.* have countenanced the entire process, but in recognition of the fact that it is unclear how to model certain of these 12 factors, only position, time, garment type and laboratory examination were included in their proposed model. A graphical modelling technique was found to offer a greater measure of consistency and reliability than other methods. These authors conclude that it is possible to derive a partial quantification of the information loss resulting from the examination of a smaller subset of the total number of particles found in the clothing of a suspect, and that in any event the probability of recovery of any given number of fragments from a suspect is very low.

6.10 Future directions for the physical examination of glass evidence

There is some evidence that fractal geometry, with its considerable power to describe the real world in terms of margins of fractional dimensions, is applicable to the issue of glass fracture. Freiman *et al.* [63] describe what appears to be a fundamental mathematical relationship between parameters determined by fracture surface analysis and those derived from fractal analysis. This relationship may ultimately affect the manner in which glass fracture processes are viewed, and may enhance our understanding of the uniqueness of fracture surfaces. This approach to glass evidence has yet to be explored from the standpoint of any benefits it may provide to the forensic sciences.

6.11 References

1. O'Hara C., Osterburg J. (1953). *An Introduction to Criminalistics*. Macmillan, New York.
2. Radley J. (1948). *Photography in Crime Detection*. Chapman & Hall, London.
3. Stapleton D. (1940). Evidence of fractured glass in the investigation of crime. *Forensic Science Circular*; No. 6, HMSO. London.
4. Nelson D. (1959). Illustrating the fit of glass fragments. *Journal of Criminal Law and Criminology* **50**:312–314.
5. Knight M. (1956). Cords in glass. *Glass Industry* Part I, **37**:491–495; Part II, **37**:553–574; Part III, **37**:613–690.
6. von Bremen U. (1975). Shadowgraphs of bulbs, bottles and panes. *Journal of Forensic Sciences* **20**:109–118.
7. Welch A., Richard R., Underhill M. (1989). The observation of banding in glass fragments and its forensic significance. *Journal of the Forensic Science Society* **29**:5–13.
8. Thornton J., Cashman P. (1979). Reconstruction of fractured glass by laser beam interferometry. *Journal of Forensic Sciences* **24**:101–108.
9. Field J. (1964). Fracture of solids. Annual report of the Smithsonian Institution, pp. 431–438.
10. Lawn B. (1993). *Fracture of Brittle Solids* (2nd edn). Cambridge, New York.
11. Gross H. (1962). In *Criminal Investigation* (ed. R. Jackson) (5th edn). Sweet & Maxwell, New York.
12. Matwejeff S. (1931). Criminal investigation of broken window panes. *American Journal of Police Science* **2**:148–157.
13. Federal Bureau of Investigation. (1936). Evidence of fractured glass in criminal investigations. *FBI Law Enforcement Bulletin* **October**:2–11.
14. Tryhorn F. (1936). The fracture of glass. *Forensic Science Circular* **2**:1–5.
15. Tryhorn F. (1939). The examination of glass. *Journal of Criminal Law and Criminology* **30**:404–406.
16. McJunkins S., Thornton J. (1973). Glass fracture analysis – a review. *Forensic Science* **2**:1–27.
17. Thornton J., Cashman P. (1986). Glass fracture mechanism – a rethinking. *Journal of Forensic Sciences* **31**:818–824.
18. Shinkai N. (1994). Fracture and fractography of flat glass. In *Fractography of Glass* (eds R. Brandt, R. Tressler). Plenum, New York.
19. Orr L. (1972). Practical analysis of fractures in glass windows. *Materials Research and Standards* **12**:21–23, 47–48.
20. Phillips C. (1972). Fracture of glass. In *Fracture* (ed. H. Liebowitz), Vol. 7. Academic Press, New York.
21. Mecholsky J. (1985). Fracture analysis of glass surfaces. In *Strength of Inorganic Glass* (ed. C. Kurkjian). Plenum, New York.
22. Mencik J. (1992). *Strength and Fracture of Glass and Ceramics*. Elsevier, New York.
23. Kepple J., Wasylyk J. (1994). The fracture of glass containers. In *Fractography of Glass* (eds R. Brandt, R. Tressler). Plenum, New York.
24. Griffith A. (1920). Phenomena of rupture and flow in solids. *Philosophical Transactions of the Royal Society of London, Series A* **221**:163–204.
25. Poncelet E. (1958). The markings of fracture surfaces. *Journal of the Society of Glass Technology* **42**:279–288.
26. Preston F. (1939). Bottle breakage – causes and types of fractures. *Bulletin of the American Ceramic Society* **18**:1–26.
27. Murgatroyd J. (1942). The significance of surface marks on fractured glass. *Journal of the Society of Glass Technology* **26**:153–171.
28. Andrews E. (1968). *Fracture in Polymers*. Oliver & Boyd, London.
29. Rice R. (1988). Perspectives on fractography. In *Fractography of Glasses and Ceramics* (eds V. Fréchette, J. Varner), Vol. 1. American Ceramic Society, Westerville, Ohio.
30. ASTM. (1986). Method C162-85a, Standard definitions of terms relating to glass and glass products, Vol. 15.02. American Society for Testing and Materials, Philadelphia.
31. Kerkhof F. (1975). Fracture mechanics analysis of damage to glassware. *Glastechnologie* **48**:112–124.

32. Thompson J. (1969). The structure of hackle lines on glass. *International Criminal Police Review* **226**:62–64.
33. Christie D., Kolsky H. (1952). The fractures produced in glass and plastics by the passage of stress waves. *Journal of the Society of Glass Technology* **36**:65–73.
34. Kolsky H., Rader D. (1972). Stress waves and fracture. In *Fracture* (ed. H. Liebowitz), Vol. 1. Academic Press, New York.
35. Wallner H. (1939). Line structures on fracture surfaces. *Zeitschrift für Physik* **114**:368–378.
36. Hopkinson B. (1914). A method of measuring the pressure produced in the detonation of high explosives or by the impact of bullets. *Philosophical Transactions of the Royal Society of London, Series A* **213**:437–443.
37. Chaudri M., Liangyi C. (1986). The catastrophic failure of thermally tempered glass caused by small-particle impact. *Nature* **320**:48–50.
38. Preston F. (1939). Bottle breakage – causes and types of fractures. *Bulletin of the American Ceramic Society* **18**:1–26.
39. Mould R. (1952). The behavior of glass bottles under impact. *Journal of the American Ceramic Society* **35**:230–235.
40. Thornton J. (1985). Interpretation of glass fractures of curved surfaces. *FBI Crime Laboratory Digest* **12**:82.
41. Rhodes E., Thornton J. (1975). The interpretation of impact fractures in glass polymers. *Journal of Forensic Sciences* **20**:274–282.
42. Andrews E. (1979). *Developments in Polymer Fracture*. Applied Science Publishers, London.
43. Manfré G. (1985). Strength of automotive window glass. In *Strength of Inorganic Glass* (ed. C. Kurkjian). Plenum Press, New York.
44. Turfitt G. (1940). The fracture of safety glass by revolver bullets. *Forensic Science Circular* **6**:12–15.
45. Kirk P. (1953). *Crime Investigation*. Wiley, New York.
46. Nelson D., Revell B. (1967). Backward fragmentation from breaking glass. *Journal of the Forensic Science Society* **7**:58–61.
47. Luce R., Buckle H., McInnis I. (1991). A study on the backward fragmentation of window glass and the transfer of glass fragments of individual's clothing. *Journal of the Canadian Society of Forensic Science* **24**:79–89.
48. Zoro J. (1983). Observations on the backward fragmentation of float glass. *Forensic Science International* **22**:213–219.
49. Underhill M. (1997). The acquisition of breaking and broken glass. *Science and Justice* **37**:121–127.
50. De Haan J. (1991). *Kirk's Fire Investigation* (3rd edn). Prentice-Hall, Englewood Cliffs, NJ.
51. Cooke R., Ide R. (1991). *Principles of Fire Investigation*. Institution of Fire Engineers, 148 New Walk, Leicester, UK.
52. Schudel D. (1996). Glass fracture analysis for fire investigators. *Fire and Arson Investigator* **46**:28–35.
53. Pearson E., May R., Dabbs M. (1971). Glass and paint fragments found in men's outer clothing. Report of a survey. *Journal of Forensic Sciences* **16**:283–300.
54. Pounds C., Smalldon K. (1978). The distribution of glass fragments in front of a broken window and the transfer of fragments to individuals standing nearby. *Journal of the Forensic Science Society* **18**:197–203.
55. Brewster F., Thorpe J., Gettinby G., Caddy B. (1985). The retention of glass particles on woven fabrics. *Journal of Forensic Sciences* **30**:798–805.
56. McQuillan J., Edgar K. (1992). A survey of the distribution of glass on clothing. *Journal of the Forensic Science Society* **32**:333–348.
57. Lambert J., Satterthwaite M., Harrison P. (1995). A survey of glass fragments recovered from clothing of persons suspected of involvement in crime. *Science and Justice* **35**:273–281.
58. Hicks T., Vanina R., Margot P. (1996). Transfer and persistence of glass fragments in garments. *Journal of the Forensic Science Society* **36**:101–107.
59. Lau L., Beveridge A., Callowhill B., Conners N., Foster K., Groves R., Ohashi K., Sumner A., Wong H. (1997). The frequency of occurrence of paint and glass on the clothing of high school students. *Canadian Society of Forensic Sciences Journal* **30**:233–240.

60. Davis R., De Haan J. (1977). A survey of men's footwear. *Journal of the Forensic Science Society* **17**:271–283.
61. Curran J., Triggs C., Buckleton J., Walsh K., Hicks T. (1998). Assessing transfer probabilities in a Bayesian interpretation of forensic glass evidence. *Science and Justice* **38**:15–21.
62. Curran J., Triggs C., Bucketon J. (1998). Sampling in forensic comparison problems. *Science and Justice* **38**:101–107.
63. Freiman S., Mecholsky J., Becher P. (1991). Fractography: a quantitative measure of the fracture process. In *Fractography of Glasses and Ceramics* (ed. F. Fréchette), Vol. 2. American Ceramic Society, Westerville, Ohio.

7

Composition, manufacture and use of paint

JOHN BENTLEY

7.1 Introduction

Paints (or surface coatings as they are termed technically) are generally recognised as materials applied to substrates ranging from wood and paper to a variety of metals, plastics and many composite assemblies. They generally have a dual role which is to protect and to decorate, the latter including an ability to disguise. The protective role is that of shielding the substrate from such environmental agents as ultraviolet radiation, moisture and oxygen. This frequently extends to more aggressive attackers such as the salt applied to roads in winter and carried by spray in coastal regions and atmospheric pollutants in industrial environments. In addition to its role in decoration, paint may disguise inferior construction materials or even prevent recognition of the object as with camouflage and infrared reflective paints. Inks are differentiated from paints by being characterised as more highly pigmented and are used as thin coatings laid down in a precise manner, where properties of colour and hiding power are paramount.

Paints consist of three principal components, namely binder, pigment and solvent. The first two are the permanent constituents, with the binder providing the adhesion and cohesion, keeping the pigment within the coating and ensuring that the paint remains attached to the substrate. Pigments provide colour and opacity. Solvents are present to aid manufacture and application, but are lost from the coating during application and the subsequent period of curing, the loss often aided by the application of heat. Varnishes are non-pigmented coatings. This chapter discusses the components, manufacture and use of paint, with emphasis on the decorative, automotive and refinish markets, the last two being similar in ways that will become apparent. See reference [1] for a general introduction to paint chemistry and technology. Sources of further information are provided in the references at the end of the chapter. These include general references [1–4], and references on resin chemistry [5, 6], pigments [7–12], paint making, pigment dispersion and colloid science [13], decorative paints [14] and automotive paints [15, 16].

7.2 Categories of coating

There are many different types of paint, which can be grouped in various ways, the principal ones being by chemistry and by usage (that is by market). In dividing by market, half of paint sales can be designated as decorative or architectural, that is used to decorate and preserve houses and other buildings. An increasingly blurred division for these paints is between 'do-it-yourself' (DIY) and trade, that is those applied by the homeowner and those applied by specialist contractors. The other half of paint sales are 'industrial', more usually split into a number of distinct sectors where general industrial holds a large difficult-to-categorise residue. A further distinction in these sectors is sometimes between paint applied on original manufacture (OEM) and that used for refurbishment.

Large industrial sectors that are reasonably self-descriptive are automotive, marine, aircraft and wood-finishing (furniture). While automotive and wood-finishing are OEM, the marine and aircraft sectors include a sizeable maintenance/refurbishment content. The so-called refinish sector, whose sales volume approaches that of automotive, embraces the repainting of cars after accidental damage, principally after road accidents but also in the factory to repair blemishes. Other sizeable sectors are coil coatings with the painting of steel or aluminium strip, can coatings for the internal protection and exterior decoration of food and beverage cans, and heavy duty coatings for protection of structural steel and concrete. The residual general industrial sector includes agricultural machinery, domestic appliances and a miscellany of finishes principally applied onto metallic surfaces.

7.3 Terms used in the coating industry

Since this chapter uses terms specific to the coating industry, it is essential to define some of these now to avoid confusion. The term 'surface coating' could be very widely embracing, but as used, its context essentially refers to organic paints or finishes. Varnishes are unpigmented coatings where the substrate at least partly shows through. They are used on wood and over pigmented coatings to provide additional protection. The term 'lacquer' may be confusing but the technologist normally uses the term for those coatings which form a film by solvent evaporation only; lacquers may be clear or pigmented. 'Drying' covers all mechanisms of film formation, which is the transformation of applied liquid paint into a solid film, and may involve physical change possibly with polymer particle coalescence and chemical change, not just the evaporation of solvent. 'Curing' usually applies to processes where applied agents such as heat and light are used to effect chemical reaction. Drying and curing are generally polymerisation processes. The terms 'resin', 'vehicle' and 'binder' are all used for the film-forming part of the paint, which will nowadays be polymeric (wholly or partly synthetic), though in early paints may have been simply oil or gum.

7.4 Paint systems

Paint systems are invariably multilayer, because adhesion to substrate, anticorrosion properties, barrier effects, attainment of the required appearance and resistance to atmospheric degradation are difficult to attain from a single layer. These properties are achieved much more readily by different paint layers. Table 7.1 lists typical systems.

Table 7.1 Typical paint layer systems

Decorative		Automotive			
		Solid colour		Metallic	
Gloss coat	50 μm	Topcoat	40 μm	Clearcoat	40 μm
				Metallic basecoat	15 μm
Undercoat	40 μm	Primer surfacer			35 μm
Primer	20 μm	Cathodic electrocoat primer			20 μm
		Pretreatment			2 μm
Substrate (typically wood)		Substrate steel (possibly galvanised)			

The table shows the different layers and their thicknesses; it also indicates some differences in nomenclature but from which one can readily deduced similarities in function. In all cases the function of a primer coat is to adhere to the substrate and make it more receptive to the paint system. Where the substrate is metallic, this is also the most important part of the corrosion protection system achieved by including anticorrosive pigments in the formulation. Undercoats and surfacers provide essentially the same function, that is they provide a thicker layer with full opacity (hiding effect) and at the same time a layer ensuring that the top coats have a smooth base. They are formulated to be sandable, though this is not necessarily carried out.

Gloss in the top layer is ensured by having just sufficient pigment present to provide colour; the layer requires good but not total opacity. Metallic and pearlescent automotive and refinish systems are now always produced by the basecoat/clearcoat system. The metallic coat contains aluminium or mica platelets which, on solvent evaporation after application, align parallel with the surface, giving the required appearance. Because of the size of these platelets, it is not possible to guarantee that they are all fully below the film surface. For this reason full-system durability can only be ensured by covering with a clear (unpigmented) 'varnish' layer.

Automotive bodies now make considerable use of either zinc alloys or galvanised steel, at least in lower areas vulnerable to corrosion. Otherwise, ductile forms of steel suited to press forming are used. To provide the levels of paint adhesion and corrosion protection required, the base metal following degreasing is treated to provide a phosphate layer consisting of zinc or mixed zinc/iron phosphate crystals.

Refinishing systems use essentially the same classes of paint coatings as in the OEM situation. One difference is that where damage or sanding is down to the metal, protection to compensate for the loss of the phosphating may be provided by an etch-primer layer. In addition, to assist in achieving smoothness over a surface which may have had panel beating and heavy abrasion operations carried out, thicker coats of 'putty' filler or sanding surfacer may have been applied. (Putty here refers to a filler which has been applied by knife rather than spray, not to the use of the traditional decorator's material.)

7.5 Formulating principles

The age-old balance required with paint formulation, and hence directed to the resin system specifically, is that of adequately achieving final properties, particularly mechanical strength, getting surface coverage and build, and of being able to apply the paint evenly, all at acceptable cost. In molecular terms, good mechanical properties require solid polymeric binders of high or infinite molecular weight (three-dimensional networks). Most application techniques demand low viscosities, and hence paint compositions always contain some solvent or diluent to make the paint fluid for application. If the binder is of high molecular weight, a proportionately higher quantity of solvent will be required for application than for a low molecular weight binder, the latter, however, having inferior mechanical properties. The effect of this can be seen with spirit varnishes and nitrocellulose paints (lacquers) which may contain 80–90 per cent solvent as they are applied. However, the objective is to build up a solid layer; the solvent is lost and maximising solid content and minimising solvent content should be the formulator's aim. The well-recognised solution to this is to use a curing or drying reaction so that lower molecular weight binder can be used which will increase in molecular weight and improve its properties following application. Lower molecular weight binders require less solvent to achieve a given viscosity.

Reaction can be assisted by component or catalyst addition just prior to application and by the use of heat or other radiation after application. Where possible, ovens are used. It will be recognised that a car body under construction is metal only and can be stoved at high temperature, whereas a car under repair will have tyres and significant amounts of rubber and plastic trim in place and can tolerate only limited raising of temperature. For this reason, traditional refinishing has been achieved with room-temperature drying. However, in recent years, and becoming dominant, 'low-bake' assisted drying systems are being used.

Paints can be applied by brush, spray, roller coating and dipping, house paints being applied mainly by brush and roller; factory systems include all of these. Spray application dominates for most painting of cars with one major exception, which is for the application of the first primer layer to bodies. In this case, a form of dip application has been developed to force the paint to deposit by the application of an electric current. So-called 'electrodeposition' of the primer is now used for all private vehicles and also many commercial bodies, by which, in a process analogous to electroplating, an even paint layer can be deposited on all surfaces, including those inside box chassis sections. Another spray technique is that used with powder coatings, where dry paint particles of size 30–40 μm are applied by electrostatic spraying. These coatings are now being used, for example, on truck cabs, particularly in the USA, and as clear coats on cars in Germany.

Solvent or diluent, as already stated, is an essential component of paints. The solvent used for oil paints has moved on from natural turpentine to the petroleum distillate, white spirit. Short oil alkyds, acrylic and epoxy resins require the use of a full range of organic solvents (aromatic, esters, ketones, etc.) to achieve full solubility. Solvent choice is determined both by the resin system and by the method of application. For brush application, the presence of solvent assists flow and lapping. In spray application, the properties of the solvent blend used crucially affect the spray appearance. Users now expect limits to the solvent present, differing for different coating types, and these limits are being enforced by legislation. Water is now the diluent of choice, with polymers often present in dispersion rather than in solution in many instances. It is not correct, however, to assume that all solvent traces are lost from the coating after application and cure. In some hard coatings, solvent can be retained in small but measurable amounts because of the crosslinked and glassy nature of the coating.

7.6 Pigments and extenders

Pigments contribute colour and opacity to the coating; in addition, they may have an effect on mechanical properties and anti-corrosion properties, and can have a considerable effect on cost. Generally there is an optimum pigment loading to achieve their colouring effect; beyond this they have little effect. Extenders are a further particulate component; they are not opacifying, and are generally cheap. However, their presence affects the mechanical and application properties of the paint, and also film appearance, that is gloss, when present in higher concentrations.

To function as pigments and to provide opacity, pigments must scatter light and be of optimum particle size. To scatter, the refractive index of the pigment must be different from that of the resin in which it is dispersed. Extenders, which do not scatter and hence do not provide opacity, have refractive indices close to or identical with the resin. Pigment particles have sizes generally of the order of a fraction of a micrometre. Many are crystalline and have regular shapes, such as cubic, rectangular, and needle-shaped. Spherical shapes are not found. Some materials are plate-like, specifically metallic pigments and micas designed to be light reflective from their flat faces. Micaceous iron oxide is another pigment that by layering enhances the resistance properties of paints by forming a physical barrier layer.

Pigments are of three types, natural inorganic, synthetic inorganic and synthetic organic. It is not useful, however, to separate the types of inorganic pigments, since those pigments that are dug from the earth will have been crushed, washed and graded by size; they are likely to have synthetic equivalents of possible higher purity and more controlled particle size. The natural varieties that are still important are principally ochres, umbers and siennas which are red, yellow and black iron oxides.

The methods used in choosing pigments are as follows:

- Brilliance and clarity of hue – the most attractive, cleanest colours can only be obtained with organic pigments.
- White and black paints – the purest white pigment is titanium dioxide and it is used in its rutile form; the most jet black is carbon (usually considered inorganic). There are no organic blacks and whites.
- Non-bleeding pigments – inorganic compounds have negligible solubilities in organic solvents. Some organics are very insoluble, though others have some solubility in stronger solvents. A first differentiation between pigment and dyestuff is that the former is insoluble whereas the latter is soluble, at least when first applied to its particular substrate.
- Lightfastness – through their chemical structure, inorganic compounds are generally more stable to ultraviolet (UV) light than organic compounds.
- Heat stability – this requires inorganic pigments. Very few organic compounds are stable at or above 300°C. Some decompose or melt at much lower temperatures.
- Anticorrosive action – all anticorrosive pigments are inorganic.
- UV absorption – titanium dioxide blocks harmful UV radiation from the binder and substrate. Fine iron oxides are visibly transparent but again block UV and give protection.

- Reflective and pearlescent effects – metallic aluminium and treated mica are available in platelet form and are used to give these effects. Aluminium flakes are generally plain (though polished) or may be tinted. Pearlescent mica pigments are coated to give interference effects to reflected light as a further enhancement.

From the above, the reader may be surprised that organic pigments find much use. However, as stated, they are the only option for many bright colours. For decorative paints, where the ultimate colour fastness and heat stability are not required, organic pigments such as toluidine reds, phthalocyanine blues and greens and hansa yellows are suitable.

Automotive and refinish require very high standards and where possible will use inorganic pigments. Vat yellows based on flavanthrone and anthrapyrimidine, phthalocyanine blues and greens, perylene, thioindigo and quinacridone reds, and dioxazine or quinacridone violets are used. The range is necessarily restricted.

Up until fairly recently, formulators in automotive and refinish would have chosen from a range of cadmium-, chrome- and lead-containing coloured pigments. On environmental grounds, their use is now either severely restricted or banned, so that the bright lead chromes, scarlet chromes and molybdate orange are little used. Similarly, anticorrosive pigments were until recently very dependent on lead and chromium. Thus, red lead primer was excellent for metal protection. Lead and chromate metal primers are no longer sold in the decorative market. Their use is very restricted in the automotive and refinish markets, so that electrocoat primers are now chromate free and may soon be lead free. Zinc phosphate-based anticorrosive pigments are taking their place.

The primary metallic effect is the so-called flip (or flip-flop) effect whereby the paint appears light in shade when viewed from above but darker when viewed at a glancing angle. The light to dark transition is fairly sudden. The metallic aluminium used is in the form of flakes or platelets which may be 20 μm across but only 0.1 μm thick. For use in aqueous basecoats, passivating treatment with phosphate surfactants is necessary. Pearlescent pigments, so called because they provide a dual-colour effect similar to that from 'mother of pearl', are principally treated mica flake, where controlled thickness layers of, for example, titanium or iron oxide are deposited onto the mica to give interference layers. The latest development has been a very expensive pigment where, in the derived paint, the colour varies across the spectrum, dependent on the angle of view. For all of these pigments, it is vital that during paint application the pigment flakes are aligned flat and as near parallel to each other as possible in the film.

Extenders variously reduce the cost, reduce gloss or affect the rheological properties of paint. Coarser calcium carbonate extenders are generally satisfactory for gloss reduction. Fine-particle silicas and treated montmorillonite clays are effective in providing thickening for rheology control for application. Talc and barytes are other extenders.

A final mention is necessary of opacifying beads, used in decorative emulsion paints. These work on the principle of having voided centres, which in the wet paint are water-filled; however, in the drying process, the moisture diffuses out, leaving an air-filled hole. Two types are in use. Ropaque® is a fine-particle acrylic bead with a single void; Spindrift® is a coarse vessiculated styrene/polyester bead containing both multiple voids and embedded titanium dioxide particles. In both cases the voids are sized to scatter light, in the latter case scatter is aided by the titanium dioxide.

7.7 Paint manufacture

The chemistry of the paint system is defined by the resins and crosslinking processes used; the pigments, extenders and many other additives are essentially 'inert' in the broader sense. Resin manufacturing is an optional activity for the paint maker, since many suppliers provide all of the ingredients of paint. Only larger manufacturers make their own resins, both to gain the economies of in-house production and frequently to include proprietary chemistries or variations in reactants. Hence there need be no chemical reactions involved in paint making within the paint factory. Chemical reaction has occurred at the component suppliers, and then occurs again only on application and curing.

Paint making can be viewed primarily as a mixing process and consists of a number of stages as shown in Figure 7.1. The first stage is millbase manufacture, which is the physical process of pigment dispersion. This is followed by millbase let-down, mixing, testing and adjustment to specification. Both pigments and extenders are supplied typically as dry powders. Although the primary particle sizes are small, these particles have clumped together in the drying process to form agglomerates or aggregates. It is necessary to disperse these in liquid resin.

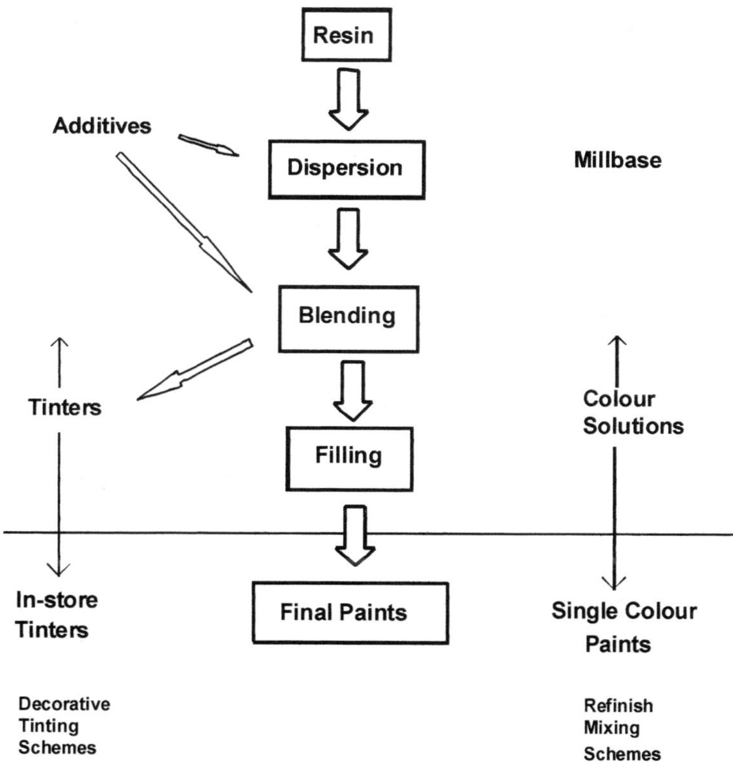

Figure 7.1 Schematic representation of the paint manufacturing process

The millbase, or pigment dispersion concentrate, can be prepared by a variety of processes, and is traditionally known as the grind stage. In this stage the particle clusters are wetted out, separated and colloidally stabilised. Generally there is negligible particle size comminution, the link with grinding being a misnomer, and the stage is one of applying shear forces sufficient only to break aggregates back into the unit sizes predetermined by the pigment manufacturer. The quantities of pigment, grind resin or dispersant, and solvent required to assemble a formulation are determined more empirically than theoretically. The object is to produce a stable, fine dispersion of pigment in the minimum of solvent and resin within a given time. Once this is achieved, further resin is added carefully to minimise colloidal shock, preferably in controlled high-shear conditions. The mixture is then transferred to other mixing vessels, where further resins, additives, solvents and tinters are added to make the paint up to its final specification. The paint is then tested, filtered and filled to containers for storage and transport.

Many pigments are treated by the manufacturer and coated to make the dispersion process easier, and hence the primary or even the sole pigment dispersion stage is now frequently carried out in high-speed dispersion (HSD) equipment. A floor-standing device is pictured in Figure 7.2, and detail of a disperser blade in Figure 7.3. The blade revolves at considerable speed with tip speeds of 20 m/s or higher encountered. The aggregate size reduction is achieved through the shearing action, when the millbase passes through the high-energy zone around the disc. The quality of dispersion achieved using HSDs is generally sufficient for emulsion paints and primers, but has its limitations for high-quality finishes.

A range of other equipment is available for dispersion, and this includes ball mills and different types of bead mills and attritors. These all work with the aid of so-called grinding media which are variously steel or porcelain balls, glass beads, zirconia or sand particles,

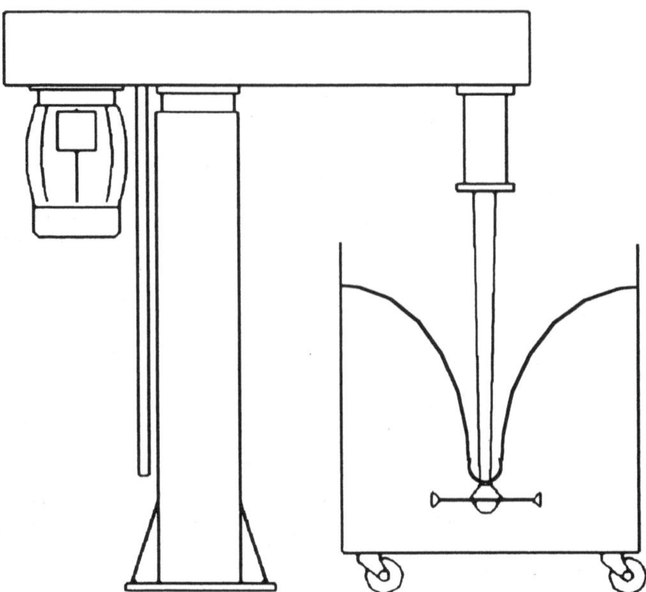

Figure 7.2 Outline of a high-speed disperser (with permission from Eiger-Torrance Ltd)

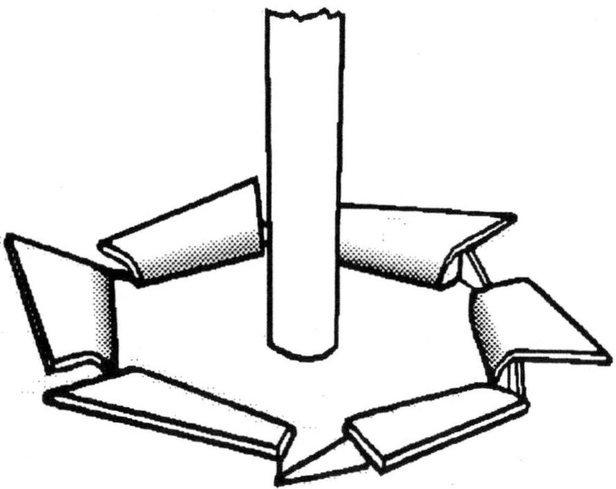

Figure 7.3 High-speed disperser blade (with permission from Eiger-Torrance Ltd)

and so on. These devices are used alone or as a second-stage refinement to obtain the ultimate fine dispersion, free from larger particles, where high gloss paint is required. Thus for decorative and especially automotive and refinish gloss paints, pigment pre-dispersed in HSD equipment will have the dispersion process completed in a bead mill. Other types of dispersion equipment are available such as roller mills and ultrasonic dispersers, but these are little used within the paint industry.

A number of references have been made to colloidal terms and to colloidal stability. Typical pigment sizes mean that paints are colloidal systems, and serious study of pigments, pigmentation processes, millbase and paint stability, and colour development involves detailed knowledge of colloid science. This will not be further discussed here and the reader is referred to Doroszkowski [13].

While the making of paint can be relatively straightforward, the logistics are complex, and several schemes exist for use both inside and outside the factory to ease the problems of supplying one or a range of colours, accurately colour matched, to the end user. The final paint is rarely made using a single pigment but requires a blend of pigments. Although large single batches of paint can be made using all of the pigments required, other methods are possible.

Paints may be made in the factory by mixing different coloured millbases, previously made and stored. It should be noted that smaller amounts of single pigment concentrates or millbases will always be required for use as tinters to adjust the colour of paints to match the required colours. Outside the factory, colour mixing has been used to ease the problems of supplying larger colour ranges. In the decorative market, base white and pastel paints can be tinted in-store to give an extended colour range that the stockist can supply to the customer.

It should be recognised that very significant differences apply between markets. Automotive and refinish paints require the highest standards of colour matching, while for decorative paints the standards are somewhat lower. The decorative market generally buys paint in 1, 2.5 and 5 litre containers. In the automotive OEM situation, large batches of single colour are required, which may be supplied to the factory in 200 litre drums, larger portable containers or even road tankers. By contrast, even the repainting of a complete car requires only a few litres of paint, while the stockist must be able to supply a large range of colours and shades for all models, even where just a single car manufacturer is involved. In this case, mixing schemes are used. For these, a range of single pigment paints of controlled colour strength are kept, the stockist having recipes to enable him to mix all the required colours and shades, including metallic and pearlescent paints.

7.8 Overview of paint resin chemistry

Binders have evolved from the initial range of naturally available vegetable oils, gums and resins, through to the present vast range of carefully designed synthetically based polymeric materials. Oils have played a significant role over the years, and latterly have been chemically combined into a range of polymer compositions; the use of most other natural resins and gums has ceased.

The paint industry uses a subset of the polymer systems available from the polymer and adhesives industries and has tailored these in often sophisticated ways to its requirements. This section will be restricted to the systems used in the decorative and automotive/refinish sectors, and will highlight those systems most encountered. For fuller detail the reader is referred to the references at the end of the chapter.

7.8.1 Basic polymer systems

While use is still made of cellulose derivatives, for example cellulose acetate butyrate (CAB) in automotive resins, nitrocellulose (NC) in refinish systems and certain others as thickeners in decorative paints, all other polymers now used are synthetic.

Both step-growth and chain-addition polymers are used. The principal step-growth polymer is the polyester, where phthalic acid isomers and adipic acid are typical acidic components. A very broad range of hydroxylic components are also used, including glycerol, trimethylol propane and pentaerythritol. The structures of some polyester components are illustrated in Figure 7.4.

Components are chosen on the basis of cost and of performance related to need, where structural variation affects susceptibility to hydrolysis and UV resistance, as well as to immediate properties of hardness/flexibility, adhesion, corrosion resistance and colour.

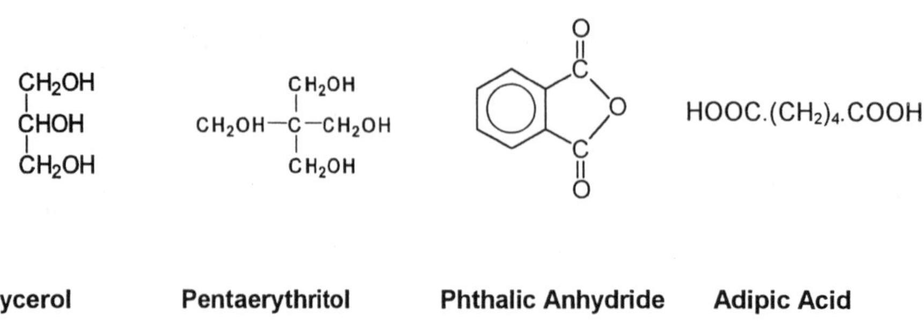

Figure 7.4 Structure of some acids and alcohols used in paint resins

Unmodified polyesters from these components are used in automotive and refinish paints, crosslinked by melamine formaldehyde resins, and by blocked and non-blocked isocyanates. Polyesters require stronger organic solvents – aromatics, esters and ketones. Polyester resins are reactive since the polymer ends will typically have hydroxyl groups, and these resins are not useful without further reaction.

Oils, whose outline structure is shown in Figure 7.5, are no longer used unmodified in coatings but are now incorporated chemically into polymer structures. Natural oils are triglycerides, where the fatty acids are always mixed, and are predominantly C_{18}, though others may be present. Principal fatty acids that are encountered in currently used oils are illustrated in Figure 7.6.

The saturated lauric acid is predominant in coconut oil, while stearic acid is present as a constituent in almost all oils. Oleic acid is singly unsaturated and is also common in nearly all oils. Linoleic acid is doubly unsaturated and is present in the semi-drying soyabean oil and tall-oil fatty acid TOFA (a by-product of the paper industry). The terms 'drying', 'semi-drying' and 'non-drying' refer to whether the oil or fatty acid supplied contains significant amounts of unsaturated components and to the extent related to this content that film-forming reaction can occur through autoxidation.

When incorporated into a polyester structure the product is termed an alkyd (coined originally from the components alcohol and acid, the latter corrupted to kyd), or more

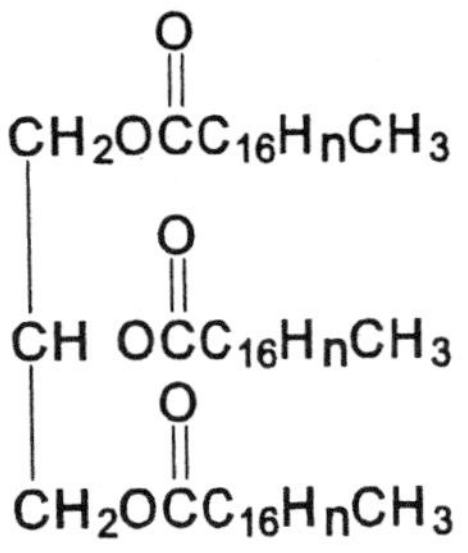

$$
\begin{array}{l}
CH_2O\overset{O}{\overset{\|}{C}}C_{16}H_nCH_3 \\
| \\
CH\ O\overset{O}{\overset{\|}{C}}C_{16}H_nCH_3 \\
| \\
CH_2O\overset{O}{\overset{\|}{C}}C_{16}H_nCH_3
\end{array}
\quad n \text{ may be} = 32, 30, 28 \text{ or } 26
$$

Figure 7.5 Structure of vegetable oils

$CH_3(CH_2)_{10}$-COOH

Lauric Acid

$CH_3(CH_2)_{16}$-COOH

Stearic Acid

$CH_3(CH_2)_7$-CH=CH-$(CH_2)_7$-COOH

Oleic Acid

$CH_3(CH_2)_4$-CH=CH-CH_2-CH=CH-$(CH_2)_7$-COOH

Linoleic Acid

Figure 7.6 Fatty acids found in paint resins

properly an oil-modified alkyd. Alkyds using soyabean oil or TOFA will be termed drying-oil alkyds.

The terms long oil, medium oil and short oil are used to categorise the oil or fatty acid content of the total resin. This concept guides the flexibility of the film and, if containing a drying oil, the rate and extent to which oxidative drying can be expected. For alkyds, elastic 'long-oil' resins contain more than 60 per cent oil, 'medium-oil' resins 40–60 per cent, and the harder 'short-oil' resins less than 40 per cent oil.

Alkyd resins will be encountered in two forms, namely in solution in more traditional decorative and industrial finishes, and now occasionally in emulsified form in decorative water-based finishes. Solution-based decorative alkyds are typically supplied in white spirit, an aliphatic hydrocarbon blend. Emulsified alkyds preserve alkyd properties while eliminating solvent to enable eco-labelled products to be formulated. This class of polymer is prepared without significant solvent in the reactor. Hence the choice of diluent may be simpler than for chain-addition polymers. However, in practice a choice will be necessary as to whether polymer modification or the use of additives (surfactants) is required to achieve the emulsified product.

Autoxidation is the reaction of atmospheric oxygen with the unsaturated fatty acid components of resins. The process is a free radical reaction, whereby hydroperoxides are first formed on susceptible carbon atoms on the fatty acid chains which are at positions adjacent to unsaturation. Under the influence of so-called driers, which are soluble salts or soaps of transition metals, hydroperoxides decompose, then react to form links between chains thereby changing an initial mixture of discrete molecules into a crosslinked mass where essentially all molecules are linked together. The process is accompanied by a number of side reactions and breakdown products which give the autoxidation process its characteristic smell. These products are shorter molecules including alkanes, aldehydes and ketones.

Drier metals used always include cobalt in combination with other metals which may be zirconium, aluminium, calcium or sometimes strontium and iron. Lead is no longer in use in decorative paints. The drying process in air is slow and is used mainly for decorative paints. The process is accelerated by heat and drying-oil modified polymers may also be present in some stoving primers and some automotive underbonnet component finishes.

Longer oil-length alkyds formulated with drying oils will be encountered in decorative formulations, and medium-oil drying-oil alkyds in a range of industrial coatings. Long-oil alkyds may be modified to enhance their properties. Thus reaction of a low molecular weight alkyd with toluene diisocyanate (see later) produces a urethane alkyd with tougher properties. A long-oil alkyd reacted with 5 per cent of polyamide resin will have its rheological properties altered to result in a thixotropic alkyd. Medium- and short-oil alkyds, most usually with non-drying oils, will be encountered used unreacted as plasticising resins, and also in formulations where, as with polyesters, chain-end hydroxyl reactivity is utilised.

Chain-addition polymers encountered are those made principally from (meth)acrylates, styrene and the vinyl monomers vinyl acetate and vinyl chloride (Figure 7.7). They will always be copolymers, with a blend of hard monomers, for example methyl methacrylate, styrene or vinyl acetate, with one or more softer or plasticising monomers such as ethyl acrylate, 2-ethyl-hexyl acrylate or butyl acrylate. The vinyl monomers familiar in bulk plastics are little encountered, ethylene has minor use in latex compositions, and propylene is not encountered.

Chain-addition polymers need to be prepared in the presence of diluent, so that very fundamental formulating differences are encountered. Products will be produced by either

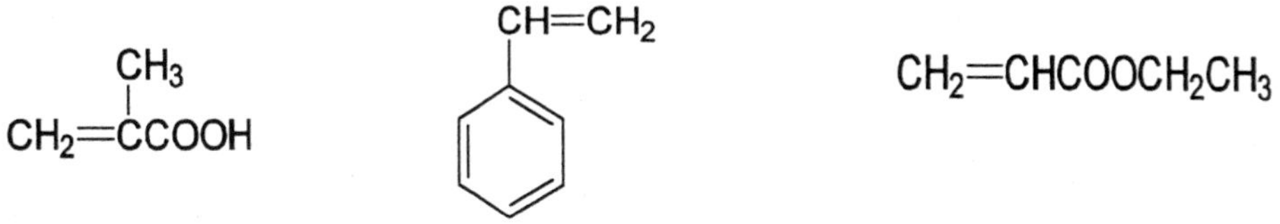

$$CH_2{=}C(CH_3)COOH$$

Methyl Methacrylate

$$C_6H_5CH{=}CH_2$$

Styrene

$$CH_2{=}CHCOOCH_2CH_3$$

Ethyl Acrylate

Figure 7.7 Examples of monomers used in chain addition polymers

solution or emulsion polymerisation. Preparation in solution requires the use of aromatic, ester, alcoholic and ketone solvents, the blend used depending on the polymer and the application. Emulsion polymerisation is carried out in water with the use of appropriate surfactants. The product then consists of polymer particles of size typically 0.2 μm.

Both types of polymer may be formulated to be reactive or non-reactive, in the former case by incorporating monomers such as hydroxy ethyl acrylate. If they are of high molecular weight, acrylic polymers in solution may find application as lacquers. Emulsion polymers are used decoratively in non-reactive form. Providing that the hardness of the polymer in particles is appropriate, particles coalesce to form a continuous film when water evaporates after application. Hydroxy reactive emulsion polymers are used in stoving applications.

The third major polymer encountered is the epoxy resin. This is a step-growth polymer, whose basic structure is illustrated in Figure 7.8. It may be purchased in its final required form, or the basic epoxy, the glycidyl ether of bisphenol-A (diphenylol propane), may be purchased and further reacted by the coatings' supplier. The basic component will always be purchased to avoid handling toxic epichlorohydrin.

This class of step-growth resin has a structure without hydrolysable groups in its backbone, unlike the polyester. Its major use is in primer formulations with anticorrosive pigments and applied to metallic substrates; it will be encountered primarily in electrocoat resins. In this case the epoxy resin is made water dispersible by attaching amine groups to the polymer which are then neutralised with salt-forming acetic or lactic acids.

While the above will have led the reader to believe that resins will be found only either in solution in solvent or in dispersion form in water, most resin types can be made partly or completely soluble in water by chemical modification. In general this means the inclusion of solubilising groups, principally salt-forming groups in the resin structure. In the case of polyester resins formulation and preparation technique will together be used to make a resin with a significant acidity from unreacted carboxyl groups. These groups, when neutralised with an amine such as dimethyl ethanolamine, will enable the polymer to be

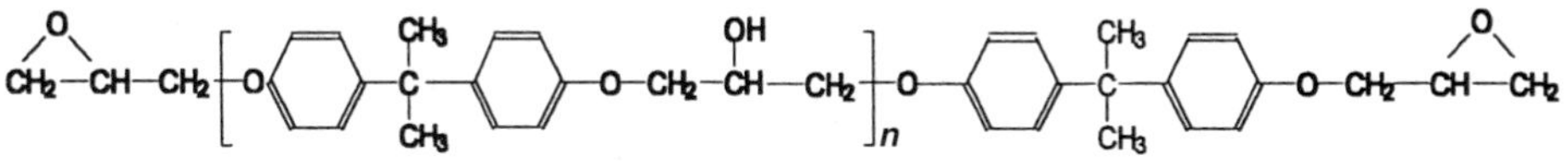

$$\overset{O}{\overbrace{CH_2{-}CH}}{-}CH_2{-}\left[O{-}C_6H_4{-}C(CH_3)_2{-}C_6H_4{-}O{-}CH_2{-}CH(OH){-}CH_2\right]_n{-}O{-}C_6H_4{-}C(CH_3)_2{-}C_6H_4{-}O{-}CH_2{-}\overset{O}{\overbrace{CH{-}CH_2}}$$

Bisphenol-A-based epoxy resin (n = 0–12)

Figure 7.8 Basic epoxy resin structure

diluted with water. This technique is also possible with acrylic resins. These may also be made with attached amine groups which could be neutralised with an organic acid. Cationic resins of this type are not generally encountered, though they are used in industrially electrodeposited resin, which may be used on automotive components. All water-soluble resins of this type will have water-miscible organic solvent present, typically butoxy ethanol. This helps both in resin preparation and in paint formulation and application.

7.8.2 Crosslinking systems

In addition to the oil-derived autoxidative system already described, a range of reactive chemical components are added to attain crosslinking of the resin types described above. The most established of these are crosslinking resins based on formaldehyde, that is the phenol/formaldehyde (PF), urea/formaldehyde (UF) and melamine/formaldehyde (MF) resins. These are based on the reaction of formaldehyde with compounds with susceptible hydrogen atoms to form methylol derivatives, these then being alkylated to modify their reactivity and solubility. The basic reaction scheme is seen in Figure 7.9. The scheme also shows one reaction by which these reactive products can link with hydroxy-reactive polymers. These reactions usually take place with the aid of heat and in the presence of acidic catalysts.

While PF and UF resins are still used, the major resins of this type encountered are the MF resins based on melamine. Formaldehyde can react with the N–H groups, and a dominant reactive MF is now hexamethoxymethyl melamine as illustrated in Figure 7.10. This has largely displaced the use of more polymeric MF resins which were etherified with alcohols such as butanol.

The other major reactive system is that involving isocyanate resins, whose reaction with hydroxy compounds gives the urethane group. The isocyanate grouping is highly reactive so that with suitable catalysis, it can form the basis of a highly effective room-temperature curing system. The links formed are hydrolysis resistant and through their polarity contribute hardness and toughness to the cured paint system. Since the components are reactive, it is essential to keep them apart until reaction is required. These systems are so-called two-pack, now referred to as 2K from the German ‘zwei komponenten’.

Isocyanates can also be used in stable one-pack systems where the isocyanate group has been blocked by another group. The link with this blocking group is reversible at higher temperatures, and since these compounds are lower molecular weight, they are also volatile

$\text{R–H} + \text{HCHO} \rightarrow \text{R–CH}_2\text{OH}$
Reaction with formaldehyde to give methylol derivative

$\text{R–CH}_2\text{OH} + \text{R}'\text{OH} \rightarrow \text{R–CH}_2\text{OR}' + \text{H}_2\text{O}$
Further reaction to give alkylated derivative

$\text{Polymer–OH} + \text{R}'\text{OCH}_2\text{–R} \rightarrow \text{Polymer–OCH}_2\text{–R} + \text{R}'\text{OH}$
First step in crosslinking reaction, with loss of alcohol

Figure 7.9 Reaction scheme for the synthesis of methylol derivatives

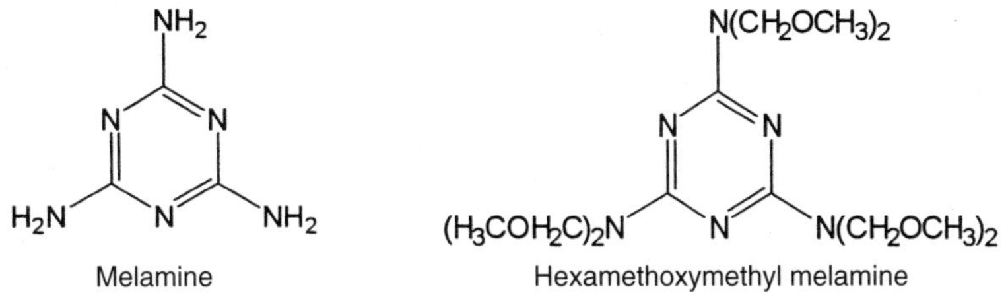

Melamine   Hexamethoxymethyl melamine

Figure 7.10 Melamine and reactive hexamethoxymethyl melamine

and lost from the film on stoving. Suitable blocking agents include low molecular weight alcohols and lactams. These reactions are shown in Figure 7.11.

Isocyanates used in coatings, including hexamethylene diisocyanate, isophorone diisocyanate and to a lesser extent toluene diisocyanate, are illustrated in Figure 7.12.

Isocyanates are irritants and sensitisers, and for this reason low molecular weight isocyanates are not used. Instead a number of reacted variants are available; these include adducts with water (allophanates) and isocyanurates, each resulting in a compound trifunctional with respect to isocyanate.

This section has given an overview of those systems most likely to be encountered, and has endeavoured to put them in some context. However, other resin systems may be met as part of decorative and automotive/refinish systems, and certainly in the wider area of industrial coating systems. The reader is referred to the list of references, particularly [15–17], for further information.

R–NCO + HO–Polymer → R–NHCOO–Polymer
Reaction of isocyanate with hydroxy-polymer to give urethane

R–NHCOOR′ + HO–Polymer → R–NHCOO–Polymer + R′OH
Reaction of alcohol-blocked isocyanate to give urethane

Figure 7.11 Polyurethane reactions used in paints

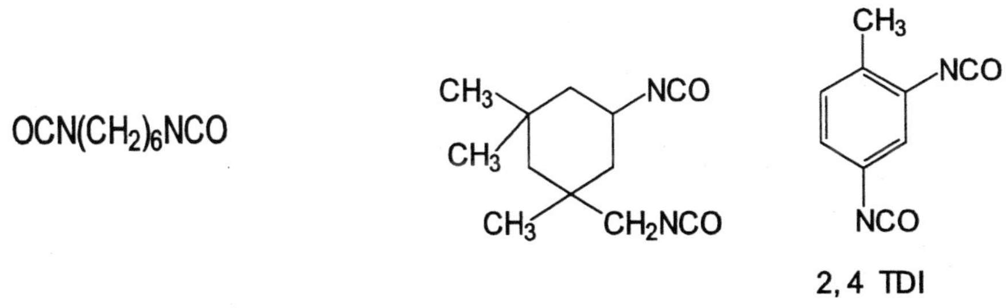

Hexamethylene diisocyanate   Isophorone diisocyanate   TDI Isomer

Figure 7.12 Some isocyanates used in paint resin systems

7.9 Decorative paints

The effects of environmental pressures coupled with enforcing legislation is driving the decorative paint area increasingly waterborne, and is eliminating or has already eliminated the use of major airborne and environmental pollutants; for example, lead is now no longer included in any decorative formulations in either pigment or drier (soluble soap) form. Two types of paint dominate decorative paints, those based on drying-oil alkyds and those based on vinyl/acrylic latexes. Decorative alkyd paints are predominantly solventborne; vinyl/acrylic-based paints are waterborne, only containing minor amounts of 'solvent' in many cases to aid particle coalescence.

Decorative painting involves the painting for decoration of interior wood and plaster, and the painting principally for protection of exterior wood and cement rendering. Paint systems have evolved from the use of simple bound whiting used in distempers for plaster, and the use of oil-based paints for woodwork. Most interior surfaces are now painted with aqueous emulsion paint where appearance can vary from a high sheen to completely matt. In the UK, white and pastel paints are the most popular colours and contain both white titanium dioxide pigment and amounts of extender such as calcium carbonate or china clay. Colour is provided mostly by inorganic oxide pigments, in addition to the white titanium dioxide. Matt latex paint frequently also contains Spindrift® beads, while higher sheen paint may include Rhopaque® in the formulation.

Interior latexes may be based on either vinyl polymer (vinyl acetate as the hard monomer) or acrylic polymer (with methyl methacrylate). The latter resins are used in the higher sheen products and where the environment may need better performance, for example in kitchens, prone to grease pick-up, and in bathrooms where moisture is more problematical. The polymer composition will comprise a hard monomer, a plasticising monomer and often a minor monomer included to improve adhesion to old gloss paint and bare wood in damp conditions. A range of additives is essential in formulating water-based paints including dispersants, anti-foam agents, can corrosion inhibitors and biocides.

Fundamentally, latex paints have poorer application properties than oil paints, making even coverage and brush application avoiding brush marks more difficult. Various thickening options are available to the formulator. Choice is from a range of water-soluble polymers, ranging from modified celluloses (hydroxy ethyl cellulose and carboxy methyl cellulose as examples), through acidic acrylic polymers to more complex materials known as associative thickeners. These latter are soluble block copolymers with components which include polar groupings, for example urethane segments. Paints may also be 'gelled' to various extents by the addition of titanium chelates, paints for roller application being more highly gelled to prevent roller-splatter.

Latex for exterior paints requires a higher hydrolysis resistance and will be formulated to resist both damp and mould. Resistance from alkaline hydrolysis is found with acrylic latex and particularly from those vinyl latexes incorporating Vinyl Versate (Shell Chemicals) as their plasticising monomer. While the use of biocidal and fungicidal additives is now kept to a minimum, the use of these additives is essential in these particular paints.

Some external timber (cladding, fences, sheds) may be treated with specially formulated latex paints containing water-based preservative. Generally, however, both interior and exterior wood is treated with solvent-based alkyd products. Full gloss paints of satisfactory performance and appearance still cannot be formulated from water-based paint. Solvent-based alkyd is hence the basis of most gloss paint, gloss varnish, woodstain and floor varnishes.

The basis of the alkyd paint is the long-oil soya or tall-oil alkyd. Where greater toughness is required, polyurethane alkyd may be blended, in greater proportion for floor varnish. Thixotropic alkyd will also be included to improve application properties, preventing runs and sagging. Pigmentation will be most likely with titanium dioxide and inorganic pigments, but the more expensive organic pigments will be required for brighter colours. Woodstains, which may be transparent or translucent, can contain transparent iron oxides to provide some red coloration along with high UV protection.

Higher durability interior wall paints, both matt and eggshell, have been formulated from alkyd resin. Even when low odour (low aromatic) hydrocarbon diluent has been used, the solvent hazard has remained, and emulsified alkyd-based paints are now being introduced. While emulsified alkyd paints have been used for decades in Europe for external wood cladding, their introduction in the UK has been recent, and formulating possibilities are still being explored.

Decorative undercoats and primers have been traditionally based on alkyd resins. However, water-based primers based on acrylic latexes designed for good wet adhesion can be formulated to give better performance on wood, and have largely replaced solvent-based primers. Solvent-based preservative, however, might be the preferred preparation coat for modern exterior woodstains. Both solvent and latex based undercoats are now available, the advantage of the former being that they brush out better and may be more sandable, the latter, however, being faster drying.

In addition to the above, a range of sundries are used, for example patent knotting, aluminium sealants, and specialist paints such as primers for galvanised metal. Floor coatings and sealants may use moisture-curing reactive or fully reacted polyurethane coatings. Describing all of these is beyond the scope of this summary chapter.

7.10 Automotive and refinishing systems

These paint systems are used on metal bodies which increasingly also have plastic parts such as bumpers which require painting to the same standards as the main body. Pressures include environmental pollution control on the factories. Demands to ensure that vehicles are recyclable affect the paint system principally in the need to avoid toxic metals. This ensures that recyclable components do not require special treatment because of the coating.

While the OEM sector is the most technically advanced, refinish is more traditional. Thus where OEM factories have discarded NC lacquers, air-drying alkyds and acrylic lacquers, these are still evident in refinish. Refinishers who individually have a wide range of facilities and a requirement for flexibility make use of all of these systems. Several contrasting features of the two sectors have already been described. Most evident is that bare car bodies can be and are stoved at up to 180°C, while many refinish paints air dry, with the rest force dried at up to 80°C. Cleanliness in refinish workshops can be a problem, and the ability to polish is a significant advantage; OEM conditions should be essentially dust free.

Automotive OEM primer is now almost always an epoxy-based cathodic electrocoat, which is crosslinked with masked aromatic isocyanate, and still contains anticorrosive pigment. Older anodic systems included epoxy/fatty acid systems and maleinised polybutadienes. The primer and phosphating pretreatment together provide the major corrosion protection for the vehicle. The use of powder primers on truck cabs has already been mentioned.

To increase protection in vulnerable areas, further protection is often found. Thus sills and wheel arches have further antichip coatings, which may be polyvinyl chloride or hydrocarbon resin based. Finally, inner areas such as chassis box sections, inner sills and doors may have a final injected wax spray.

Solvent-borne surfacers are of a number of types with polyester/MF, epoxy/polyester/MF and epoxy/polyester/MF/blocked isocyanate all used. The MF resin is likely to be hexamethoxy methyl melamine. Both aliphatic and aromatic blocked isocyanates are used. Very similar waterborne systems are being introduced. Surfacers are more highly pigmented than topcoats and a typical off-white surfacer will contain titanium dioxide pigment and barytes extender, providing gloss and opacity, and improved filling and cost reduction respectively.

Solids colour topcoats are now acrylic/MF though with some alkyd/MF used. The acrylic resins will contain methyl methacrylate, possibly styrene, and other acrylate and methacrylates including hydroxy monomer, prepared by a solution polymerisation process. Alkyd resins will be medium or short oil, based on a range of both semi- and non-drying oils. The best colour is obtained from coconut or synthetic fatty acid-based alkyds. To reduce factory pollution, high-solid topcoats in which solvent content is reduced are increasingly used.

Metallic paints all now use the basecoat/clearcoat process. Since high film shrinkage of the basecoat is required to align metallic platelets, low solids is essential. Solvent-borne basecoats are likely to be based on polyester or acrylic resin containing CAB. A polyethylene wax dispersion may also be present. In OEM, basecoats are crosslinked with MF, but in refinish remain uncrosslinked. To reduce pollution problems, waterborne basecoats are available in both OEM and refinish sectors.

Clearcoats provide both the final appearance and protection. For the latter, since UV absorption from pigment is absent, additives are essential to absorb UV and counter the effect of any free radicals produced by radiation. Both thermosetting acrylic/MF and acrylic/PU systems using blocked isocyanates are used. Polyester blends may also be used. There is also some use of silane curing systems, and a development is the use of powder resin (solvent free) and slurry (redispersion of powder resin in water) clearcoats.

Refinishing uses systems similar to OEM. However, 2K PU will be used in place of single-pack using masked isocyanate, with aliphatic isocyanates rather than aromatic. Primers, surfacers, solid colours and clearcoats may all be acrylic/PU or polyester/PU. Because of pollution legislation application solids are steadily increasing.

In smaller operations, where low-bake ovens and operator protection against isocyanates cannot be provided, both NC and acrylic lacquer systems are still in use. Putties and surfacers may be based on styrene-thinned unsaturated polyester. Air-drying alkyd systems have poorer performance and cannot be polished; for this reason only a minority of refinishers and the commercial transport sector still use these.

Where refinishers expose bare metal, some protection can be restored by the use of polyvinyl butyral-based etch primers.

Solvents for OEM systems are often simple blends. However, for refinish, to cope with a range of finishes, varying degrees of coverage and application conditions, solvent thinner blends are often complex. These are designed to assist the spray operator, who must be highly skilled, especially with regard to metallic systems.

It has already been noted that refinish-type repair systems will be used in the factory where line damage occurs. With anticorrosion guarantees now standard, refinish systems now seek manufacturers' approval so that repair work will not invalidate warranties.

This section has described current practice in refinish and OEM. However, this has allowed only a glimpse of older systems, which will still be on vehicles in use today. Variations in scale mean that the relatively low volume output from smaller sports car and quality manufacturers and commercial vehicle producers use different systems than those of the large producers.

7.11 References

1. Bentley J., Turner G. P. A. (1997). *Introduction to Paint Chemistry and Principles of Paint Technology* (4th edn). Nelson Thornes.
2. Lambourne R., Strivens T. A. (eds) (1999). *Paint and Surface Coatings, Theory and Practice.* (2nd edn). Woodhead Publishing Ltd, Cambridge.
3. Surface Coatings Association of Australia (1993). *Surface Coatings, Volume 1, Raw Materials and their Usage* (3rd edn). Kluwer Academic Publishers.
4. Wicks Z. W., Jones F. N., Pappas S. P. (1999). *Organic Coatings Science and Technology* (2nd edn). John Wiley and Sons.
5. Bentley J. (1999). Organic Film Formers. In Lambourne R., Strivens T. A. (eds), *Paint and Surface Coatings, Theory and Practice* (2nd edn), pp. 19–90. Woodhead Publishing Ltd, Cambridge.
6. Stoye D., Freitag W. (1996). *Resins for Coatings.* Hanser, Munich.
7. Abel R. G. (1999). Pigments for Paint. In Lambourne R., Strivens T. A. (eds), *Paint and Surface Coatings, Theory and Practice* (2nd edn), pp. 91–165. Woodhead Publishing Ltd, Cambridge.
8. Buxbaum G. (ed.) (1998). *Industrial Inorganic Pigments.* Wiley-VCH, Weinheim.
9. Herbst W. & Hunger K. (eds) (1997). *Industrial Organic Pigments* (2nd edn). Wiley-VCH, Weinheim.
10. Walton R. E. (1993). Titanium Dioxide Pigments. In *Surface Coatings Volume 1, Raw Materials and their Usage.* Surface Coatings Association of Australia (3rd edn), pp. 435–448. Kluwer Academic Publishers.
11. Austin M. J. (1993). Anticorrosive Inorganic Pigments. In *Surface Coatings Volume 1, Raw Materials and their Usage.* Surface Coatings Association of Australia (3rd edn), pp. 409–434. Kluwer Academic Publishers.
12. Broad R., Power G., Sorego A. (1993). Extender Pigments. In *Surface Coatings Volume 1, Raw Materials and their Usage.* Surface Coatings Association of Australia (3rd edn), pp. 514–529. Kluwer Academic Publishers.
13. Doroszkowski A. (1994). Paint. In *Technological Applications of Dispersions*, McKay R. B. (ed.), pp. 1–67. Marcel Dekker New York.
14. Graystone J. A. (1999). Coatings for Buildings. In Lambourne R., Strivens T. A. (eds), *Paint and Surface Coatings, Theory and Practice* (2nd edn), pp. 330–410. Woodhead Publishing Ltd, Cambridge.
15. Fettis G. (ed.) (1995). *Automotive Paint and Coatings.* Wiley-VCH, Cambridge.
16. McBane B. N. (1987). *Automotive Coatings.* Federation of Societies of Paint Technology. Bluebell, PA.

8

The role of colour and microscopic techniques for the characterisation of paint fragments

WILFRIED STOECKLEIN

8.1 Ways in which colour can be described

Colour is one of the most eye-catching and important characteristics of opaque objects such as paints. The determination and comparison of colour is therefore the first step to be taken in forensic paint analysis. The visual comparison of paint fragments with or without a microscope is subjective. The different colour sensitivities of the respective observers frequently lead to non-reproducible evaluations. Although colour must be measured under conditions that are as constant as possible, up until now there have been no uniform guidelines to clearly define the examination of microscopic samples with respect to, for example, illuminant, magnification, sample preparation, etc. Some discussion of this subject can be found in the works of Hudson *et al.* [1] and Audette and Percy [2]. A recommended method for the examination of microscopic samples is to use a stereo microscope with fibre optic tungsten filament illumination (reflected light) against a neutral grey background. With larger paint chips a macroscopic visual examination under different light sources (average daylight, fluorescent and incandescent) against a neutral grey background is also useful. Visual colour description/colour-order systems are still used today in many laboratories. Among the most important of these systems are the Munsell colour coordinate system [3], the Standard colour chart DIN 6164 [4], the Natural colour system [5] and the Methuen Handbook of Colours [6]. Nearly all have been designed to classify colours in a three-dimension colour space which enables them to be given labels or coordinates. For the spatial representation of their hue (colour group), saturation (chroma) and lightness (value), spheres, cones, cubes or other less-regular solids are used. The much used Munsell system [7] arranges the colours by visual interpretation but lacks the support of objective measurement. The new Acoat colour codification system (Akzo Coatings bv, The Netherlands) is based on objective colorimetric procedures. None of the systems mentioned is suitable for the colour determination of special-effect paints that form a high percentage of automotive paints but are also seeing increasing use in other areas, for example for furniture and printed colours. For describing and classifying car paint colours the Colour Collections distributed by the main paint producers have proved successful. These collections contain touch-up paints of standard car colours applied on a paper base. Colour nuances caused by weathering, which may be found on second-hand vehicles as a result of damage, are often included in addition to the original standard shades. In the case of the leading paint producers Du Pont, ICI, Nippon Paint and Herberts, the colour charts/chips are arranged according to car manufacturers and the time periods over which they have been used. A fundamentally different concept was applied by Sikkens (Color map, Akzo Coatings bv, Sassenheim, The Netherlands) and by Glasurit (Color-Profi-System, BASF, Germany). Here closely related colour shades, and their variant colours, of all car manufacturers are combined either on one page of the collection or as swatches where the colour cards form a loose stack, separated according to solid colours and metallic/pearlescent paints. These compilations, which include a total of 3800 (Sikkens) and 3300 (BASF) shades of solid colours plus 1979 (Sikkens) and 5600 (BASF) shades of metallic paints in use since 1955, are particularly suitable for forensic purposes.

It is a prerequisite that comparative colour examinations are carried out by people with perfect colour vision. In this context it is important to know that 8 per cent of men and 0.6 per cent of women are deficient in this respect, some to the extent of being colour blind. Even among those with perfect colour vision, the visual evaluation of colour is subjective and therefore full of uncertainties. This is why at a very early stage of a paint analysis, instrumental tests are carried out to acquire more objective results in connection with colour measurement. In objective colour measurement and colorimetry the subjective single observer is replaced by the colour measurement 'Standard Observer'. The basis for the proper use and understanding of methods for colour measurement is directly associated with the processes of colour vision which take place in the human eye.

Colour perception is a visual sensation caused by a stimulus of the retina by visible light of an appropriate luminance (>3 cd m^{-2} (candela per square metre)) so that there is full light cone vision. For colour vision only electromagnetic radiation between the wavelengths of approximately 380 and 780 nm is of importance. The colour of a paint registered by an observer is the complex interaction between the radiation from this narrow wave region with the pigments inside a paint film and on its surface in the form of absorption, refraction, diffraction and reflection. A paint sample that reflects all of the radiation from the visible spectrum without any absorption is white. A paint that is totally absorbing is black. A paint that always reflects a constant part of the radiation over the whole of the visible range, whereas the remainder is absorbed, appears grey. If certain wavelengths of the white radiation are absorbed more than others the paint appears coloured. If a curve is plotted of the reflectance against wavelength, the characteristic reflectance curve for a paint is obtained. The reflectance curve of a paint sample is an exact objective description of physical characteristics free from subjective psychophysical influences that occur when the human eye observes colour. A spectrophotometer records this spectral curve by comparing reflectance or transmittance from a standard of known spectral response with light reflected or transmitted from a coloured sample, the wavelength being changed step by step through the visible spectrum. With this kind of colour measurement the physical results can be interpreted in a new way such that statements on colour perception may be made. Empirical results gained from the investigation of the colour vision of a large number of subjects serve as a base for colour measurement [8].

In order to understand colour vision one needs to know some facts about the principles and function of the human eye [9]. Light enters through the lens which can alter its focal length so that the image is focused on the retina, the inner wall of the eyeball. Over most of this part of the eye some hundred million rod-shaped nerve endings are found. These rods are only sensitive to light variations, movement, shape and texture, that is, they can only distinguish between black and white. In the centre of the retina there is a small pit called the fovea. In this area are packed some seven million nerve endings which are called cones and provide for colour vision.

Experimental findings show that the visual perception of a certain colour can be satisfactorily matched by an appropriate mixture of three monochromatic radiations. This so-called trichromatic theory of Young and Helmholtz [10] postulates three different types of cones, which are sensitive to different ranges of wavelengths. Outside a 4° field from the fovea, the cone population diminishes rapidly. The varying distributions of cones and rods cause individuals to show different spectral responses of the eye to colour, and also cause differences in colour discrimination between observers.

In order to attain independence of subjective observer judgement and to establish the identity of a colour stimulus on a quantitative scale, the responses by the colour-sensing cones in the retina to different wavelengths of light have been standardised.

8.1.1 Colour space and the use of tristimulus values and chromaticity coordinates

The CIE (Commission International de l'Éclairage) has established three colour-matching response functions of the 2° Standard Observer in 1931, using three imaginary red, green and blue lights derived from additive light-mixing experiments [11, 12]. They specify the relative response of the CIE Standard Observer to each wavelength (Figure 8.1).

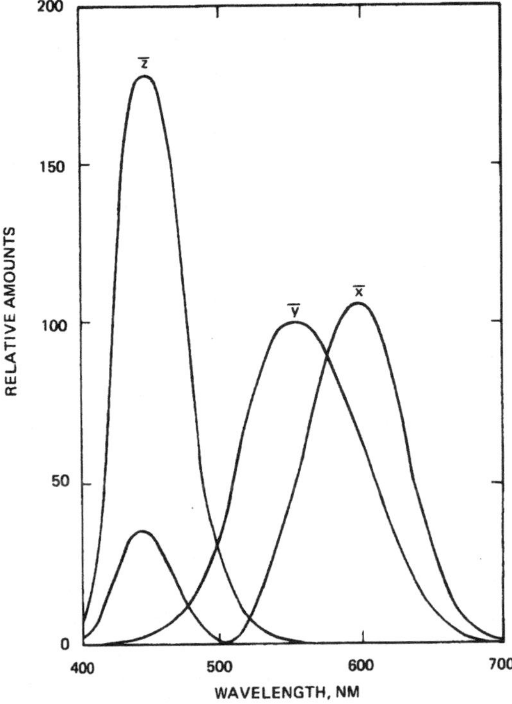

Figure 8.1 The colour matching response function of the 1931 CIE 2° Standard Observer

The $\bar{x}$, $\bar{y}$ and $\bar{z}$ values of each wavelength are called the CIE tristimulus values for that wavelength. The 2° observer was decided upon for technical reasons connected with the equipment used for colour measurement. In practice the human eye sees colour from a much greater viewing angle. Because the spectral sensitivity of the eye is not the same in all positions, in 1964 the CIE established the tristimulus values for a 10° standard observer to conform to reality. They are known as $\bar{x}_{10}$ (λ), $\bar{y}_{10}$ (λ) and $\bar{z}_{10}$ (λ). Since the spectral distribution of the light source is of paramount importance to the colour seen by the eye, it is necessary to define 'standard illuminants'. The CIE has established different artificial light sources, the spectral distribution of which corresponds to natural light sources and which are reproducible in the laboratory. Illuminant A defines light from a typical tungsten incandescent lamp for a colour temperature of 2854 K. Illuminant C represents average daylight from an overcast sky in the 400–700 nm range. Illuminant D 65 corresponds to typical daylight for a colour temperature of 6500 K, where the spectral energy distribution in the 300–400 nm ultraviolet range contributes to the colour appearance.

Just as the illuminant has to be defined, standardised conditions must be established for the geometric relationships of observer, light source and object. Light can be reflected to the eye from the smooth surface of an opaque object such as paint in two ways. Most of the light penetrates the surface, and after selective absorption re-emerges from the surface at varying angles. We call this kind of reflection diffuse because the light is reflected in all directions, and even if we move the light source off the eye the surface colour remains essentially the same. About 4 per cent of light is reflected without interaction from smooth paint surfaces. This component is reflected at the mirror angle (the angle of incidence equals the angle of reflection) and is called specular reflection. Specular reflection is directional and gives rise to the phenomenon of gloss. When observing or measuring colour, one should avoid any specular reflection since gloss highlights will mask the colour. The CIE has defined two geometries for colour measurement (Figure 8.2):

- Illumination at 45° and 0° viewing or vice versa.
- Diffuse illumination and 0° viewing or vice versa.

These measuring conditions are only valid for the colour measurement of paints whose optical properties are determined by absorption pigments.

If the paints contain metallic and/or interference pigments other measurement parameters must be used, not only for visual assessments but also for instrumental measurements. Only by doing this can the 'two-tone effect' as well as the 'colour flop' (also known as goniochromatism), which appear in special-effect pigments, be measured. Until now no standard measuring parameters have been set. Illumination under 45° is recommended for large areas or 'bulk' paint films, with an observer angle of 25°, 45° or 70° from gloss [13, 14]. The American Society for Testing and Materials (ASTM) suggested the parameters 15°/45°/110° (ASTM E-12.03). It should be noted here that for microscopical colour measurements only the measuring parameters of 45°/0° are realisable. Finally, one further condition remains to be standardised, namely the reference material for diffuse measurements. All colours are assessed relative to other colours in the field of view. Therefore, colour measurements are always made relative to the luminosity of a perfectly white object (reflectance equal to 100 per cent at each wavelength). The reference normally used for colour-measuring instruments is the surface of a tablet pressed from pure barium sulphate powder. This reference is calibrated at 10-nm intervals. Another standard that is more stable than $BaSO_4$ over long periods is a polished opal glass – type MS 20 (distributor:

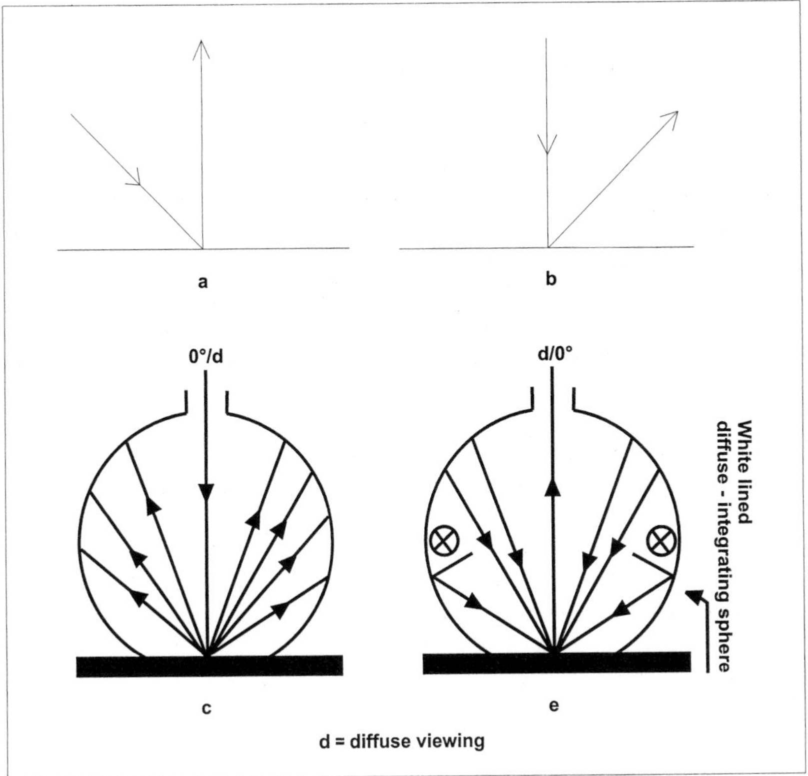

Figure 8.2 Schematic diagram of optical systems designed to measure diffuse reflection: (a) 45°/0°; (b) 0°/45°; (c) 0°/d; (e) d/0°

Hemmendinger, USA). Using the Standard Observer and the standardised light source energy distribution it is possible to convert any percent reflectance or transmittance curve into three numbers, called tristimulus values X, Y, Z.

For reflecting objects these values are integrals of $\bar{x}$, $\bar{y}$, $\bar{z}$, each multiplied by the corresponding energy of the standard illuminant (S_λ) and the measured percentage of reflectance (R_λ) of the object at a specific wavelength

$$X = \int_{380}^{700} S_\lambda \times R_\lambda \times \bar{x}_\lambda \times d_\lambda$$

$$Y = \int_{380}^{700} S_\lambda \times R_\lambda \times \bar{y}_\lambda \times d_\lambda$$

$$Z = \int_{380}^{700} S_\lambda \times R_\lambda \times \bar{z}_\lambda \times d_\lambda$$

Colour-matching response functions and illuminant energy distributions are published as tables. To calculate tristimulus values for a specific illuminant, the tabulated values of the colour-matching functions simply have to be multiplied by the tabulated values of the source and by the reflectance or transmittance values at each specific wavelength (10-nm intervals). Then, each of the three products has to be summed separately throughout the visible range at the wavelength intervals chosen to obtain X, Y and Z respectively. These calculations can be done by computers in a split second. An example of calculation of tristimulus values can be found in the literature [15]. The tristimulus values identify the colour of an object in terms of the mixture of the three imaginary primary lights that match the colour visually. It is important to note that two objects with different reflectance or transmittance curves can have identical tristimulus values.

Objects that appear visually to be of the same colour under one illuminant but have different spectral curves are called metameric. Such colours, which are indistinguishable to the eye under a number of illuminants, can be discriminated by a comparison of the shapes of their reflectance or transmittance curves (Figure 8.3).

The tristimulus values X, Y and Z have no clear-cut colour specifications because they do not correlate with visual attributes such as hue, saturation, etc. Only Y corresponds to perceived lightness, being zero for black and 100 for white. (Y is equivalent to the CIE luminosity function; luminosity refers to the relative visual intensity response to different wavelengths of light at normal levels of illumination.) To develop a relationship between the tristimulus values with colour appearance, the CIE recommends a chromaticity system. In this system chromaticity coordinates are defined as

$$x = \frac{X}{X+Y+Z} \quad y = \frac{Y}{X+Y+Z} \quad z = \frac{Z}{X+Y+Z}$$

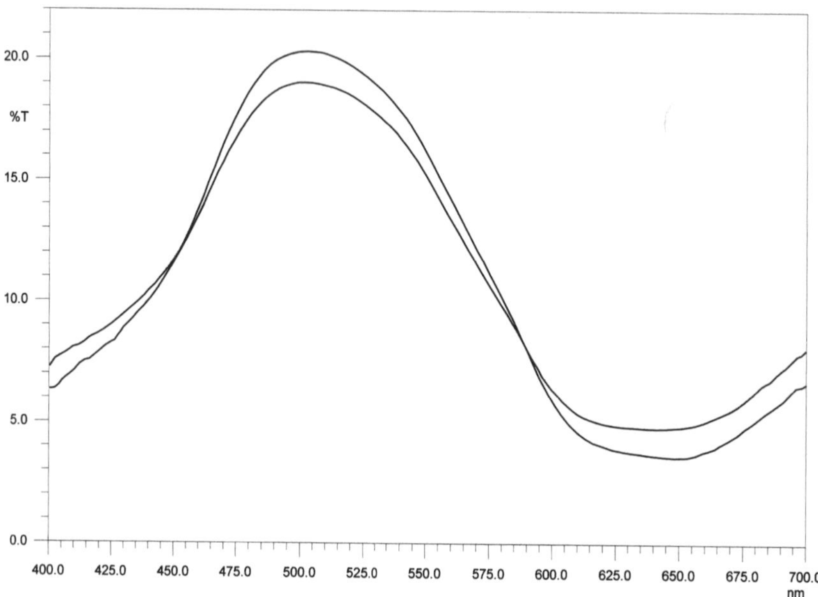

Figure 8.3 Reflectance spectra of two moderately metameric green two-coat metallic paints. (Top coat colour electronic green of VW AG from two different paint suppliers)

As the sum of *x*, *y* and *z* will always amount to 1, only two of the coordinates, usually *x* and *y*, are needed to describe chromaticity. The *x* and *y* coordinates can be used as the axes of a rectangular graph to derive the CIE chromaticity diagram (Figure 8.4).

In such a graph, chromaticity coordinates of all pure spectral colours yield a horseshoe-shaped curve. The non-spectral purple line connects both ends of the spectrum and closes the diagram. All real colours fall within the enclosed area. It must be remembered that such a diagram is two-dimensional and does not take the lightness value (or luminance value, *Y*) into account.

8.1.2 The CIELAB colour space and colour differences

The CIE system is based on experiments in which observers were asked only to match one test light with a mixture of three other coloured lights, not to estimate differences. While the tristimulus values and chromaticity coordinates exactly describe a colour in terms of the human visual response, the CIE scales are non-linear in the spacing of colours as related to visual differences. Equal numerical differences between colours as plotted in

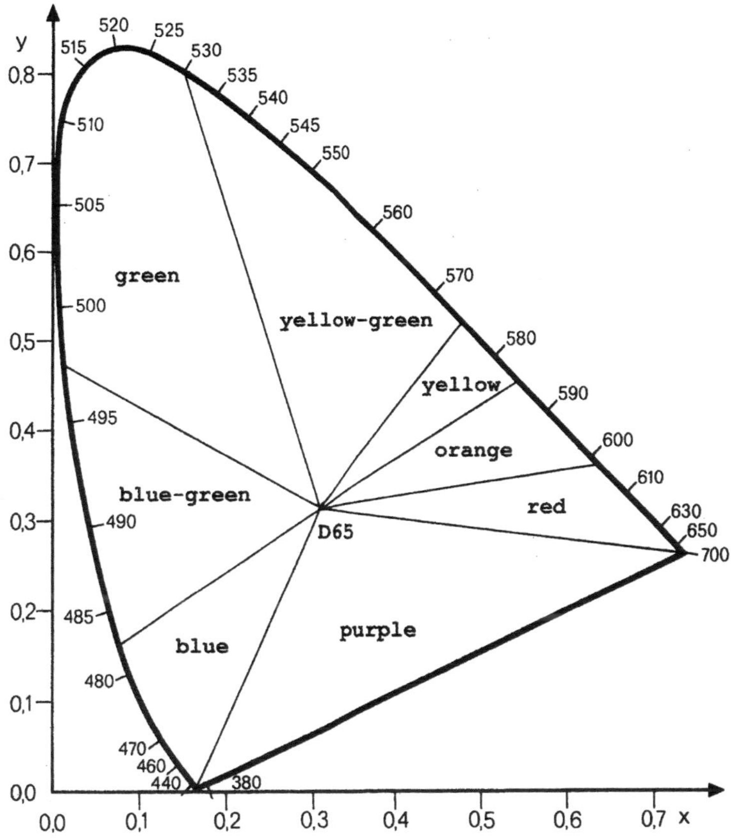

Figure 8.4 The 1931 CIE *x*, *y* chromaticity diagram

various parts of the CIE x, y, Y colour space do not correspond to the size of the differences as seen by the human eye. Thus differences between colours cannot be calculated simply from differences between the respective tristimulus values. The CIE system requires transformation so that equal numerical steps are also visually equal. None of the numerous equations to form equal colour spacing from CIE coordinates is completely successful. As a compromise two transformation equations are employed wherever colour analysis is used for industrial purposes: the CIELAB equation [16] and the new modification of the CIELAB equation CIE 94 [17, 18]. In both equations, the magnitude of the total colour difference calculated from the tristimulus values is represented by a single number. The CIELAB colour difference equation is the most widespread and generally recognised today. It is based on a transformed colour space which is composed of the colour scale values L^*, a^*, b^* (Figure 8.5).

The following mathematical interrelation ($Y/Y_n > 0.01$) exists between the tristimulus values X, Y, Z in the CIE system and L^*, a^*, b^* in the CIELAB system, respectively, these latter being defined by the equations:

$$L^* = 116(Y/Y_n)^{1/3} - 16$$

$$a^* = 500[(X/X_n)^{1/3} - (Y/Y_n)^{1/3}]$$

$$b^* = 200[(Y/Y_n)^{1/3} - (Z/Z_n)^{1/3}]$$

In these transformation equations, X_n, Y_n, Z_n are the tristimulus values of the entirely off-white surface for both the standard illuminant and the standard observer which have been made the basis of the tristimulus values. These values have been tabulated. The total colour difference in the CIELAB colour space is calculated according to the following formulae:

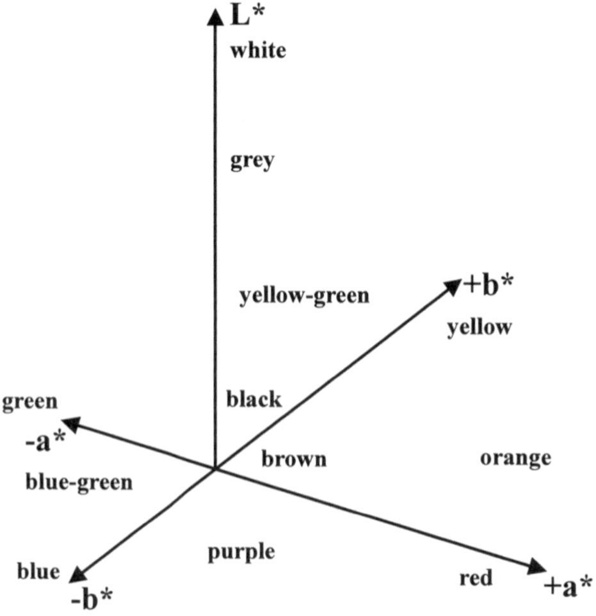

Figure 8.5 CIELAB colour space

$$\Delta E^*_{ab} = [(\Delta L^*)^2 + (\Delta a^*)^2 + (\Delta b^*)^2]^{1/2}$$

$$\Delta E^*_{ab} = [(\Delta L^*)^2 + (\Delta C^*_{ab})^2 + (\Delta H^*_{ab})^2]^{1/2}$$

Both formulae give the same result.

In the formulae the meaning of the symbols is as follows:

ΔL^* = lightness difference, where $\Delta L = L^*_O - L^*_S$ and S = standard sample and O = object sample.
Δa^* = difference in the direction of the red–green axis, where $\Delta a^* = a^*_O - a^*_S$
Δb^* = difference in the direction of the yellow–blue axis, where $\Delta b^* = b^*_O - b^*_S$
ΔH^*_{ab} = hue difference, where $\Delta H^*_{ab} = [(\Delta E^*_{ab})^2 - (\Delta L^*)^2 - (\Delta C^*_{ab})^2]^{1/2}$
ΔC^*_{ab} = chroma difference, where $\Delta C^*_{ab} = (a^{*2}_O + b^{*2}_O)^{1/2} - (a^{*2}_S + b^{*2}_S)^{1/2}$

Figure 8.6 shows an example of two yellow–orange samples with different values of L^*, a^* and b^*, illustrating the differentiating characteristics that have been included in the colour difference formula of the CIELAB system. The CIELAB colour space is also not completely uniform. Even in this system colour differences between two samples which appear visually to be the same are not represented in all locations of the three-dimensional colour space as spheres which have the same diameter. The surfaces having the same visual colour difference form ellipsoids which differ in their form and orientation depending on their position in the colour space. The colour difference formula ΔE^*_{94} introduced by CIE in 1994 improves the situation by using factors to correct the values calculated by using the

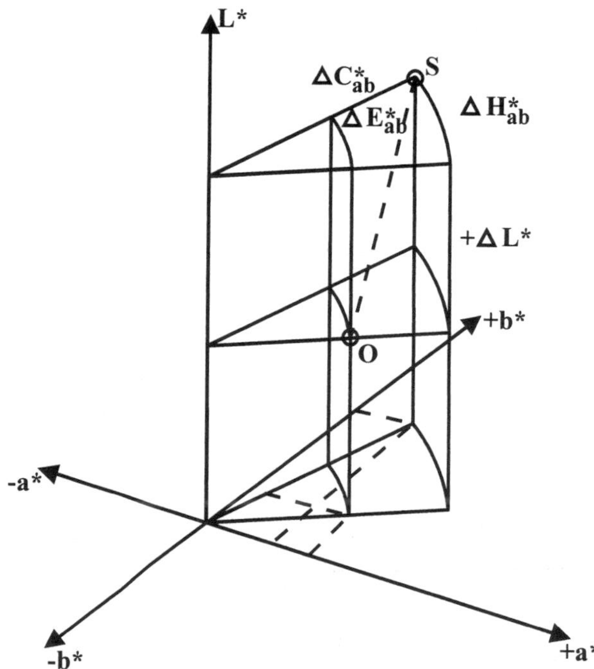

Figure 8.6 Colour difference between standard (S) and object (O) colour in the CIELAB colour space

CIELAB equations. The human eye can detect colour differences more accurately in the white range (ΔL) than in the coloured range (ΔH). According to the German Standard DIN 6175 [19, 20], during automobile repairs colour differences up to 0.3 CIELAB units are permitted. In the blue–turquoise range differences up to 0.5 CIELAB units are allowed, in the green area up to 0.7, and in the red area up to 0.9 CIELAB units.

8.1.3 Instrumentation for colour measurement

Although Williams [21] had previously attempted to make colour measurements on microscopic samples, the microspectrophotometer only became available to the forensic scientist in the 1970s thanks to the integration of modern electronics and personal computers which made them fast and easy to operate. User-friendly software with graphic user interface minimised the time between measurement and obtaining useful results. Reviews about the early use of microspectrophotometers for colour measurement have been written by Stoecklein [22] and Cousins [23]. Modern microspectrophotometric systems usually consist of a research-grade microscope incorporating all the methods of illumination and image formation with both transmitted and incident light.

Systems recommended for colour measurement are today offered by Leica, Zeiss and SEE. The well-known Nanometrics instruments are no longer in production. All microspectrophotometers are single-beam instruments. This means that the instruments measure first the standard or blank, store the results and then measure the sample. To ensure accurate and reproducible results high-stability xenon or halogen lamps and a precise mechanical monochromator design must be available. The system which most lends itself to expansion is currently the MPM 800-system (Zeiss), as shown in Figure 8.7.

The basis of this system is a research microscope equipped with infinity-corrected high-performance glass or quartz optics for the spectral range 240–2100 nm. The photometer unit and the detector form the nucleus of the microspectrophotometer. The measuring diaphragm can be observed or photographed in the binocular or TV tube, regardless of the measuring technique used. The smallest diaphragm corresponds to a measuring field diameter of 0.5 μm in the object plane. The detector unit comprises a photomultiplier (PMT for $\lambda = 240$–800 nm) and an infrared detector (PbS for $\lambda = 800$–2100 nm). Holographic dispersion gratings on both the image-side and illumination-side monochromators ensure optimum resolution and high efficiency over the entire spectral range. High-stability xenon, mercury and halogen lamps are integrated into the system. A microscope system A to D and D to A processor is a link between the photometer and the control and data processing computer (Figure 8.7). The computer is used for instrumental control, measurement and data analysis. The software package contains a library of specialised processing routines including spectral arithmetic and integration. Colorimetry software can be used to calculate tristimulus or chromaticity values, and CIE colour coordinates. Measured values are documented by printer or plotter. In addition, data can be stored on hard or floppy disks. Colour measurements in this system are carried out with an image-side monochromator with a measuring geometry of 45°/0°.

For rapid measurement of colour values of limited dynamic range, the MPM system can be equipped with a diode array spectrometer (MMS or MCS). The light emerging from the microscope is conducted by fibre optics to the diode array detector. At the entrance of a fibre-bundle cross-section changer is a monochromator comprised of a holographic, blazed,

Figure 8.7 The Zeiss MPM 800 microspectrophotometer equipped with the diode array spectrometer MMS

flat-field grating with a fixed entrance slit. The monochromator projects the spectrum onto the fixed diode array. The individual diodes are therefore always aligned to one wavelength and their charges can be read out by the computer and assigned spectrophotometrically. All of the optical components including the grating, the fibre-bundle cross-section changer working as an optical entrance slit and the diode array working as an opto-electronic exit slit are positioned and cemented to a central lens-like glass body. This holds them in place and also offers optimal protection against dust and gases. The spectrometer has a wavelength accuracy of 0.3 nm and a spectral resolution better than 10 nm. Because of the monolithic design the instrument has a high wavelength accuracy and reproducibility. The instrument does not require servicing. The weekly calibration necessary for photometers with mechanical gratings is not necessary for this instrument.

The optical bench of the SEE equipment also consists of a sensitive charged coupled device (CCD) spectrometer combined with a research-grade microscope and the latest infinity-corrected optics. The system takes high-resolution spectra (optional 0.8 nm) from the ultraviolet to the near-infrared region.

Instruments for colour measurement of large samples should not be forgotten. These instruments were used and are still operated by workers in the UK (Forensic Science Service) and Germany (BKA, Wiesbaden) for the building up of colorimetric databases from large-area samples. These instruments, in addition to the measuring geometry of 45°/0°, have geometries available for measuring pearlescent and metallic paints (e.g. instruments from Kollmorgan and Zeiss-Datacolor). The illumination geometry 45°/0° method of signal output detection (diode arrays) and output format of these devices are comparable to that of the Zeiss MPM/MMS diode array microspectrophotometer.

However, they differ in that they employ the double-beam technique, the source of illumination (xenon flash tube) and the sample areas (10 to 15 mm diameter) [22, 24, 25].

8.2 Recovery and preparation of trace paint samples for analysis

In industry, colour measurements are made on clean, undamaged large-sample areas. The area used for the colour measurement is not standardised or defined as a norm, but most colour measurement instruments have measuring areas of 10 mm or more. In contrast, the forensic scientist normally has at his disposal micro traces with an edge length of less than 10 mm in the form of multilayered chips, single-layer fragments or only smears with particle sizes of less than 1 mm. The top-coat surfaces of the samples are frequently dirty or scratched. Under these conditions, sample preparation plays a decisive role in the production of correct, reproducible and comparable results. Two different situations that arise will now be discussed.

In carrying out colour comparisons the weathered areas near the surface of the suspect and control samples must definitely remain intact as they contain valuable discriminatory features. Therefore, in these cases the surfaces of the samples should be cleaned carefully with a cotton plug soaked in alcohol to make them dust free. Following a cleaning procedure developed by Taylor *et al.* [26] and Cousins *et al.* [27] the paint flake is attached with a quick-setting resin or glue to a small, flat platform cut on the tip of a sharpened wooden stick. However, before making colour measurements in connection with the determination of origin (e.g. hit-and-run cases) it is necessary to remove these weathered areas to ensure that a comparison with standard colours does not lead to false results. The removal is accomplished by polishing the surfaces attached on the platform of the wooden stick as described above using either a diamond paste, an aluminium/magnesium oxide mixture [28], cerium oxide [22] or by applying an ultra-milling cutter [22].

8.2.1 *Embedding technique*

In order to obtain accurate results during colour measurement the upper surfaces must not only be free of scratches, dirt and blemishes but they must also be mounted with the surface at a right angle to the optical axis of the microscope. Even with larger samples this can be difficult to achieve if, for example, they are bent or curled up because they have sprung off a curved surface. Such samples should, as with all small samples, be embedded so that a cross-section of the paint chip can be examined. Before measurements the surplus resin must be removed by grinding and polishing. However, polished specimens are only useful for examination of solid paints. With very small samples of special-effect paints top-coat measurements are not possible because the limited area does not allow serial measurements to obtain an average. In these cases, thin sections should be prepared and examined by UV-visible spectroscopy in transmission. Various embedding techniques have been described in the literature [28, 29]. The following technique [30] has proved useful for all sample sizes. A wooden stick of 2 mm diameter with a platform cut on the tip is positioned in a vice and viewed under a stereo microscope. A small piece of double-sided adhesive tape is fixed to the platform of the stick using superglue. The paint chips to be compared are aligned on the adhesive tape side by side (Figure 8.8) either flat to the surface of the adhesive tape (for planar sections) or on edge (for cross-sections).

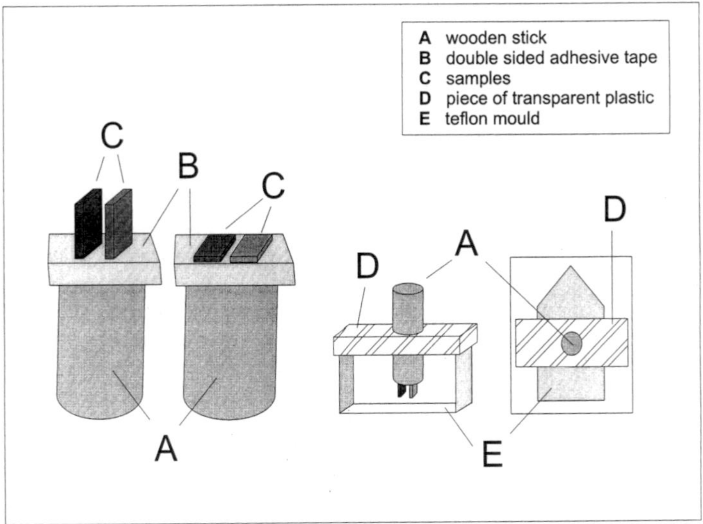

Figure 8.8 Technique for preparing thin sections

In the next step, the prepared samples are positioned and mounted in a special wedge-shaped casting mould made of Teflon (Histoform, Heraeus Kulzer GmbH, Wehrheim, Germany). This is done with the help of a piece of transparent plastic (10 mm × 10 mm × 0.5 mm), through which a 2-mm-diameter hole has been drilled in the centre, bearing a strip of double-sided adhesive tape. The plastic strip is fixed on the mould, after the stick has been pulled through the hole. The cavity in the Teflon mould is then filled with a one-component resin based on a mono- and bifunctional methacrylate *Technovit 2002 LC* (Heraeus Kulzer, GmbH). After curing (30 minutes under blue light) the resin block is removed and then ground and polished. Diamond paste, cerium oxide applied with a polishing cloth, or an ultra-milling cutter can be used for polishing.

8.2.2 Microtome sectioning

For optical and for UV-visible microspectrophotometric analyses, the sample material should be in the form of microtomed thin sections (suspect and control samples embedded side by side). Wilkinson and co-workers [31, 32], Stoecklein and Gloger [30] and Allen [29] have described some suitable methods for microtoming thin sections of paint flakes. Completely extended flat sections can be obtained by using Technovit 2000 LC. This material cures quickly and does not react with paints. The sections of paint are prepared using a tungsten steel blade (3-μm sections for UV-visible analyses as well as for optical microscopy; 20-μm for microspectrophotometric determination of UV-absorbers in clear coats). To stretch the section out flat, it is placed on a microscope slide and treated with a drop of alcohol. Embedded sections of this type can be examined using microscopy and microspectrophotometry at magnifications of up to ×1000.

8.3 Microscopic examination

Stereo and optical research microscopes are the most important tools in the forensic analysis of paint. They are used for the determination of the number, thickness and colour sequence of layers in paint fragments, and for the recognition of textures as well as the fundamental features of pigment and extender mixtures. Polarised light microscopy (incident and transmitted light) up to a magnification of ×1000 [33–35] and fluorescence microscopy are necessary for paint examinations. The determination of particle form, size class distribution and the number of special-effect pigments (metal and pearl lustre pigments) can also be accomplished very quickly with plane-polarised light in incident light/bright field without prior sample preparation. On the other hand, all analyses for the determination of paint structures and for the most complete description of the morphological and physico-chemical features of the coloured pigments and extenders used in the paints (colour, size, crystal habit, crystallinity and refractive index) must be carried out at the highest magnification possible (×250 to ×1000) in transmitted light/bright field or in transmitted light/dark field (crossed polarisers) respectively. The most suitable specimen preparations are 3-μm sections of paint fragments embedded parallel or perpendicular to the surface of the resin. After polarised light microscopy, the microtome sections should be subjected to fluorescence microscopy at an excitation wavelength of 365 nm. Using this technique further differences are often apparent. In addition to pigments, additives and film-forming components also contribute to fluorescence.

Cathodoluminescence (CL) examinations are among the special procedures that can be conducted using optical microscopy. This phenomenon is the excitation of visible light or of radiation of adjacent wavelengths in semiconductors and insulators (e.g. inorganic pigments and extenders) when these are bombarded by electrons emitted from a cathode and accelerated in an electric field. This procedure is used for the comparative analysis of multilayered white or beige paint fragments [35, 36]. Polished sections are analysed in special sample chambers with observation windows or gem chambers with colour video camera observation accessories.

8.4 Microspectrophotometry

The aim of microspectrophotometrical colour measurement of microscopic paint particles is first to measure the reflectance or transmittance spectra from the suspect and control samples and to calculate the tristimulus values or the CIE colour coordinates. Afterwards the results of the examination are evaluated by a spectral comparison or by using the colour differences in CIELAB units. First, the reliability of the results must be checked. This is no easy task with colour measurement. The accuracy of the results depends on systematic errors in the system, for example the photometric accuracy, the spectral accuracy, the measuring geometry and stray light as well as statistical errors such as electronic noise and sample heterogeneity. The calibration of 0 per cent and 100 per cent for an ideal matt reflector and the linearity of the photometric scale are of paramount importance for the photometric accuracy of the reflectance values. Zero reflectance is measured by blocking out the light in the measuring beam so that in reality the dark current of the photomultiplier is measured. Determination of the 100 per cent value is made using a calibrated white standard ($BaSO_4$). The Opal Glass standard MS 20 can be checked against the $BaSO_4$ and then used as a working standard. However, it is necessary to remember that with opal glasses the values measured depend on the size of the measuring field. Suitable coloured standards (Labsphere, Minolta) are used for checking the overall operation of the equipment. Calibration should not take place until at least two hours after the lamp has been switched on so that the photometer has become thermally stabilised. The spectral accuracy of a grating monochromator should be checked weekly with a wavelength calibration standard (such as didynium glass). Inaccuracies of ±0.5 nm are not significant. The measuring geometry of 45°/0° can only be achieved with the help of a dark field illuminator (e.g. bright field–dark field reflector slider) fitted with dark field objectives. The short time stability of a microspectrophotometer is normally excellent. If, for example, solid paints are measured with a measuring slide of 50 μm × 50 μm several times on the same spot colour, differences in CIELAB units of <0.05 are normal. The heterogeneity of the sample material can be a problem, particularly with special-effect paints. The statistical uncertainty of the results increases considerably as the size of the measuring area decreases. Therefore, an evaluation of mean values from repeated measurements (5 to 10, at least) in different areas is necessary. All measurements should be carried out under identical conditions of temperature (e.g. time of illumination). Because of the thermochromy of pigments a difference in temperature of only a few degrees can result in distinct colour differences [37]. As already stated, to obtain correct values it is essential that the sample is mounted with the surface perpendicular to the optical axis of the microscope [22]. In addition, the upper surface of the carefully cleaned paint sample must be exactly in the focal plane of the objective [38]. As paints may show the characteristics of fluorescence, colour measurement should be carried out exclusively using an image-side monochromator [39]. If all measuring conditions are strictly followed the mean colour difference for replicates of solid colours measured in cross-section is 1.86 CIELAB units with a standard deviation of 1.23 CIELAB units [40]. To evaluate the characteristics of reflectance spectra (peaks, shoulders, inflections) a visual comparison is normally sufficient [28]. Laing *et al.* [41] suggested a simple mathematical approach (the square root of the sum of the squares of the numerical differences between reflectance values at 10 nm intervals over the range 390–740 nm) to evaluate the differences in the reflectance curves from household gloss paints. Using this calculation Laing *et al.* found a discrimination power of 0.97 for surface measurements of white paints and 0.89 for cross-sectional measurements.

For comparative colour examinations a general ground rule must be complied with. The control sample must originate from as near as possible to the suspect sample. If this is not the case it is probable that with top-surface measurements such great differences will arise that it is not possible to determine whether the samples came from the same source. Light, weathering and ageing in the upper zones of a paint sample can lead to such strong variations within one object (e.g. the covered part of a painted window frame) that, for example, in surface measurements of car paints maximum colour differences up to 6–7 CIELAB units can occur [28, 42] in different areas of the same vehicle. In areas of paint adjacent to each other in new cars, provided the paint is original, colour differences are normally not greater than 3 CIELAB units [22]. A colour difference that exceeds this value shows that the two paint layers being compared have a different colour. If surface measurements are made on several areas from the same coloured surface of individual older cars, differences of up to 14.5 CIELAB units can occur. This might be a consequence of weathering and fading or of batch-to-batch differences from one paint supplier or it might be due to metamerism if OEM (Original Equipment Material) or repair paints from different suppliers are used. If the same samples are examined as cross-sections the differences can decrease to 3.0 CIELAB units, especially when the differences are due to weathering effects

[22]. Identical chromaticity coordinates or small colour differences do not in fact imply that the paints also contain the same pigments. They just mean that the colour of the samples cannot be distinguished by the human eye. Paints showing the same tristimulus values may be made up of entirely different pigment formulations. If that is the case we are dealing with metameric colours, the reflectance spectra of which cannot be congruent. If two samples come from the same object the reflectance spectra should show no significant differences.

The main use of chromaticity coordinates or CIE colour coordinates is in clearing up hit-and-run accidents. Recently [24–27], data banks have been established in the UK and Germany which include CIE colour coordinates of all car paints produced in the past 20 years. In an actual case, the reflectance spectrum will be measured from the polished top coat of the sample from the crime scene and the CIE colour coordinates calculated. This is followed by a comparison of this number triplet with the colour coordinates filed in the data bank on the basis of the colour difference in CIELAB units. All paint colours with a colour difference smaller than or equal to 8 CIELAB units are assessed and indicated as samples which cannot be differentiated from the suspect sample. Since the CIE colour coordinates of different OEM car paints are often closer to one another than the values of the samples of one OEM colour from different vehicles, paints of different manufacturers are generally quoted when a search is being made. CIE colour coordinates do not therefore produce unequivocal differentiations. They are suitable, however, for a quick presorting. From a very large collection a manageable small subset is selected which must be further subdivided with the help of other methods. The reflectance spectra can also be used for the identification of pigments.

The basis for identification of pigments in colour mixtures is the theory of Kubelka and Munk [43], who established a connection between the reflectance spectrum and the absorption and scatter coefficients [15]. This method of pigment identification is very laborious for a forensic laboratory [44] and the process is therefore seldom used. If pigments are soluble after separation using thin-layer chromatography (TLC) their identification can be accomplished by reflectance measurements in conjunction with a comprehensive library of reference spectra [45, 46]. Special pigments such as the phthalocyanines can readily be identified by dissolving them in concentrated sulphuric acid and recording the absorption curves from such solutions [47, 48].

In the examination of paints of similar shades of colour, results with significantly higher evidential value can be achieved if UV-visible spectra instead of the reflectance spectra of paint samples measured in transmission are compared. As Zeiss and SEE microspectrophotometers can be fitted with quartz or Cassegrainian optical components, an extended range (230–1500 nm) is available for the colour measurements made in transmission. Quantifiable results are obtained from 3-μm thin sections fixed on quartz specimen slides [35]. To prevent discrepancies in measurement results caused by differences in layer thickness from section to section (resulting from microtomy), the suspect and control samples must be embedded together side by side. The reproducibility of absorbance spectra is very high for solid as well as for metallic paints even when a small measuring aperture size ($10\,\mu m \times 10\,\mu m$ to $12\,\mu m \times 120\,\mu m$) is used. To protect the sample from overheating and UV degradation in UV-visible measurements, the monochromator should be located in the illumination beam path. This can only be realised with Zeiss instruments. Suitable light sources are xenon high-pressure lamps (XBO 75). Because light is not only absorbed but also scattered in microtome sections, poor sample focusing can also lead to errors in transmission measurements. Sample heterogeneities can also be the reason for statistical variations in colour measurement by transmission. It is therefore necessary, with transmission as well as with reflectance measurements, to take the average of 5–10 measurements made on different areas.

Because pigments in paint films do not build a molecular solution but occur as particles of crystalline solid with dimensions up to a few micrometres, scattering effects (multiple scattering and dependent scattering) are significant [15]. The Lambert–Beer law and the concept of complementary chromaticity coordinates (CCCs) x' and y' [49] are not strictly valid for microtomed thin sections of paint. In fact, when measuring in transmission there is a linear relationship between absorbance and concentration (according to Beer's law). But this does not apply to scattering [15]. The CCC values will therefore vary depending on the section thickness. However, the errors in measurement that occur within a thin section are small in comparison to those that are normally found in replicate measurements from fibres [50]. The absorption spectra measured using UV-visible spectroscopy primarily provide information on the pigments used in the paint film. This allows not only the acquisition of qualitative data but also the detection of quantitative differences. For instance, paints that cannot be distinguished in reflectance measurements can be clearly differentiated after their UV-visible spectra have been recorded, as some of the paint films contain the same main pigments, but in different concentrations [35]. Individual organic pigments are identifiable when, for example, they are present in the form of aggregates and reagglomerates in the basecoats of two-coat metallic paints (Figure 8.9).

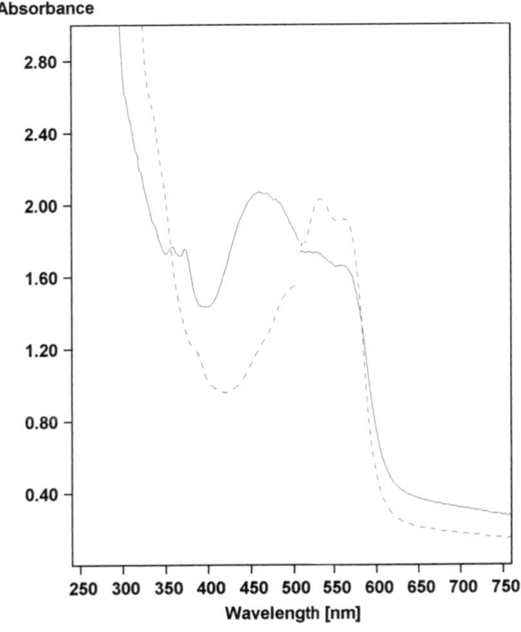

Figure 8.9 UV-visible absorption spectra of two individual red organic pigments present in the form of aggregates in a 3-μm thin section of two-coat metallic top coat burgundy red (BMW 199): solid line, Pigment Red 179 [51]; broken line, Pigment Red 122 [51]

The identification and differentiation of pearlescent pigments of similar composition is also possible (Figure 8.10).

Suitable preparations are 3-μm thin sections of paint chips aligned parallel to the surface. The recording of UV-visible spectra from pigments is only possible with high-magnification quartz objectives (e.g. Zeiss ×100/1.20 Ultrafluar). For the identification of pigments a spectral library must be available for comparison [51, 52]. Finally, batch-to-batch discrimination is also possible using UV-visible spectroscopy (Figure 8.11).

UV-absorbers can also be identified in 20-μm thin sections of the clear coats of two-coat metallic paints (Figure 8.12).

Discrimination between different vehicles which were originally painted in an identical manner is based on the varying decrease of the UV-absorber concentration with time, depending on the substance class of light stabilisers and the environmental conditions [53].

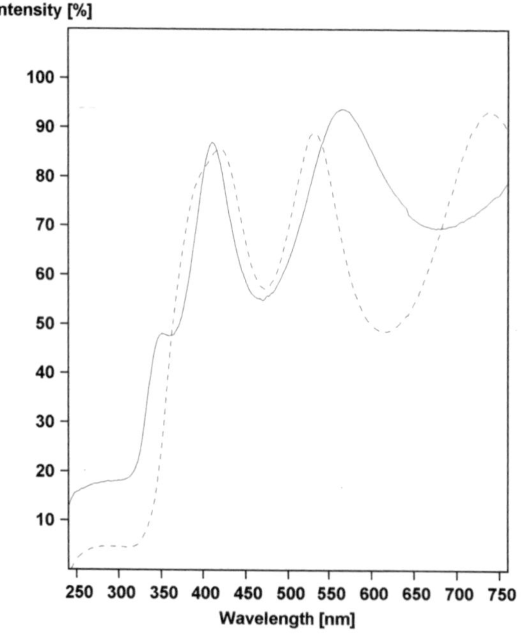

Figure 8.10 UV-visible absorption spectra of two individual pearl lustre pigment flakes present in a 3-μm thin section: solid line, 105 Rutile Brilliant Silver, EM Industries Afflair®; broken line, 207 White Gold Pearl, EM Industries Afflair®

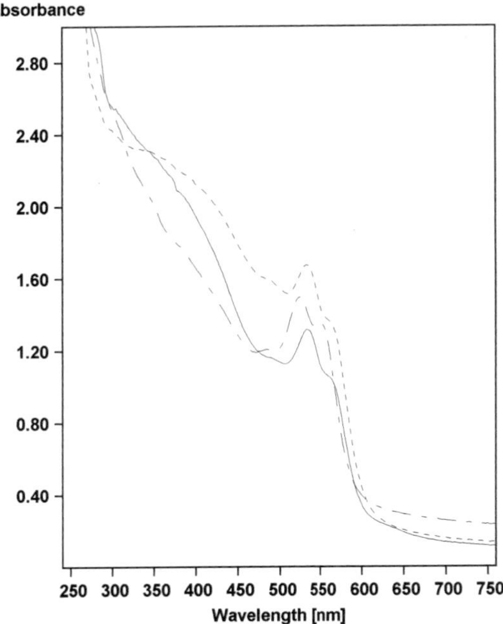

Figure 8.11 UV-visible spectra of the basecoats and the relevant clear coats of three vehicles painted with the OEM two-coat metallic paint carnelian red (OP 534) by Adam Opel AG. All three chips were embedded together in one synthetic resin block. The analyses could therefore be conducted on the same 3-μm thin section. All three basecoats can be clearly differentiated because of pigment differences in the batches

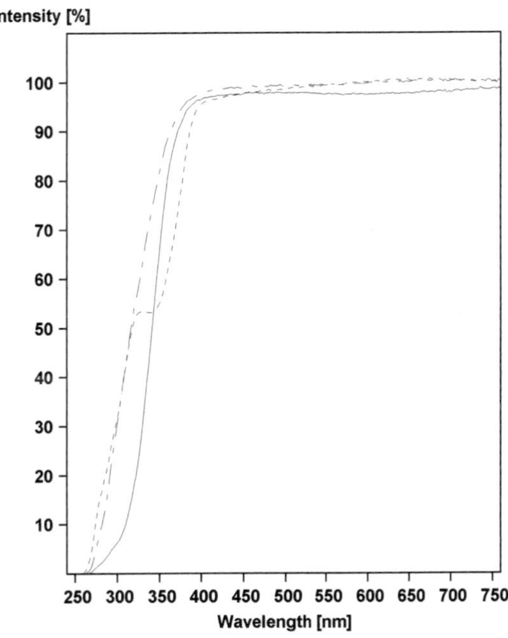

Figure 8.12 The relevant clear coats (Figure 8.11) also display differences on the basis of different UV-absorbers and their concentrations

8.5 References

1. Hudson G. D., Andahl R. O., Butcher S. J. (1977). The paint index – the colour classification and use of a collection of paint samples taken from scenes of crime. *Journal of the Forensic Science Society* **17**:27–32.
2. Audette R. J., Percy R. F. E. (1979). A rapid, systematic and comprehensive system for the identification and comparison of motor vehicle paint samples. 1. The nature and scope of the classification system. *Journal of Forensic Sciences* **24**:790–807.
3. *Munsell Book of Color* (1976). McBeth Division of Kollmorgan Corporation, Baltimore.
4. Richter M. (1974). Umstellung des Farbsystems DIN 6164 auf Normlichtart D 65. *Die Farbe* **23**:108–130.
5. Natural Colour System. SIS Colour Atlas NCS (1979). Swedish Standard No. 019102. Scandinavian Colour Institute, Stockholm.
6. Kornerup A., Wanscher J. H. (1963). *Methuen Handbook of Colours.* Methuen, London.
7. Cartwright L. J., Cartwright N. S., Norman E. W. W., Cameron R., MacDougall D. A., Clark W. H. (1984). The classification of automotive paint primers using the Munsell color coordinate system. A collaborative study. *Canadian Society Forensic Science Journal* **17**:14–18.
8. Hunter R. S., Harold R. R. (1987). *The Measurement of Appearance.* John Wiley and Sons, New York.
9. Richter M. (1976). *Einführung in die Farbmetrik.* Sammlung Göschen, 2608, Walter de Gruyter, Berlin.
10. Helmholtz H. v. (1911). In *Handbuch der physiologischen Optik.* Voss, Leipzig.
11. CIE Publication No. 15. (1971). Colorimetry. Official recommendations of the International Commission on Illumination. Bureau Central de la CIE, Paris.
12. Judd D. B., Wyszecki G. (1975). *Color in Business, Science, and Industry* (3rd edn). John Wiley and Sons, New York.
13. Hofmeister F., Pieper H. (1989). Remissionsmessungen von Interferenz-, Aluminium- und Körperfarbenpigmenten. *Farbe + Lack* **95**:557–560.
14. Hofmeister F., Pieper H. (1990). Einfluß einer variablen Meßgeometrie auf die Farbwerte von Interferenzpigmenten. *Farbe + Lack* **96**:773–775.
15. Völz H. G. (1990). *Industrielle Farbprüfung.* VCH Verlagsgesellschaft mbH, Weinheim.
16. CIE Publication No. 15.2. (1986). *Colorimetry* (2nd edn). Bureau Central de la CIE, Paris.
17. CIE Publication No. 116. (1995). *Industrial Colour-difference Evaluation.* Bureau Central de la CIE, Paris.
18. Witt K. (1995). CIE – Empfehlungen zur industriellen Farbabstandsbewertung. *Farbe + Lack* **101**:937–939.
19. Terstiege H. (1986). Colour tolerances in automotive paints. *Journal of Coatings Technology* **58**:37–41.
20. DIN 6175. (1985). Farbtoleranzen für Automobil-Lackierungen. Deutsche Normen, Beuth-Vertrieb GmbH, Berlin.
21. Williams J. F. (1953). Examination of paint chips and scrapings with the spectrophotometer. *Police Science* **44**:647–651.
22. Stoecklein W. (1985). The examination of small opaque objects by microspectrophotometry. *Presentation at the 6th Nordic Forensic Science Meeting, Helsinki, May.*
23. Cousins D. R. (1989). The use of microspectrophotometry in the examination of paints. In *Forensic Science Review* (ed. R. H. Liu), Vol. 1, pp. 141–162. Central Police University Press, Taipei.
24. Locke J., Cousins D. R., Russel L. W., Jenkins C. M., Wilkinson J. M. (1987). A data collection of vehicle topcoat colours. 1. Instrumentation for colour measurement. *Forensic Science International* **34**:131–142.
25. Locke J., Wilkinson J. F., Hanford T. J. (1988). A data collection of vehicle topcoat colours. 2. The measurement of colour samples used in the vehicle refinishing industry. *Forensic Science International* **37**:177–187.
26. Taylor M. C., Cousins D. R., Holding R. H., Locke J., Wilkinson J. M. (1989). A data collection of vehicle topcoat colours. 3. Practical considerations for using a national database. *Forensic Science International* **40**:131–141.
27. Cousins D. R., Holding R. H., Locke J., Wilkinson J. M. (1989). A data collection of vehicle

topcoat colours. 4. A trial to assess the effectiveness of colour identification. *Forensic Science International* **24**:183–197.
28. Cousins D. R., Platoni C. R., Russel L. W. (1984). The variation in the colour of paint on individual vehicles. *Forensic Science International* **24**:197–208.
29. Allen T. J. (1992). Paint sample presentation for Fourier transform infrared microscopy. *Vibrational Spectroscopy* **3**:217–237.
30. Stoecklein W., Gloger M. (1988). FT-infrared spectroscopy of automobile paints using infrared microscopy. *Nicolet FT-IR Spectral Lines* **Spring–Summer**:6–12.
31. Wilkinson J. M., Rickard R. A., Locke J., Laing D. K. (1987). The examination of paint films and fibres in thin section. *Microscope* **35**:233–248.
32. Wilkinson J. M., Locke J., Laing D. K. (1988). The examination of paints as thin sections using visible microspectrophotometry and Fourier transform infrared microscopy. *Forensic Science International* **38**:43–52.
33. Hamer P. S. (1982). Pigment analysis in the forensic examination of paints. III. A guide to motor vehicle paint examination by transmitted light microscopy. *Journal of the Forensic Science Society* **22**:187–192.
34. Kilbourn J. H., Marx R. B. (1994). Polarized light microscopy of extenders in structural paints – forensic applications. *Microscope* **42**:167–175.
35. Stoecklein W., Tuente J. (1994). Using the light microscope for analytical procedures – aids for solving cases involving hit-and-run offences. *Zeiss Information with Jena Review* **3**:19–22.
36. Stoecklein W., Goebel R. (1992). Application of cathodoluminescence in paint analysis. *Scanning Microscopy* **6**:669–684.
37. Hoffmann K. (1977). Temperaturabhängigkeit der Remission von Pigmenten. *Farbe + Lack* **83**:1067–1071.
38. Döring G. (1988). Probleme bei der mikroskopischen Farbmessung. *Farbe + Lack* **94**:610–615.
39. Hoffmann K. (1982). Zur Problematik von Spektralmessungen an fluoreszierenden Proben. *Farbe + Lack* **88**:365–370.
40. Laing D. K., Dudley R. J., Isaacs M. D. J. (1980). Colorimetric measurements on small paint fragments using microspectrophotometry. *Forensic Science International* **16**:159–171.
41. Laing D. K., Dudley R. J., Home J. M., Isaacs M. D. J. (1982). The discrimination of small fragments of household gloss paint by microspectrophotometry. *Forensic Science International* **20**:191–200.
42. Cousins D. R., Russel L. W. (1989). Maximum differences of colours on the surfaces of relatively new vehicles. Unpublished observations.
43. Kubelka P., Munk F. (1931). Ein Beitrag zur Optik der Farbanstriche. *Zeitschrift für technische Physik* **12**:593–609.
44. Cousins D. R., Platoni C. R., Russel L. W. (1984). The use of microspectrophotometry for the identification of pigments in small paint samples. *Forensic Science International* **24**:183–196.
45. Fuller N. A. (1985). Analysis of thin-layer-chromatograms of paint pigments and dyes by direct microspectrophotometry. *Forensic Science International* **27**:189–204.
46. Massonnet G., Stoecklein W. (1999). Identification of organic pigments in coatings: Application to red automotive topcoats. Part I: Thin layer chromatography (TLC) with direct visible microspectrophotometric (MSP) detection. *Science and Justice* **39**:128–134.
47. Schweppe H. (1977). Identification of dyes on textile fibres. In *The Analytical Chemistry of Synthetic Dyes* (ed. K. Venkataraman), pp. 149–196. John Wiley and Sons, New York.
48. Voskertchian G. P. (1995). Quantitative analysis of organic pigments in forensic paint examination. *Journal of Forensic Sciences* **40**:823–825.
49. Garland C. E. (1977). Solution coloristics. In *The Analytical Chemistry of Synthetic Dyes* (ed. K. Venkataraman), pp. 149–196. John Wiley and Sons, New York.
50. Hartshorne A. W., Laing D. K. (1987). The definition of colour for single textile fibres by microspectrophotometry. *Forensic Science International* **34**:107–129.
51. *Colour Index* (3rd edn). (1975). The Society of Dyers and Colourists, Bradford, UK.
52. Stoecklein W. (1995). The UV-Vis spectra of organic pigments in automobile paints. Unpublished results.
53. Stoecklein W., Fujiwara H. (1999). The examination of UV-absorbers in 2-coat metallic and non-metallic automotive paints. *Science and Justice* **39**:188–195.

9

Pyrolysis techniques for the characterisation and discrimination of paint

JOHN M. CHALLINOR

9.1 Introduction

In a recent survey of the results of laboratory proficiency trials involving forensic evidence types [1], it was concluded that paint, glass, textile fibres and body fluid mixtures presented the greatest difficulties for the forensic examiner. In paint trials, the failure to perform pyrolysis gas chromatography (PyGC) examinations and the misinterpretation of pyrolysis data were the causes of erroneous results. The Collaborative Testing Services Proficiency Advisory Committee concluded that PyGC was an indispensable tool for performing many of the examinations but proficiency with this technique varied widely.

One objective of this chapter is to encourage forensic examination of trace evidence, namely paint, plastics, textile fibres, rubbers and other macromolecular material, by pyrolysis methods. Once the technique has been fully appreciated, the advantages of the method for forensic examinations are both absorbing and rewarding. Some recent reviews of the general application of PyGC have been published [2, 3], while pyrolysis techniques and their application to the identification of textile fibres have been described by the author [4]. Similarly, the pyrolysis of paint and other types of forensic evidence were described in a review of other applications of analytical pyrolysis [5].

This chapter outlines the principles and practice of PyGC and pyrolysis gas chromatography–mass spectrometry (PyGC/MS). The applications of PyGC pertinent to forensic science examinations of automotive, architectural and industrial paint are illustrated. A brief account of the thermal degradation mechanisms relevant to different paint resin polymeric systems assist in an understanding of thermolysis processes. The principles and applications of pyrolysis mass spectrometry (PMS) and pyrolysis infrared (PIR) spectroscopy for the identification of paint are also described. The advantages and disadvantages of the pyrolysis techniques compared to other paint identification and comparison methods are outlined and some potential future developments in pyrolysis techniques are also discussed.

Pyrolysis is the thermal fragmentation of a substance in an inert atmosphere. The pyrolytic process produces molecular fragments which are usually characteristic of the composition of the original macromolecular material. Forensic scientists have used analytical pyrolysis since the early 1970s to characterise a wide range of crime scene evidence. The pyrolysis products have been detected and identified by coupling the pyrolysis unit to a gas chromatograph or a mass spectrometer or a combination of both instruments as in PyGC/MS. The infrared spectrometer may also be used as a detector for GC or it may be used directly to identify a pyrolysate as in PIR spectroscopy. PyGC/MS is usually the preferred method of analysis because the data provides a reliable identification and comparison of analytes, they are usually easy to interpret and the technique is more cost-effective than other methods. Generally, PyGC gives good discrimination, involves minimal sample manipulation and generates reliable and reproducible results, and is able to detect very small quantities of material. Despite these advantages, it is a technique which is often underutilised in paint examinations. Criticisms of poor reproducibility were made in the developing years of the PyGC method. Experience has shown that these misgivings are unfounded when modern instrumentation and correct analytical techniques are employed. If differences in results exist between laboratories, then it is probably because of the differences in the type of pyrolyser used to conduct the thermal fragmentation process. The variables in the pyrolysis process include temperature-rise-time, pyrolysis temperature, sample mass, the dimensions of the pyrolysis chamber, carrier gas type and flow rate. These factors influence the proportion of primary pyrolysis products, the production of secondary products from recombination processes and the introduction of catalytic effects. A system which avoids secondary products and minimises catalytic effects should be adopted [6].

9.2 Pyrolysis gas chromatography

PyGC techniques have progressed considerably from the early days of custom-built pyrolysers with unwieldy sampling handling arrangements and the use of packed GC columns. Modern pyrolysis instruments are 'off the shelf' units which usually require little skill in sample loading. There has been a move away from packed chromatography columns to the more versatile silica capillary columns. These columns give improved results in terms of better peak shape and the efficiency of separation of the compounds. They are also capable of chromatographing a wider range of compounds of greater polarity. Recently, pyrolysis derivatisation techniques have been applied to the identification of a wide range of polymers [7].

9.2.1 Modes of pyrolysis

Essentially three different types of pyrolyser have emerged since the introduction of commercial units for analytical pyrolysis applications. These are the pulse mode filament and the pulse mode Curie point types and the continuous mode furnace types. The filament type employs a resistivity heated platinum coil or ribbon. The coil houses a quartz tube to hold the sample and the ribbon acts as a surface on which to evaporate the sample. The advantage of this type is that the pyrolysis temperature can be varied continuously from about 200°C to 1400°C. The Curie point system depends on the inductive heating of a metal or an alloy wire to its Curie point using a high-frequency oscillating current in a coil surrounding the wire. The wire can be flattened and bent over to hold the sample and has the advantages of low dead volume and ease of sample loading. The furnace-type pyrolyser involves introducing the sample into an oven unit by a gravity feed, a magnetic push rod or a plunger arrangement. Care must be taken to ensure that the pyrolysis zone is free of contaminants from previous experiments before the system is used.

Walker and Wolf [8] reported a study of the three different pyrolysis systems using a branched-chain alkane as a model compound. The filament pyrolyser gave a higher proportion of lower molecular weight products than the Curie point instrument, suggesting that the 'true' pyrolysis temperature may have been much higher as a result of a faster temperature-rise-time for the particular filament instrument used. A more recent study [9], in which the performance of a modern Curie point pyrolyser, an injection port mounted filament pyrolyser and an oven-type unit were all compared, indicated that the Curie point unit gave the best overall performance. However, the oven-type pyrolyser gave superior reproducibility. The sensitivity of the Curie point system was greater in the conventional mode of operation using a hypodermic needle pyrolysate transfer arrangement and had the potential for even better performance by connecting the GC capillary column directly to the pyrolysis chamber. Discrimination due to activity and the effects of dead volume affecting peak broadening were not apparent in any of the systems. The Curie point system appears to have the advantage of an easier loading arrangement for microgram-sized specimens.

Consideration should be given to true final pyrolysis temperatures, especially when comparing results from different systems. Other factors, such as rapid temperature-rise-time, low dead volume and ease of sample loading, should be considered when choosing a pyrolysis system. A rapid temperature-rise-time is necessary to avoid secondary reactions which give spurious pyrolysis products. Small pyrolysis chamber geometry is also preferred. High dead volumes also result in secondary reactions and permit band spreading leading to peak broadening of pyrolysis products and consequently reduced sensitivity. Where very small specimens are to be pyrolysed, the design of the pyrolysis probe should facilitate sample loading.

Pyrolysers equipped with laser beams have been developed, but they do not appear to have been commercially exploited for forensic organic analysis to any appreciable extent. A new apparatus based on laser micropyrolysis GC-MS has been developed [10] which has the potential for the chemical characterisation of microgram-sized specimens in forensic examinations.

9.2.2 Gas chromatography

Contemporary PyGC techniques require the use of capillary columns and temperature programming [11–15]. By contrast, early PyGC work was carried out with packed columns for the chromatographic separation. Numerous applications to paint, plastics, adhesives, rubbers and many other materials have been reported [16, 17]. Early instrumentation and applications have been described by May *et al.* [6] while forensic applications of PyGC using capillary column GC has been described and its advantages compared and critically assessed [11].

One criticism of PyGC as a viable technique is the lack of standardisation of analytical conditions with the consequence that interlaboratory comparison of data may not be possible. The type of pyrolyser will be the choice of the user, but the GC conditions could be standardised. A common base for paint examinations could be the adoption of capillary columns, flame ionisation detection and temperature programming from ambient to maximum column temperature at an intermediate ramp rate, e.g. 8–10°C min^{-1}. A mid-polarity phase column (e.g. OV 17 type or equivalent – cyanopropyl phenyl methyl silicone) would be favoured for the identification of paint resins. This type of phase gives good peak shape for polar pyrolysis products without baseline drift or the relatively poor phase stability experienced with the more polar polyester (e.g. Carbowax) phases.

9.2.3 Practical aspects

To achieve optimum performance, attention must be paid to certain practical aspects of the pyrolysis and GC systems. Capillary columns, though now very stable compared to the early types, will become active or adsorbing after a period of time with the result that peak tailing or complete loss of product may be experienced. Test mixtures should be used regularly to monitor column performance. Mixtures of a selected range of compounds, recommended by Grob *et al.* [18], have been found to be satisfactory for this purpose. If the column performance deteriorates, then a coil of the column at the inlet end can be cut off to return the chromatography to an acceptable standard. The flow rate value should be maintained by reducing inlet system pressures to compensate for the shorter column, in order to retain column efficiency and retention time values. Alternatively, a replacement section can be used at the front of the column. Column 'washing' using organic solvents does not usually improve performance to any appreciable extent.

The pyrolyser unit should be maintained in a clean condition to avoid carry-over to subsequent samples. Carry-over problems could be experienced when small quantities of an analyte are pyrolysed subsequent to larger quantities of comparison material. Blank pyrolysis runs should be conducted prior to subsequent experiments. Glass/quartz inserts should be washed regularly in a suitable solvent (e.g. dichloromethane). Curie point pyrolysis wires should be replaced at regular intervals particularly if they appear soiled. Resinous pyrolysis products may periodically cause blockage problems at the vent line from the injection system and this area should be checked periodically by observing deviations from normal behaviour of carrier gas flows and pressures.

9.2.4 Detection systems

The flame ionisation detector (FID) is the most commonly used detection system for the gas chromatograph. Pyrolysis products are identified by their retention times compared to those of known standard compounds. Other detectors provide more specific responses for compounds eluting from the GC column. The alkali flame ionisation detector (AFID) specifically measures nitrogen- and phosphorus-containing compounds. The flame photometric detector (FPD) is used for monitoring sulphur- and phosphorus-containing compounds and the electron capture detector (ECD) is specific for halogenated hydrocarbons and other electron-accepting organic compounds; these two detectors are not normally used for paint pyrolysis work.

There can be no doubt that the unequivocal identification of pyrolysis products by GC is necessary for an interpretation of the polymer composition of resins in paint. The detection system most favoured is the mass spectrometer. Electron-impact (EI) mass spectral libraries make it possible to rapidly identify many of the pyrolysis products facilitating interpretation of polymer composition. Chemical ionisation mass spectrometry techniques may also be employed and these have the advantage that molecular ions are produced in greater abundance to give molecular weight/molecular formula information on the pyrolysis products.

A Fourier-transform infrared (FTIR) spectrometer employed as a detector provides a means of identification of compounds eluted by GC. For this technique there are two methods of collection of the pyrolysis products, the 'light pipe' and 'matrix isolation'. Wide-bore capillary columns are usually required for these operations to provide sufficient sensitivity, which is of the order of low nanogram levels. The technique has been used for the characterisation of epoxy and phenolic resins and their combinations [19]. The topic of GC detection systems has been discussed in some depth by Leibmann and Levy [20].

9.2.5 Pyrolysis derivatisation

Another development involves the flash heating in a pyrolyser of an intimate mixture of an aqueous solution of a tetraalkylammonium hydroxide, particularly tetramethylammonium hydroxide (TMAH), and the material under examination. The high-temperature reaction is a concerted hydrolysis and methylation process, which produces methyl derivatives of groups such as esters and ethers. This reaction was initially described as simultaneous pyrolysis methylation (SPM) [21] but the term was later modified to thermally assisted hydrolysis and methylation (THM) to avoid misunderstandings of the reaction mechanism [22]. The method is rapid, sensitive and gives data unobtainable by conventional pyrolysis methods. This procedure is more rapid when compared to more complex wet chemical degradation methods and is cost-effective and applicable to the minute samples typically encountered in forensic casework.

Figure 9.1 shows an example of the significant change in pyrolysis profiles achievable by the THM reaction of an alkyd paint, when compared to conventional packed column PyGC and capillary column PyGC. The interpretation of the THM-GC data from the pyrogram shown in Figure 9.1 is that the alkyd enamel is a pentaerythritol *o*-phthalic alkyd consistent with having a high proportion of a soya bean drying oil (from the C16.0:C18.0 peak height ratio). The total absence of unsaturated C18 acid methyl esters and the significant peak for the methyl ester of azelaic acid indicates that the paint has a high degree of cure. No rosin or other modification is detected. An in-depth study of the identification of alkyd resins by the THM technique has been reported [23].

Although PyGC is useful for discriminating between different resins, THM more clearly identifies the components of coating resins susceptible to hydrolysis, such as polyesters.

9.2.6 Applications

There have been numerous applications of analytical pyrolysis reported in the past two decades. A selected bibliography of applications was reported by Wampler [17] while Wheals [14] has described the forensic applications of the technique, particularly with respect to paint. Wheals [24] also reviewed analytical pyrolysis techniques applied to paint and other polymeric materials encountered in crime scene evidence. Challinor [25] has described the use of PyGC and pyrolysis derivatisation methods in forensic applications. A

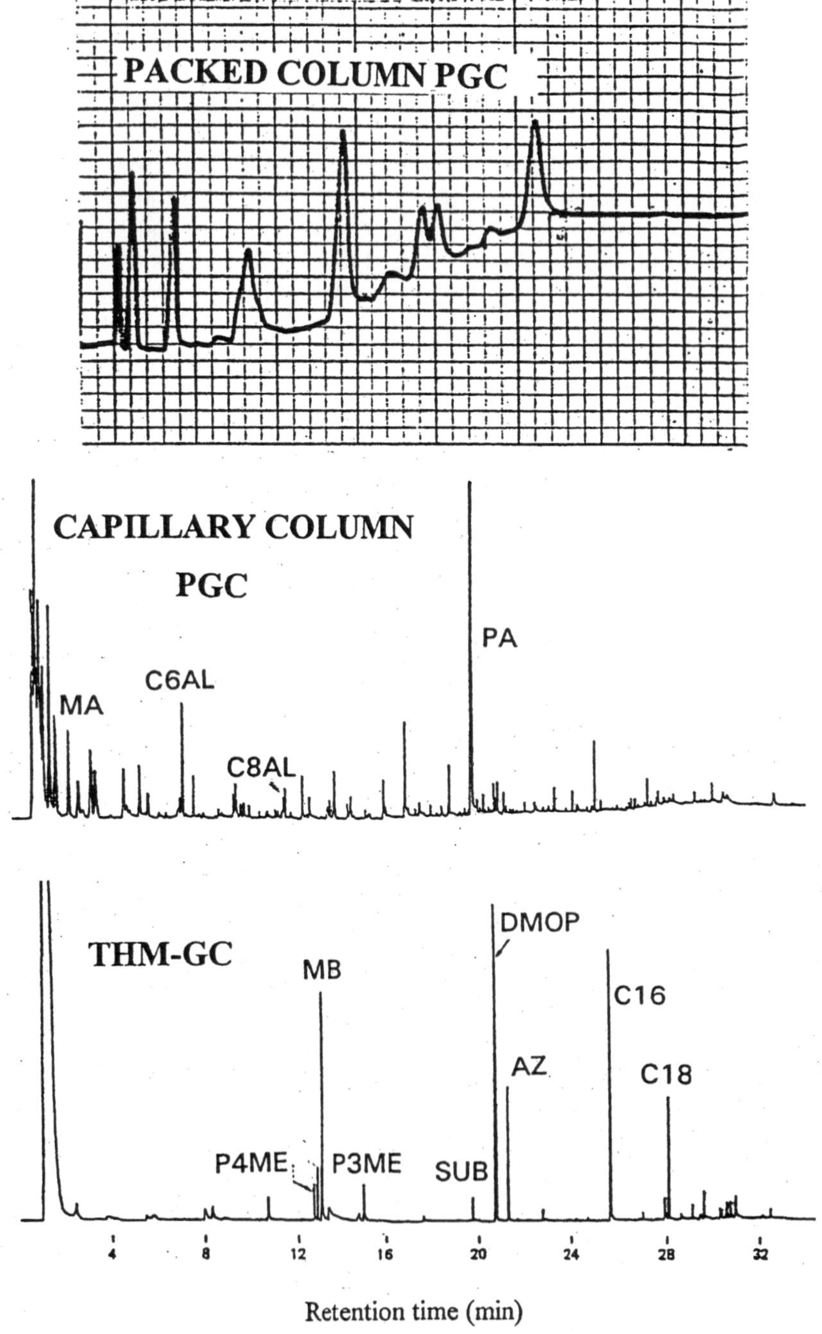

Figure 9.1 Pyrograms obtained by packed column PyGC, capillary column PyGC and THM-GC of a pentaerythritol *ortho*-phthalic alkyd resin paint. Key: MA, methacrolein; C6AL, hexanal; C8AL, octanal; PA, phthalic anhydride; P4ME and P3ME, pentaerythritol termethyl and trimethylmethyl ethers; MB, methyl benzoate; DMOP, dimethyl *ortho*-phthalate; SUB, AZ, C16 and C18, suberic, azelaic and long-chain fatty acid methyl esters, respectively

combination of the two techniques has been used for end-group analysis of polystyrene having methacryloyl end groups [26]. The number-average sequence length in the microstructure of emulsion polymers was determined by PyGC using data from consideration of the trimer region of the pyrograms of styrene-butyl acrylate [27].

The variability and complexity of paint physical and chemical characteristics can be such that the material provides strong evidence in a forensic investigation [14], in particular the resin or binder composition may vary widely within an individual class. PyGC is an effective method for identifying and differentiating the organic binder of a paint and, in some cases, additives and organic pigments may also be detected and identified. The PyGC examination of paint, in the context of an overall forensic analysis of this type of evidence, has been described Challinor [28] but the PyGC identification of organic pigments in a resin matrix continues to be a challenging prospect for the future.

The most common paint types occurring in forensic casework include alkyd enamel, acrylic latex, solution resins and thermosetting polymers, and vinyl acetate latex types. Other paint types occurring less frequently are epoxy, urethane and chlorinated rubber latex.

9.2.6.1 Architectural paint

Break-in tools used to gain access to premises could carry traces of paint which have abraded from painted surfaces at the point of entry. Commonly encountered household paint types include polyvinyl acetate (PVA), acrylics and alkyd enamels, while epoxies and chlorinated rubber coatings are occasionally found as evidence. These different classes are distinguished in PyGC by the detection of one or two major pyrolysis products. PVA resins are recognised by the detection of acetic acid, acrylics may have methylmethacrylate as a major product and alkyd enamels are recognised by the major phthalic anhydride peak. Epoxy resins are indicated by the detection of phenol, isopropenyl phenol and bisphenol A and chlorinated rubber by xylenes and trimethylbenzenes [5]. These different paint binder classes can also be identified by FTIR spectroscopy without difficulty (see Chapter 10 of this book). Within-class differentiation, however, may be more of a problem in FTIR analyses. By contrast, subtle differences in composition can be readily determined by PyGC. Pyrograms of four acrylic latex paints having different monomer compositions are shown in Figure 9.2.

Although in some cases it is possible to identify paint additives, the presence of a latex coalescing agent, trimethylpentanediol monoisobutyrate (Texanol), may also be detected. The relative proportion of the isomers of this coalescing agent diminishes with time due to evaporation and this may sometimes be used to assess the age of the coating.

The differentiation of alkyd enamels by PyGC has been attempted [29, 30] and, as discussed previously, more structural information about alkyd enamels may be obtained by the pyrolysis derivatisation THM technique [23]. Polybasic acid, polyhydric alcohol, drying oil composition, oil length, degree of cure and rosin modification can all be determined using the THM technique, thus improving discrimination.

Epoxy resins are usually found more frequently in industrial applications, notably tools, hardware and equipment, and pyrolysis methods have been used for their characterisation. PyGC/MS was used for the determination of phenols and heterocyclic nitrogen compounds in the identification of thermal degradation of epoxy powder paint [31]. The same method has been used for the analysis of cured polyfunctional epoxy resins [32], where it was

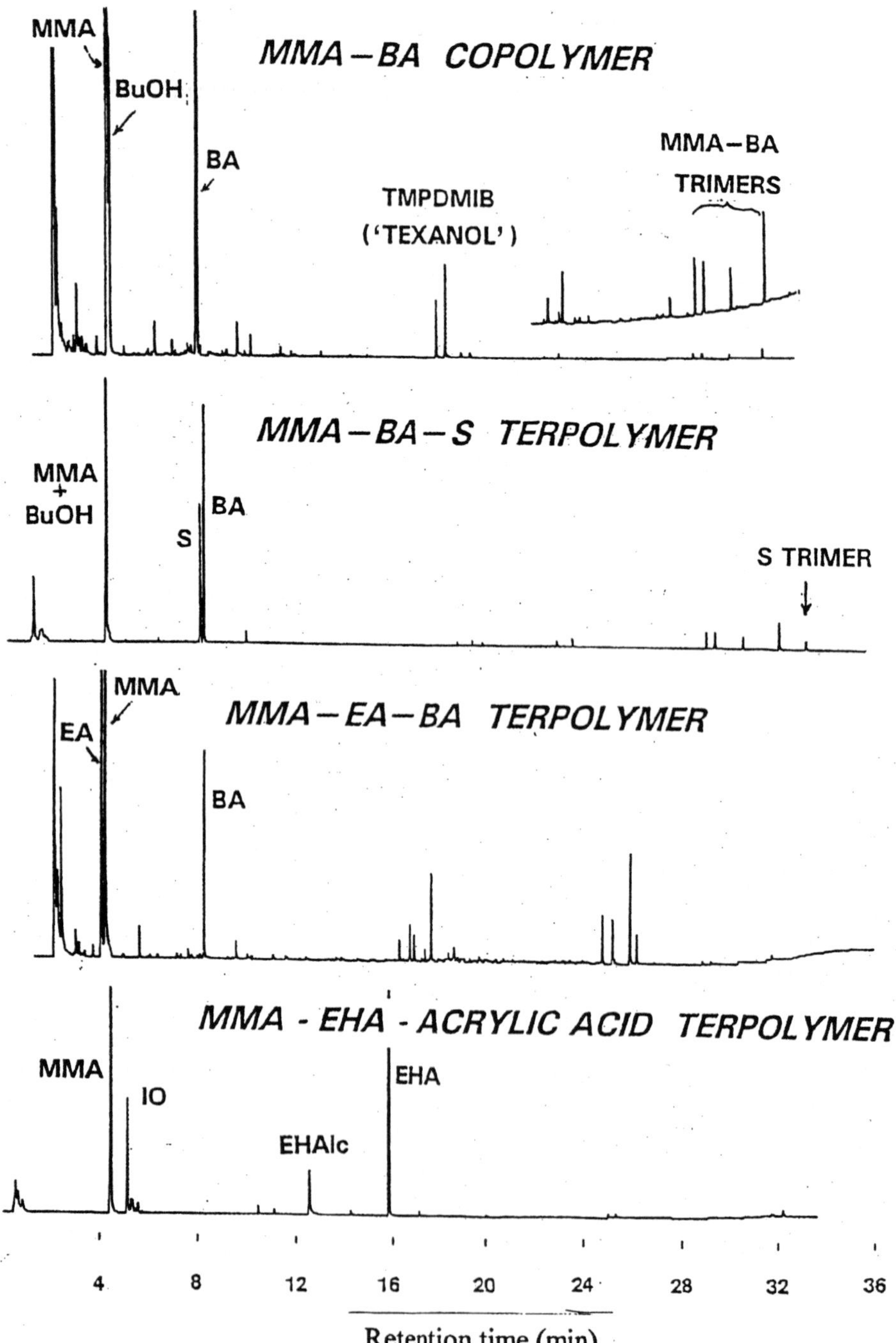

Figure 9.2 Pyrograms of four acrylic latex architectural paints having different monomer compositions. Key: MMA, methylmethacrylate; BuOH, *n*-butanol; BA, butylacrylate; TMPDMIB, 2,2,4-trimethyl-3-pentanediol monoisobutyrate; S, styrene; EA, ethylacrylate; IO, iso-octene; EHAlc, 2-ethylhexyl alcohol; EHA, 2-ethylhexylacrylate

found that it was possible to identify the curing agents, diaminodiphenylsulphone, dicyandiamide and substituted ureas. The same authors [33] examined the pyrolysis products of epoxy binders in carbon fibre composites by using the same technique. They concluded that PyGC made it possible for the method to be used for quality control purposes. However, the absence of characteristic pyrolysis products of hardeners prevented their unambiguous identification by the technique, because of the thermal lability of polysulphones, and characteristic recoverable pyrolysis products of the hardeners were not produced.

9.2.6.2 Automotive paint

Motor vehicle-related crime may include hit-and-run, wilful damage and homicide incidents. Paint from criminal activity can often provide useful evidence in such incidents because paint colour, layer sequence and resin composition vary widely between different makes of vehicle, and also as a result of refinishing treatments. Automotive paint binder types have been identified on microgram-sized samples of topcoat by PyGC [15, 34, 35]. Once again, PyGC can readily distinguish between different classes of organic binder in paint found in original (new vehicle) paint systems, such as acrylic lacquer, acrylic enamel and alkyd enamel [5]. The technique has an advantage over other identification techniques through the determination of intra-class differences. Two pyrograms of acrylic enamels used as original coatings for motor vehicles are shown in Figure 9.3.

There is a clear difference in the composition of the two enamels. The acrylic enamel represented by the upper pyrogram is an *n*-butylmethacrylate/ethylhexyl acrylate/methylmethacrylate acrylic system containing hydroxypropylmethacrylate for crosslinking functionality. Pyrolysis derivatisation using the THM technique also revealed that the resin contained hexahydrophthalic acid, adipic acid and polyols. The acrylic enamel represented by the lower pyrogram is a styrene/butylmethacrylate/butylacrylate/ethylhexylacrylate/methylmethacrylate acrylic system containing crosslinking agents, hydroxyethylmethacrylate and hydroxypropylmethacrylate. The THM-GC-MS analysis also revealed melamine residues indicating the presence of the amino resin crosslinking component.

Acrylic lacquer coatings have been frequently used for refinishing vehicles after accident damage. There is now a trend towards replacing these lacquers with polyester resins based on phthalic acids as the polybasic acid components. The acrylic lacquer formulations are almost solely based on methyl methacrylate monomer and are plasticised by the incorporation of monomers which produce ‘softer’ polymers such as butyl methacrylate and/or external plasticisers such as the phthalates. PyGC has also facilitated the determination of within-class differences of automotive acrylic lacquer refinishing paints [5].

Automotive alkyd enamels, occurring as original baked enamels or spraying enamels, may be identified by the THM-GC method [23]. These alkyd polyesters are hydrolysed and derivatised to methyl derivatives of their polyol, polybasic acid and drying oil. In this case, more chemical structure information can be gained by THM-GC than from conventional PyGC. Figure 9.4 shows the pyrogram of a paint smear found on a baseball bat which had been allegedly used to damage a white Honda vehicle. The paint is a baked alkyd enamel.

The detection of the methyl derivatives of the polyhydric alcohols, neopentyl glycol (NPG) and trimethylolethane (TME), indicate the polyhydric alcohol components. The methyl esters of *ortho*-phthalic and isophthalic acids, and the fatty acids, having a composition typical of a coconut oil, identify the polybasic acid and non-drying vegetable oil constituents. Together with the detection of *n*-butanol and melamine residues, the

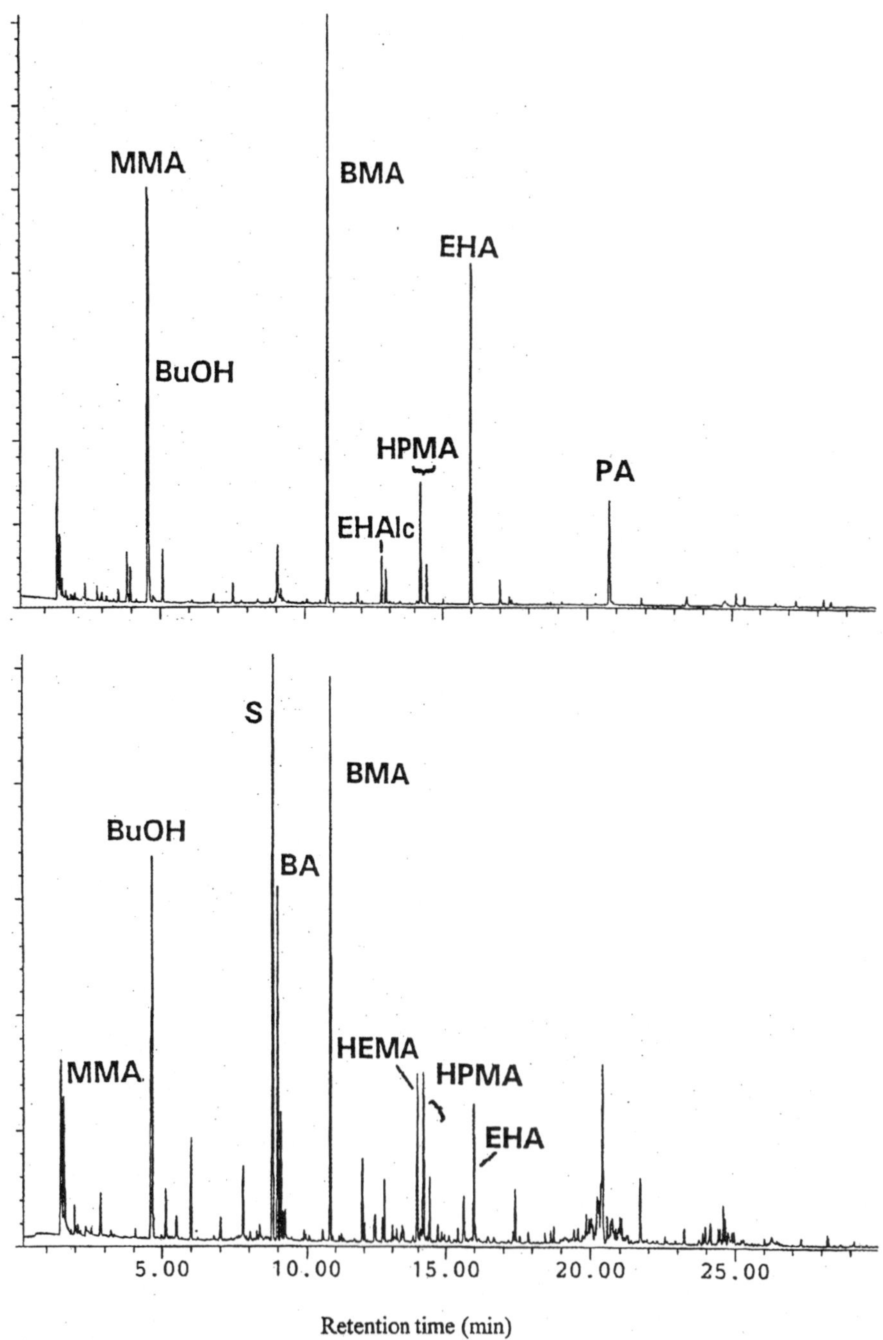

Figure 9.3 Pyrograms of two acrylic enamel automotive original topcoat finishes as determined by PyGC/MS. Key: MMA, methylmethacrylate; BuOH, *n*-butanol; BMA, butylmethacrylate; BA, butylacrylate; S, styrene; IO, iso-octene; EHAlc, 2-ethylhexyl alcohol; HPMA, hydroxypropylmethacrylate isomers; HEMA, hydroxyethylmethacrylate isomers; EHA, 2-ethylhexylacrylate; PA, phthalic anhydride

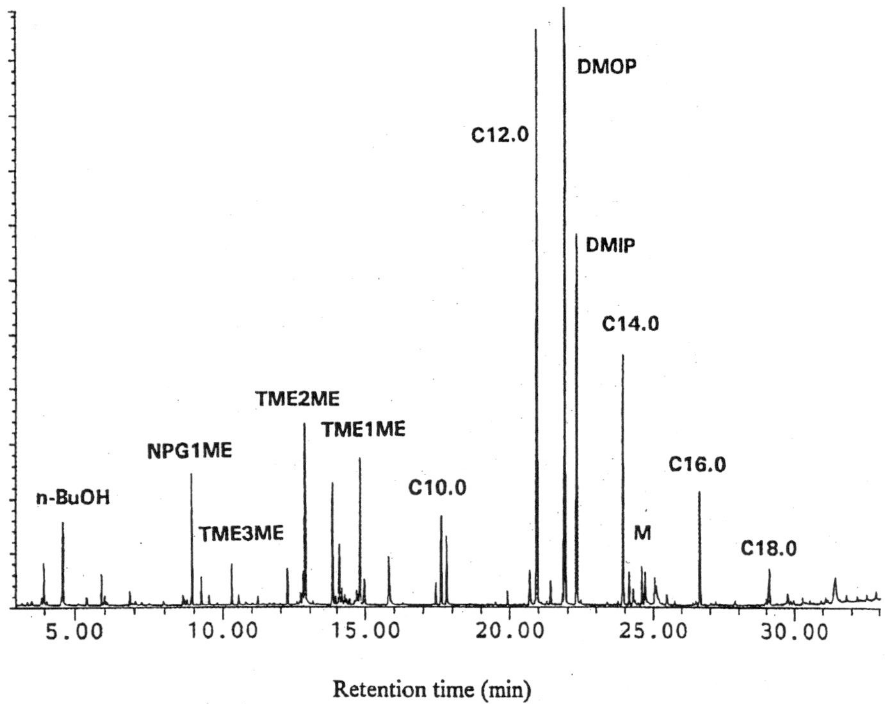

Figure 9.4 THM profile of the baked alkyd enamel paint smear found on a baseball bat. Key: *n*-BuOH, *n*-butanol; NPG1ME, neopentylglycol monomethyl ether; TME3ME, trimethylolethane trimethyl ether; TME2ME, trimethylolethane dimethyl ether; TME1ME, trimethylolethane monomethyl ether; DMOP, dimethyl *ortho*-phthalate; DMIP, dimethyl isophthalate; M, triazines from melamine; C10.0, C12.0, C14.0, C16.0 and C18.0, saturated fatty acid methyl esters

pyrogram indicates that the paint found on the baseball bat is an NPG/TME-*ortho*-phthalic acid/*iso*phthalic acid baked alkyd enamel having a coconut non-drying oil crosslinked with a butylated melamine formaldehyde resin. It is consistent with the paint having originated from a Honda vehicle.

Epoxy resins are also found in automotive undercoats and primers of original paint systems. Topcoats are usually targeted in automotive paint comparisons, but the identification of lower coats should not be ignored as these can provide further discrimination.

9.2.6.3 Motor vehicle body fillers

Body fillers are often found in the layer structure of paint from repainted vehicles. Their identification and subsequent comparison with material from the vehicle of the offender may be required to establish that the vehicle was at the scene of a traffic accident. Body fillers are usually a mixture of easily sanded inorganic fillers and styrenated unsaturated polyester resins. The composition of the resins may vary according to the application, resin supplier and the relative cost of raw materials at the time of manufacture. The resins may

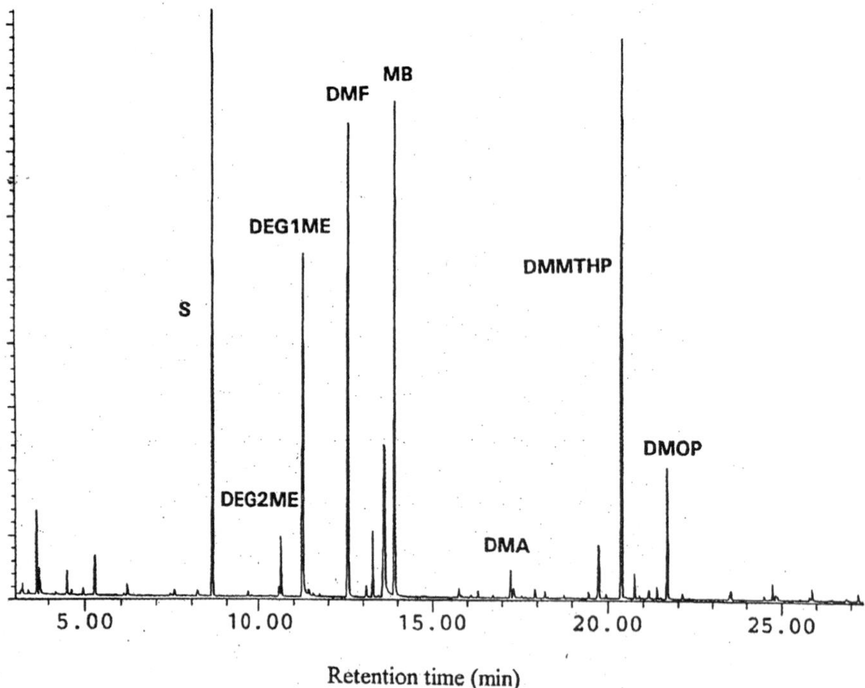

Figure 9.5 THM-GC-MS pyrogram of an unsaturated polyester automotive body filler. KEY: S, styrene; DEG2ME, diethylene glycol dimethyl ether; DEG1ME, diethylene glycol monomethyl ether; DMF, dimethyl fumarate; MB, methyl benzoate; DMA, dimethyl adipate; DMMTHP, dimethyl *endo*-methylenetetrahydrophthalate; DMOP, dimethyl *ortho*-phthalate

vary in polyhydric alcohol, polybasic acid or additive composition. The THM-GC procedure provides the most diagnostic information about the chemical composition of the resin polymer. A total ion pyrogram of an unsaturated polyester body filler used for spot repairs on painted surfaces is shown in Figure 9.5.

From these pyrolysis data it may be deduced that the resin in the body filler is a styrenated diethylene glycol-*endo*-methylenetetrahydrophthalic/*ortho*-phthalic acid polyester crosslinked with maleic anhydride/fumaric acid and modified with adipic acid to give the resin flexibility. The degree of cure may be deduced by monitoring the concentration of residual maleic anhydride/fumaric acid, detected as dimethyl fumarate in the THM procedure.

In a survey of the composition of body fillers (J. M. Challinor, unpublished results), it was found that the resin composition varies between products in the proportion of styrene, the type of phthalic acid isomer and the absence or presence of adipic acid. Most unsaturated polyester resins contained diethylene glycol and many used propylene glycol as a polyhydric alcohol.

9.2.6.4 Industrial paints

Products manufactured on a factory production line such as domestic appliances, furniture, tools and products derived from sheet metal are typical examples where industrial coatings are applied. The surface coating organic binders are either heat cured, air dried, catalysed air dried or radiation cured. Heat-cured coatings or baking or stoving enamels are the most commonly found industrial coatings. These binders may be alkyd enamels, thermosetting acrylics, polyesters, epoxies, silicone acrylics, vinyls and heat-resistant polyimides and fluorocarbons. The air-dried coatings include acrylic, vinyl and nitrocellulose lacquers, alkyd resins, urethanes and latex types. Polyurethanes and epoxy resins are examples of catalysed air-dried binders. Powder coatings are also used in many applications and they may be based on epoxies and polyesters.

Circumstances where these coatings occur as evidence involve items such as crowbars, tyre levers, baseball bats, domestic appliances and furniture. The organic coatings may be chemically identified by PyGC [5].

9.2.6.5 Art objects

The forensic scientist may be called upon to examine ancient works of art such as paintings and sculptures to establish authenticity or compare stolen items or even assist art conservationists in art galleries and museums.

Pyrolysis techniques may be applicable to the analysis of older organic coatings, which include drying oils, waxes, gums, resins and proteinaceous materials such as egg, milk and animal glues. These substances are inherently complex natural products and, therefore, their characterisation can be difficult, especially when only a small amount of material can be used, more than one medium has been used in the work and the material has deteriorated with age [36].

Shredrinsky *et al.* [37] have reviewed the application of PyGC to the study of art materials. PyGC has been used for the identification of animal glue, egg yolk, linseed oil and casein [38]. Protein pyrolysis products have been found to be diagnostic for glue and casein. The pyrolysis derivatisation technique (THM) was useful for identifying free fatty acids in yolk and linseed oil.

Other surface coating applications of PyGC to synthetic resins include the identification of photocopy toners and surface coatings on currency notes.

9.3 Pyrolysis mechanisms

There are at least four major pathways for the thermal degradation of polymers. In many cases, more than one mechanism may occur during pyrolysis of the polymer.

1 Random chain cleavage occurs in olefinic and vinyl polymers which have a polymethylene 'backbone' structure. Usually a series of oligomers is produced by random fragmentation at sites along the polymer chain. In the simplest case, polyethylene fragments give a homologous series of alkanes, mono-olefins and di-olefins.

In some vinyl polymers, for example polystyrene, the benzene side chain does not fragment but remains attached to segments of the polymethylene chain and head-to-head and head-to-tail oligomers are produced.

2 Side-chain scission takes place in some vinyl polymers where a pendant group to the polymethylene chain is expelled. The backbone chain then fragments and cyclisation takes place to yield aromatic compounds. Polyvinyl chloride and polyvinyl acetate undergo this

type of thermolysis mechanism. Chlorine and acetate radicals, which split off from the polymethylene chain, combine with hydrogen radicals generated by thermolysis of the backbone chain. This produces hydrogen chloride and acetic acid, respectively, and aromatic compounds including benzene, toluene and polyaromatics are also formed.

3 Chain depropagation takes place in some polymers, which give thermally stable monomer units on pyrolysis. This mechanism usually applies to polymers which have polymethylene backbone chains. Depolymerisation or 'unzipping' takes place to give high yields of monomer. Examples of this are polymethyl- and polybutylmethacrylates. These methacrylates are used as co-monomers in many acrylic surface coatings.

4 Directed chain cleavage occurs when thermolysis takes place at sites of comparatively weak bond strength in condensation polymers. Polyamides and polyesters undergo this type of cleavage at CO—NH and CO—O bonds, respectively. The pyrolysis products of polyamides are usually dibasic acids and diamines. The dibasic acid in nylon 6.6, adipic acid, suffers a loss of carbon monoxide to give cyclopentanone [39]. Polyethylene terephthalate cleaves at the CO—O bond in addition to other sites in the polymer chain to give decarboxylation and recombination products [40] and proceeds via a cyclic transition state.

The mechanism of the THM reaction of some paint organic binder resins is very different to the thermolysis processes described above. In the THM reaction of a quaternary alkylammonium hydroxide with hydrolysable structures in a polymer, the hydrolysed products are converted to the corresponding tetraalkylammonium salts, which are then thermally degraded in the flash heating process to alkyl derivatives. This is a rapid concerted reaction which only occurs with those polymers or molecules which can undergo hydrolysis. The thermolytic mechanisms may occur concurrently with the THM reaction in those polymers which have both hydrolysable moieties and non-hydrolysable structures in the polymer.

9.4 Pyrolysis mass spectrometry

Pyrolysis mass spectrometry (PyMS) involves the direct introduction of the pyrolysis products into the ionisation source of the mass spectrometer. Optimum reproducibility and sensitivity have been achieved by using a custom-made interface between the two [41, 42]. Hughes *et al.* [43] achieved reasonably satisfactory results by connecting the pyrolyser to the mass spectrometer via an empty column in the GC oven. They used this system to examine a range of textile fibres and compared the results with infrared (IR) spectroscopy. The IR spectroscopic method gave better discrimination, particularly with copolymer fibres. However, PMS was capable of characterising smaller samples and analysis times were shorter. Poor reproducibility with certain polymer types and difficulties with interpretation of the composition of the polymer and additives were experienced. This problem of reproducibility was studied by Hickman and Jane [44]. The cause was claimed to be the pyrolysis process rather than the electron-impact fragmentation process in the mass spectrometer. Irwin [45] has reviewed the progress of PMS and discussed the advantages of softer ionisation processes. These included chemical ionisation [46], high-resolution field ionisation mass spectrometry [47], and field desorption mass spectrometry [48]. Probe pyrolysis mass spectrometry is also useful for characterisation and the study of thermal degradation mechanisms. The sample is distilled from the direct insertion probe using a controlled temperature–time profile and mass spectra are determined with respect to temperature. Williamson *et al.* [49] examined polyurethanes and epoxy resins by PMS. The raw mass spectrometric data were stored in an on-line data system and processed to produce individual spectra, total ionisation current profiles, and selected ion profiles characteristic of specific components of the formulation.

Burke *et al.* [50] carried out a comparison of PMS, PyGC and IR for the analysis of paint resins and emphasised the importance of using more than one technique in forensic comparisons.

Wheals [14, 24] has reviewed pyrolysis methods including PMS. The advantages of PMS compared with PyGC are its speed, ease of data handling and high sensitivity. The limitations are its higher equipment cost, preferred use of dedicated instrumentation and difficulty with data interpretation. By comparison, PyGC profiles permit identification of a polymer and assessment of relative concentrations of pyrolysis products which assist in understanding differences in composition of polymers within a class. For these reasons it is likely that PyGC would be the method of choice for most forensic science laboratories. It is surprising that there was only one reported user of PMS for paint organic binder identification in recent paint collaborative trials conducted by Collaborative Testing Services Inc.

9.5 Pyrolysis infrared spectroscopy

This technique has been used for the identification of paint, plastics and rubbers in forensic casework [51]. Milligram samples were pyrolysed in ignition tubes and the pyrolysate transferred to a halide disc which was then analysed in an infrared spectrometer. Comparable results matching the IR spectrum of the polymeric material were claimed. The method is particularly suitable for heavily filled, opaque samples which are difficult to analyse by conventional IR techniques. An IR spectrum of the mixture rather than individual pyrolysis products is obtained, but the method has been largely superseded by the techniques described previously. The method is not suitable for trace paint examinations because of the relatively large quantity of material required for the analysis. A pyrolysis cell using a filament unit which fits into the optical bench of an FTIR spectrometer could have potential for the rapid identification of polymers.

9.6 Advantages and disadvantages of pyrolysis gas chromatography

Pyrolysis gas chromatography is a powerful technique for the identification and discrimination of polymers in surface coatings. The compositional data of paint within a class are clearly indicated in the pyrograms shown previously. The method also has the potential for the identification of additives, such as coalescing agents, flow promoters and organic pigments.

Some of the more negative aspects may be resolved by developments in instrumental design and pyrolysis technique. Although the method is destructive, for many forensic investigations there is often sufficient material for subsequent examination. The removal of a polymeric matrix from material may, indeed, simplify the examination of other components. For example, some pigments and dyes can be recovered unchanged from the pyrolysis zone and they can then be subsequently identified by other means.

Low microgram levels of detection are usually attained in PyGC experiments. This limit of detection is adequate for many forensic examinations. Pyrolyser design, derivatisation methods and the use of more sophisticated MS techniques may improve sensitivity levels. Standardised conditions have not yet been adopted universally, although Australian and New Zealand forensic science laboratories have moved towards standardising on mid-polarity (DB1701 type) phase columns and temperature-programming conditions, following the UK Home Office forensic laboratories practice of using Carbowax packed columns in the 1970s. Many regular users of PyGC have adopted one particular GC column phase type as their standard to assist in profile comparisons. Analysis times are of the order of 30 minutes. Sample handling and data interpretation occupy a short period of time. Therefore, the actual time required to perform the examination is comparable to other techniques. Sample automation, available on some commercial pyrolysers, would improve throughput by utilising time outside normal laboratory hours.

9.7 Future developments

PyGC has developed in the past two decades from the pioneering work of De Forest [52] and others in the 1970s to the present where researchers now carry out sophisticated work on pyrolysis mechanisms. Although pattern recognition techniques have a place in the identification and comparison of polymers, the identification of pyrolysis products and study of pyrolysis mechanisms will lead to a more thorough interpretation of polymer composition. With the development and greater use of MS and other identification methods, it is expected that there will be further advances in this direction.

Some improvements in sensitivity can probably be achieved by improvements in instrumental design and modifications such as relocating the analytical column to the pyrolysis chamber of the pyrolyser. Developments in the design of pyrolysis chamber, GC injector system, capillary columns and detectors will also give improvements in sensitivity. The selected ion monitoring mode in mass spectrometry for diagnostic pyrolysis products may also be used to identify trace quantities of polymeric material.

Modifications to the pyrolysis process are being developed. An example of this is the use of pyrolysis derivatisation techniques [21, 23, 53–55]. Pyrolysis alkylation tends to produce simpler pyrograms and at the same time gives more compositional data. The direct results of simpler pyrolysis profiles are the greater sensitivity and ease of interpretation of pyrolysis profiles. The procedure is pertinent to polymers which can be hydrolysed and derivatised and includes the polyesters and polyurethanes. Variations on this theme may be anticipated.

It may be predicted that PMS techniques for the characterisation of paint and other polymers will be favoured in the future with the increased availability of improved data-handling facilities. The advantages of speed and the elimination of GC separations with inherent column deficiencies and loss of sensitivity by analyte splitting may outweigh the relative ease of interpretation of the data.

9.8 References

1. Peterson J. L., Markham P. N. (1995). Crime laboratory proficiency testing results, 1978–1991. II. Resolving questions of common origin. *Journal of Forensic Sciences* **40**:1009–1029.
2. De Forest P. R., Tebbett I. R. (1992). Pyrolysis gas chromatography in forensic science. In *Gas Chromatography in Forensic Science* (ed. I. R. Tebbett), pp. 165–185. Ellis Horwood, Chichester, W. Sussex, UK.
3. Blackledge R. D. (1992). Applications of pyrolysis gas chromatography in forensic science. *Forensic Science Review* **4**:1–15.

4. Challinor J. M. (1993). Pyrolysis gas chromatography. In *The Forensic Examination of Fibres* (ed. J. Robertson), pp. 219–237. Ellis Horwood, Chichester, W. Sussex, UK.
5. Challinor J. M. (1995). Examination of forensic evidence. In *Applied Pyrolysis Handbook* (ed. T. Wampler), pp. 207–241. Marcel Dekker, New York.
6. May R. W., Pearson E. F., Scothern D. (1977). *Pyrolysis Gas Chromatography*. Analytical Science Monographs No. 3. The Chemical Society.
7. Challinor J. M. (1989). A pyrolysis derivatisation gas chromatography technique for the structural elucidation of some polymers. *Journal of Analytical and Applied Pyrolysis* **16**:323–333.
8. Walker J. Q., Wolf C. J. (1970). Pyrolysis gas chromatography: a comparison of different pyrolysers. *Journal of Chromatographic Science* **8**:513–518.
9. Audino M. (1994). On the evaluation of pyrolyser units for trace pyrolysis gas chromatography investigations. In *Identification of Trace Quantities of Synthetic Fibres Found as Contact Evidence by Pyrolysis Gas Chromatography Techniques* (ed. J. M. Challinor). A report to the National Institute of Forensic Science, Melbourne, Australia.
10. Greenwood P. J., George S. C., Wilson M. A., Hall K. J. (1996). A new apparatus for laser micropyrolysis–gas-chromatography/mass spectrometry. *Journal of Analytical and Applied Pyrolysis* **38**:101–118.
11. Challinor J. M. (1983). Forensic applications of pyrolysis capillary gas chromatography. *Forensic Science International* **21**:269–285.
12. Saferstein R. (1985). Forensic aspects of analytical pyrolysis. In *Pyrolysis and Gas Chromatography in Polymer Analysis* (eds S. A. Liebmann, E. J. Levy). Marcel Dekker, New York.
13. Wampler T. P., Levy E. J. (1985). Analytical pyrolysis in the forensic science laboratory. *Crime Laboratory Digest* **12**:25–28.
14. Wheals B. B. (1985). The practical application of pyrolytic methods in forensic science during the last decade. *Journal of Analytical and Applied Pyrolysis* **8**:503–514.
15. McMinn D., Carlson T. L., Munson T. O. (1985). Pyrolysis capillary gas chromatography/mass spectrometry for analysis of automotive paints. *Journal of Forensic Sciences* **30**:1064–1073.
16. Irwin W. J. (1979). Analytical pyrolysis – an overview. *Journal of Analytical and Applied Pyrolysis* **1**:3–25.
17. Wampler T. (1989). A selected bibliography of analytical pyrolysis applications 1980–1989. *Journal of Analytical and Applied Pyrolysis* **16**:291–322.
18. Grob K., Grob G., Grob K. (1978). Comprehensive standardised quality test for glass capillary columns. *Journal of Chromatography* **156**:1–20.
19. Foelster U., Harres W. (1986). Pyrolysis capillary GC–FTIR for the analysis of macromolecules. Application to epoxy and phenolic resins and their combinations. *Farbe + Lack* **92**:13–17.
20. Liebmann S. A., Levy E. J. (1985). *Pyrolysis and Gas Chromatography in Polymer Analysis*. Marcel Dekker, New York.
21. Challinor J. M. (1989). A pyrolysis derivatisation gas chromatography technique for the structural elucidation of some polymers. *Journal of Analytical and Applied Pyrolysis* **16**:323–333.
22. Challinor J. M. (1994). On the mechanism of high temperature reactions of quaternary ammonium hydroxides with polymers. *Journal of Analytical and Applied Pyrolysis* **29**:223–224.
23. Challinor J. M. (1991). Structure determination of alkyd resins by simultaneous pyrolysis methylation. *Journal of Analytical and Applied Pyrolysis* **18**:233–244.
24. Wheals B. B. (1980/1981). Analytical pyrolysis techniques in forensic science. *Journal of Analytical and Applied Pyrolysis* **2**:277–292.
25. Challinor J. M. (1990). Pyrolysis gas chromatography – some forensic applications. *Chemistry in Australia* **April**:90–92.
26. Ohtani H., Ueda S., Tsuhahara Y., Watanabe C., Tsuge S. (1993). Pyrolysis-gas chromatography for end group analysis of polystyrene macromonomers using stepwise pyrolysis combined with on-line methylation. *Journal of Analytical and Applied Pyrolysis* **25**:1–10.
27. Wang F. C.-Y., Gerhart B. B., Smith P. B. (1995). Structure determination of polymeric materials by pyrolysis gas chromatography. *Analytical Chemistry* **67**:3536–3540.
28. Challinor J. M. (1993). Paint analysis. In *Expert Evidence: Advocacy and Practice* (eds I. Freckleton, H. Selby). Law Book Company, North Ryde, NSW.
29. Challinor J. M. (1984). Pyrolysis capillary gas chromatographic examination of alkyd paints. *Journal of the Forensic Science Society* **24**:451.

30. Bates J. W., Allinson T., Bal T. S. (1989). Capillary pyrolysis gas chromatography: a system employing a Curie point pyrolyser and a stationary phase of intermediate polarity for the analysis of paint resins and polymers. *Forensic Science International* **40**:25–43.
31. Peltonen K. (1986). Gas chromatographic–mass spectrometric determination of phenols and heterocyclic nitrogen compounds in the thermal degradation products of epoxy powder paint. *Journal of Analytical and Applied Pyrolysis* **10**:51–57.
32. Bradna P., Zima J. (1991). Use of pyrolysis-gas chromatography–mass spectroscopy in the analysis of cured polyfunctional epoxy-resins. *Journal of Analytical and Applied Pyrolysis* **21**:207–220.
33. Bradna P., Zima J. (1992). Compositional analysis of epoxy matrices of carbon fibre composites by pyrolysis-gas chromatography–mass spectrometry. *Journal of Analytical and Applied Pyrolysis* **24**:75–85.
34. Cardosi P. J. (1982). Pyrolysis-gas chromatographic examination of paints. *Journal of Forensic Sciences* **27**:695–703.
35. Fukuda K. (1985). The pyrolysis gas chromatographic examination of Japanese car paint flakes. *Forensic Science International* **29**:227–236.
36. Mills J. S., White R. (1994). *The Organic Chemistry of Museum Objects.* Butterworth-Heinemann, Oxford.
37. Shredrinsky A. M., Wampler T. P., Indictor N., Baer N. S. (1989). Application of analytical pyrolysis to problems in art and archaeology: a review. *Journal of Analytical and Applied Pyrolysis* **20**:393–412.
38. Chiavara G., Galletti G. C., Lanterna G., Mazzeo R. (1993). The potential of pyrolysis-gas chromatography/mass spectrometry in the recognition of ancient painting media. *Journal of Analytical and Applied Pyrolysis* **24**:227–242.
39. Senoo H., Tsuge S., Takeuchi T. (1971). Pyrolysis gas chromatographic analysis of 6-66 nylon copolymers. *Journal of Chromatographic Science* **9**:315–318.
40. Sugimura Y., Tsuge S. (1979). Studies on the thermal degradation of aromatic polyesters by pyrolysis gas chromatography. *Journal Chromatographic Science* **17**:34–37.
41. Meuzelaar H. L. C., Kistemaker P. G. (1973). A technique for fast and reproducible fingerprinting of bacteria by pyrolysis mass spectrometry. *Analytical Chemistry* **45**:587–588.
42. Meuzelaar H. L. C., Posthumus M. A., Kistemaker P. G., Kistemaker J. (1973). Curie point pyrolysis in low voltage electron impact ionisation mass spectrometry. *Analytical Chemistry* **45**:1546–1549.
43. Hughes J. C., Wheals B. B., Whitehouse M. J. (1978). Pyrolysis–mass spectrometry of textile fibres. *Analyst* **103**:482–491.
44. Hickman D. A., Jane I. (1979). Reproducibility of pyrolysis–mass spectrometry using three different pyrolysis systems. *Analyst* **104**:334–347.
45. Irwin W. J. (1982). *Analytical Pyrolysis – A Comprehensive Guide.* Marcel Dekker, New York.
46. Saferstein R., Manura B. S. (1977). Pyrolysis mass spectrometry – a new forensic science technique. *Journal of Forensic Sciences* **22**:748–756.
47. Jones C. E. R., Cramers C. A. (1977). *Analytical Pyrolysis.* Elsevier, Amsterdam.
48. Vorhees K. J. (1984). *Analytical Pyrolysis. Techniques and Applications.* Butterworths, London.
49. Williamson J. E., Cocksedge M. J., Evans N. (1980). Analysis of polyurethane and epoxy-resin based materials by pyrolysis mass spectrometry. *Journal of Analytical and Applied Pyrolysis* **2**:195–205.
50. Burke P., Curry C. J., Davies L. M., Cousins D. R. (1985). A comparison of pyrolysis mass spectrometry, pyrolysis gas chromatography and infrared spectroscopy for the analysis of paint resins. *Forensic Science International* **28**:201–219.
51. Smalldon K. W. (1969). The identification of paint resins and other polymeric materials from the infrared spectra of their pyrolysis products. *Journal of the Forensic Science Society* **9**:135–140.
52. De Forest P. R. (1974). The potential of pyrolysis gas chromatography for the pattern individualisation of macromolecular materials. *Journal of Forensic Sciences* **19**:113–120.
53. Challinor J. M. (1991). The scope of pyrolysis methylation reactions. *Journal of Analytical and Applied Pyrolysis* **20**:15–24.
54. Challinor J. M. (1990). Latest developments in the use of simultaneous pyrolysis alkylation for the identification of crime scene evidence. Abstracts from the *12th Meeting of the International Association of Forensic Sciences*, Adelaide, SA. 24–29 October.
55. Challinor J. M. (1990). Pyrolysis gas chromatography – source forensic applications. *Chemistry in Australia* **April**:90–92.

10

Use of infrared spectroscopy for the characterisation of paint fragments

ALEXANDER BEVERIDGE, TONY FUNG and DONALD MACDOUGALL

10.1 Introduction

Paint samples received at a forensic laboratory usually consist of small chips or smears. Analytical methods, therefore, must be capable of producing significant information from microsamples. Infrared (IR) spectroscopy is one of the most useful and powerful tools for this purpose. Its microsampling capabilities are well established and it provides molecular structure information on both organic and inorganic constituents of multicomponent mixtures such as paint layers. IR spectroscopy can be applied both to identifying probable sources of paint through use of databases and to comparing samples from different sources to determine if they could have had a common origin. When comparing single layers of paint, IR data are often combined with data from other techniques such as pyrolysis gas chromatography (PGC), solvent tests and elemental analysis.

This chapter focuses on sampling and analysis techniques, interpretation of IR spectra and application of IR data to identification and comparison of paint samples.

10.2 Infrared spectroscopy

The theory of IR spectroscopy is well covered in several publications. The most useful for the forensic paint chemist is *An Infrared Spectroscopy Atlas for the Coatings Industry* published by the Federation of Societies for Coatings Technology [1] which not only describes theory but also provides an extensive two-volume library of IR spectra. This publication is referred to hereafter as the *Atlas*. This section covers only very basic theory.

Infrared spectroscopy can usefully be described as an instrumental method of analysis in which light falling in the wavelength range of IR radiation is passed through or reflected off a sample. Light of various wavelengths may be absorbed depending on the chemical composition of the sample. The instrument output is an IR spectrum in which the intensity of IR radiation absorbed or transmitted is plotted on the *y*-axis against the frequency or wavelength of the radiation on the *x*-axis. The spectrum may be interpreted to determine functional groups of elements or compared to spectra from other samples or standards collections to establish identity, similarity or differences.

Wavelength is measured in subunits of the metre – most commonly the nanometre (nm) ($1\,\text{nm} = 1 \times 10^{-9}\,\text{m}$) and the micrometre or micron (μm) ($1\,\mu\text{m} = 1 \times 10^{-6}\,\text{m}$). Spectra recorded in the 1960s and early 1970s often used wavelength measured on the micrometre scale on the *x*-axis, but in current practice, wavelength has been superseded by a frequency unit, the wavenumber (cm^{-1}). The wavenumber is the number of waves in a length of 1 cm and is the reciprocal of wavelength in cm. The spectra in this chapter have percentage transmission on the *y*-axis and wavenumber on the *x*-axis.

The infrared region of the electromagnetic spectrum extends from *c.* 12,500 to $50\,\text{cm}^{-1}$. Only the range of $4000–200\,\text{cm}^{-1}$ is of interest to the forensic chemist because the energy of that IR radiation lies in the same range as the energy of vibrational transitions or fundamental modes of vibration within molecules. Molecular vibration changes the interatomic distances between atoms to produce an alternating dipolar electric field. When electromagnetic energy passes through the sample, and the energy frequency (ν) corresponds to the natural frequency of vibration, the electrical component of the energy is absorbed. The IR spectrum provides energy/frequency information about the IR radiation absorbed or transmitted which can be related to molecular structure. Absorption occurs in all but homonuclear diatomic molecules.

To apply the theory to a polyatomic molecule with n atoms it is necessary to consider the ways in which a molecule can move relative to the classical x, y and z axes, and the ways in which the atoms move relative to each other. These movements are translation, vibration and rotation. This means that a molecule with 'n' atoms has $3n$ degrees of freedom of motion. Three of these are translations for the molecule as a whole and three (two in the case of linear molecules) are rotations of the molecule about its principal axes. The remaining $3n - 6$ ($3n - 5$ for linear molecules) degrees of freedom are for vibrational transitions. These vibrational transitions consist of stretching modes ($n - 1$) and deformation modes ($2n - 5$ for non-linear molecules and $2n - 4$ for linear molecules) and give rise to absorption bands in the IR spectrum.

The position (wavelength) of a particular absorption band can be estimated from the theory of harmonic oscillators by knowing the masses of the two atoms involved and the force constant of the bond. The intensity of the absorption band is proportional to the square of the rate of dipole moment change during the vibration. These calculated spectral positions and intensities will vary slightly in the actual spectrum because of influences of the surrounding atomic environments and the physical states of the material. The number of bands observed may also vary from the predicted number of transitions – it may be increased by overtone, combination and difference bands or be reduced by inactive absorptions and by overlapping doubly or triply degenerate absorptions due to molecular symmetry.

It follows that it is difficult to predict exactly what the IR spectrum of a complex molecule such as a paint polymer will be. However, if the structure is known, then reasonable estimates can be made of where bands corresponding to specific functional groups will appear. This is one of the fundamental underpinnings of interpretation of an unknown paint spectrum. In practice, however, identity is usually most efficiently achieved by comparison to standards.

The IR spectra of most organic molecules, including paint resins, contain many sharp bands spread throughout the spectral range of $3600–500\,\text{cm}^{-1}$, whereas ionic inorganic compounds, which include many paint pigments, generally have relatively few strong absorptions which occur primarily in the lower wavenumber region (1500–200 cm^{-1}). Infrared spectra of different compounds can often be differentiated by absorptions in the 'fingerprint' region of the spectrum at 1800–300 cm^{-1}. This makes IR a powerful tool for elucidating the chemical structure of pure compounds as well as the identification and comparison of complex mixtures such as paint.

10.3 Infrared instrumentation

The two primary instrumental methods for analysis of paint fragments by IR are dispersive and Fourier-transform infrared (FTIR) analysis.

10.3.1 Dispersive IR instruments

An IR spectrometer records the selective absorption of characteristic frequencies of radiation. The resultant IR spectrum can be interpreted to determine the molecular structure of the sample being analysed. All dispersive IR spectrometers contain a source of IR radiation, a monochromator, a sample area, a detector and a recorder.

In dispersive IR instruments, IR radiation of all wavelengths is passed through a sample and then dispersed into its individual wavelength components by a monochromator containing a prism or diffraction grating. The most commonly used prisms are alkali halides. These are transparent to IR radiation down to about 650 cm^{-1} (NaCl), 400 cm^{-1} (KBr) or 200 cm^{-1} (CsI). Diffraction gratings, however, supply better resolution. The energy throughput of the system is controlled by the apertures of the entrance and exit slits on the monochromator. The higher the resolution sought, the narrower the slits must be with consequent reduction in energy and signal-to-noise ratio.

Almost all dispersive instruments used in forensic laboratories are 'double beam', that is a beam of IR radiation produced from a source is split such that half of the beam passes through a sample and the other half passes through air. The sample and reference beams are sent alternately by the chopper via the dispersive prism or grating onto a detector, usually a thermocouple, as alternate pulses. The difference in energy between the two beams is measured and plotted as a ratio for each wavenumber and, in this way, IR absorption or transmission spectra are produced. In earlier instruments, ratioing of the sample and reference beams was achieved using mechanical detectors, but in current instruments, solid-state electronics and computers do the work. An optical diagram is shown in Figure 10.1 and individual components are discussed below.

In practice, dispersive instruments have been largely superseded by FTIR instruments which have the advantages of shorter analysis times and better resolution.

10.3.1.1 Infrared source

The infrared source is essentially a body that is electrically heated to a temperature (1200°C) high enough to emit infrared radiation throughout the range of interest (4000 to 200 cm^{-1}). The energy distribution of the emitted radiation depends on the nature of the material being heated and the temperature to which it is heated. The most common sources are the Nernst filament, the nichrome wire and the globar.

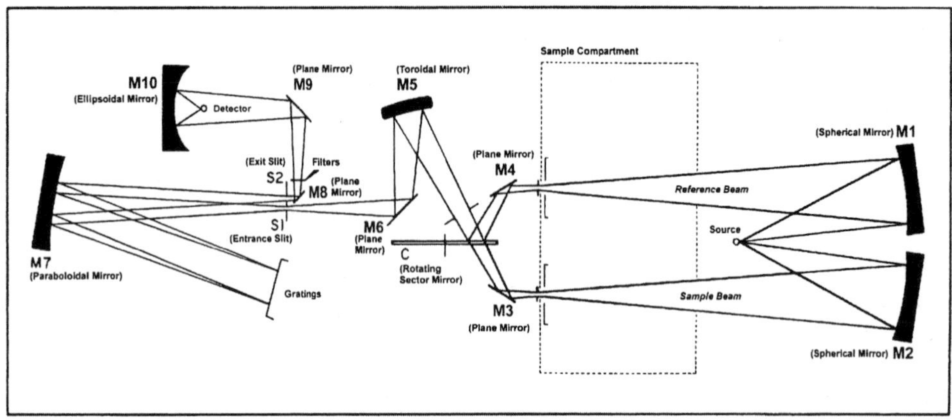

Figure 10.1 Optical diagram of a dispersive IR instrument

The Nernst filament is a small rod composed of rare earth oxides that is preheated to the conducting state and glows when an electrical current is applied to it. The nichrome wire is spirally wound and is sometimes ceramic coated. Nichrome wire is an alloy of nickel, iron, chromium and carbon that emits infrared radiation when an electric current is passed through it. The Globar consists of a silicon carbide rod that emits radiation similar to the Nernst filament. It must be mounted in either a water-cooled or an air-cooled tower.

10.3.1.2 Monochromator

Although the source emits polychromatic radiation, a dispersive spectrometer looks at the wavelength absorption monochromatically. To do this, a grating monochromator (or prism on older instruments) is employed to spatially separate the energy into its component frequencies. By rotating the grating, selected bands from the spectrum are allowed to exit the monochromator sequentially through a set of adjustable-width slits which, along with a set of filters, screen out unwanted frequencies. The slits determine the resolving power of the instrument and maintain a constant energy level reaching the detector, but they also limit the optical throughput.

10.3.1.3 Detector

The dispersed radiation is focused onto the detector which converts the thermal response into an electrical signal. The magnitude of the electrical signal is proportional to the intensity of the radiation hitting the detector. The most commonly used detectors are the thermocouple, the bolometer and the Golay cell.

Thermocouples consist of two different metals that react to the radiant energy striking their juncture by producing an electrical signal proportional to the intensity of the incident radiation. A thermistor bolometer employs a small thermal conductor that experiences a change in its resistance as a function of the temperature change caused by the radiant energy striking it. The Golay cell consists of a gas chamber with an infrared absorbing film on one wall and a flexible mirror on the other. The deformation caused in the flexible mirror as infrared radiation is absorbed and is then translated into an electrical signal.

10.3.2 FTIR instruments

Unlike dispersive instruments, the FTIR spectrometer simultaneously examines all of the frequencies of radiation emitted by the source. This is accomplished by replacing the conventional monochromator with an interferometer. Most interferometers are based on a variation of the Michelson design which consists of two perpendicular plane mirrors, one fixed and one moving, and a beamsplitter. The beamsplitter splits the radiation emitted by the source in two and reflects part to the fixed mirror and part to the moving mirror. The path difference caused by the movement of the moving mirror produces constructive and destructive interference between the two beams as they recombine in the plane of the beamsplitter. It is this recombined beam that exits the interferometer and passes through the sample and on to the detector. The signal reaching the detector undergoes a series of maxima and minima as a function of the path difference. An optical diagram is shown in Figure 10.2.

These interference patterns can be converted to infrared spectra by a complex mathematical process known as a Fourier transform. As described by Noble [2]: "Lord Raleigh realised that by using a Fourier transform method to break down the interference pattern into sine waves that represented the contributions from individual wavelengths of light, the interferogram could be converted back into wavelength absorbances. The method filtered out much of the noise that plagued dispersive spectroscopy in the far-IR range and made it easier to obtain spectra from the weak-intensity sources that were available."

By digitising the detector signal, spectra can be stored and retrieved, subtracted and compiled into searchable libraries.

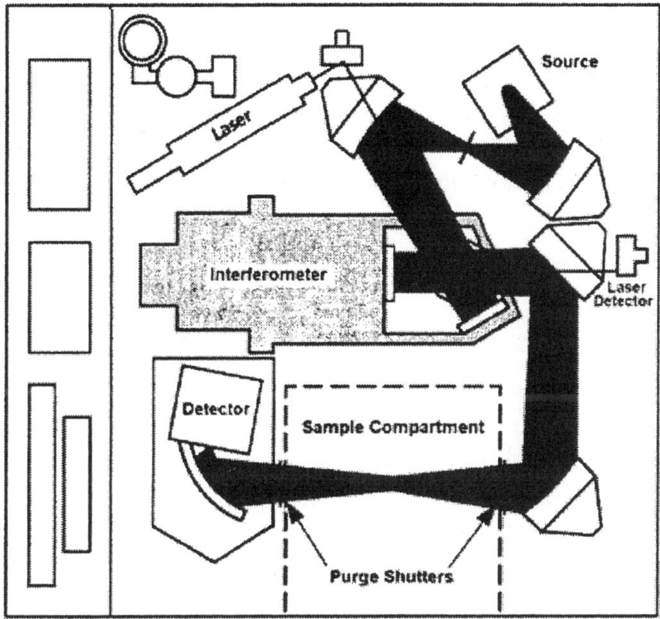

Figure 10.2 Optical diagram of an FTIR spectrometer

10.3.2.1 Other interferometer designs

A number of other interferometer designs have incorporated:

- a modified Michelson interferometer using corner reflectors or cat's-eye retro-reflectors rather than plane mirrors
- Genzel interferometers using a double-sided moving mirror located midway between two collimating mirrors
- refractively scanned interferometers that use fixed mirrors and a moving beamsplitting wedge
- a rotating stage containing the beamsplitting element and two plane mirrors.

10.3.2.2 Scanning a sample

FTIR spectrometers are normally single-beam instruments or contain two single beams that operate slightly differently from double-beam dispersive IR instruments. The analyst normally runs the sample and ratios it against a previously recorded and stored background of the empty sample cell. This enables the removal of absorptions due to the cell material (e.g. diamond absorptions) or the atmosphere (i.e. carbon dioxide and water vapour).

10.3.2.3 Beamsplitters

The beamsplitter is the heart of the interferometer. Beamsplitters must be extremely flat and have highly polished and parallel surfaces. The nature of the beamsplitter and its efficiency determine the usable wavelength range and sensitivity of the instrument. The ideal beamsplitter transmits and reflects 50 per cent of the source radiation reaching it but this is never quite achieved. Mid-range (4000 to 200 cm^{-1}) infrared instruments use alkali halide beamsplitters that are made from either KBr or CsI. KBr beamsplitters have a higher throughput than CsI beamsplitters but they are more hygroscopic and provide a much narrower (less useful) spectral range.

The throughput of an instrument, measured as the magnitude of the signal reaching the detector, is affected not only by the nature and efficiency of the beamsplitter but also by the detector response and amplifier characteristics.

10.3.2.4 Sources

The most common infrared source used in FTIR spectrometers is the Globar (see Section 10.3.1.1).

10.3.2.5 Detectors

The two most common infrared detectors used in mid-range FTIR spectrophotometry are the deuterated triglycine sulphate (DTGS) detector and the mercury–cadmium–telluride (MCT) detector.

The DTGS detector is a type of pyroelectric bolometer. DTGS detectors operate at room temperature and provide quick response to the high modulation frequencies of a mid-infrared range FTIR. The DTGS detector, in conjunction with CsI optics, allows for a wide spectral range (4000 to 200 cm^{-1}) which enables both the organic and inorganic components to be observed at the same time using one instrumental method. However, this detector is less sensitive than the MCT detector.

The MCT detector has a more limited range than the DTGS detector, but has a much higher sensitivity and speed of response. MCT detectors must be cooled by liquid nitrogen to achieve their optimum sensitivity. They are alloys of mercury and cadmium tellurides. By varying the ratio of these two alloys, the bandwidth and sensitivity of the detector can be altered. MCT detectors are available in several configurations; as a narrow band with a cutoff at 700 cm^{-1}, as a mid-range band with a cutoff at 600 cm^{-1} and as a wide band with a cutoff at 450 cm^{-1}. MCT detectors limit the amount of inorganic information observable and are best suited, because of their higher sensitivity, to time-resolved studies where large numbers of consecutive spectra are measured over time or coupled with techniques such as GC-FTIR and infrared microscopy (see below).

For analysis of paint and similar materials, the most commonly used detector is the DTGS detector which detects IR radiation as heat and converts this to an electrical signal. It may be operated at room temperature.

Figure 10.3 shows a comparison of effects of the cutoff wavelength of KBr (narrow-band MCT, wide-band MCT) and CsI (DTGS) optics on inorganic data observed in the lower IR region. The 700 cm^{-1} cutoff of the narrow-band MCT makes barium sulphate, talc and red iron oxide peaks unobservable. The 450 cm^{-1} cutoff of the wide MCT enables barium sulphate 630 and 610 cm^{-1} peaks as well as talc 670, 466 and 452 cm^{-1} peaks to be observed. The presence of these peaks identifies both pigments. The 551 cm^{-1} peak and the overlapping peak at 460 cm^{-1} suggest but do not absolutely confirm the presence of red iron oxide. The 200–250 cm^{-1} cutoff of the DTGS/CsI combination reveals all three inorganic components.

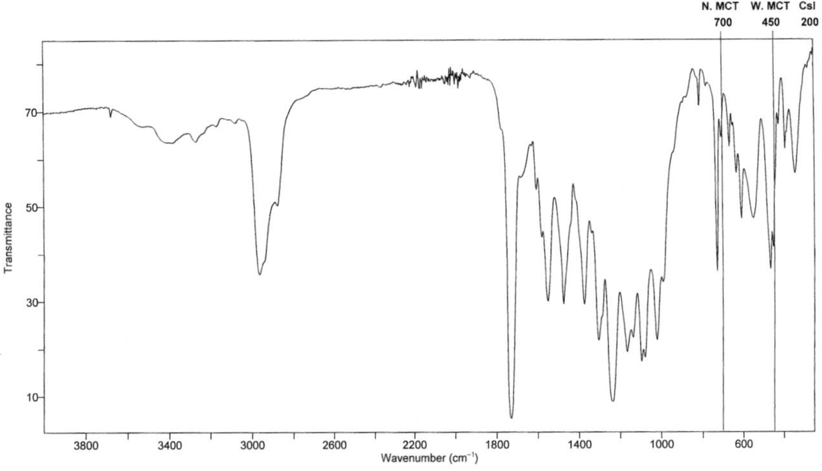

Figure 10.3 Effects of the cutoff wavelength of KBr (narrow-band MCT, wide-band MCT) and CsI (DTGS) optics on the inorganic data observed in the lower IR region

10.3.2.6 Advantages of FTIR over dispersive spectroscopy

FTIR offers three main advantages over dispersive instruments. The first is that all frequencies are detected simultaneously rather than sequentially which enhances speed and sensitivity, the second is that there is a greater energy throughput which improves the signal-to-noise ratio, and the third is use of a laser for calibration of the position of the moving mirror and combining multiple interferograms. These advantages are referred to as Fellgett's, Jacquinot's and Coone's advantages respectively and are discussed in more detail below. Other advantages are the ability to use computing power to subtract out of a spectrum absorptions due to a known component and also the fact that costs are in the same range as dispersive instruments.

10.3.2.6.1 Fellgett's advantage

Fellgett's advantage (also called multiplex) deals with the fact that all frequencies are measured simultaneously rather than one at a time. This results in a huge saving in time, since an FTIR spectrometer can measure the entire spectrum in the same amount of time it would take a dispersive instrument to measure one resolution element. For the same resolution, an FTIR instrument with a one-second scan would produce the same signal-to-noise ratio as a one-hour scan on a dispersive instrument.

10.3.2.6.2 Jacquinot's advantage

Jacquinot's advantage, although its magnitude is of some debate, refers to the increased energy throughput and therefore improved spectral signal-to-noise ratio, since there are no slits to limit the energy reaching the detector.

10.3.2.6.3 Coone's advantage

Coone's advantage (spectral frequency precision) results from the fact that a helium–neon laser, with a single known wavelength, is used to determine the exact position of the moving mirror throughout a scan. Because a laser's output is very stable it can be used as an internal reference to calibrate peak locations very precisely and accurately.

10.3.2.6.4 Other advantages

Other advantages of FTIR instruments are negligible stray light, constant resolution across the spectral range, and a lack of discontinuities in the spectrum; all of which are encountered in dispersive infrared instruments. Very few dispersive instruments are currently being manufactured because of the increased capabilities and diminished cost of FTIR instruments.

10.3.2.7 The FTIR microscope

The FTIR microscope combines the power of microscopy with the FTIR technique. A microscope designed for IR radiation has the same components as a classical light microscope with the exception that glass lenses are unsuitable since they readily absorb IR radiation. Metal-coated reflecting optics are used instead.

By coupling an optical microscope with an infrared spectrometer, minute areas of a sample can be isolated and analysed non-destructively. Aperturing devices (either adjustable knife edges or irises) are used to restrict the field of view to a particular area or layer of interest. Samples as small as 20 μm can be analysed with a minimal amount of sample preparation. Most microscopes are mounted to the spectrometer bench, use an externally directed beam and are capable of operating in either the transmittance or reflectance mode. An optical diagram is shown in Figure 10.4.

Samples for transmission studies are usually sectioned with a microtome or scalpel, flattened with a rolling device or diamond anvil cell and then placed on a KBr or BaF_2 window. A sample thickness of the order of 5–50 μm is usually required. The sample is first observed with visible light and the area to be scanned is centred in the field of view and delineated using the aperture controls. The infrared light is switched on and data collection is started. The sample is then moved out of the field of view and the background immediately scanned. The ratio between the two spectra gives the IR spectrum of the sample. The analyst must ensure that the same aperture size is used for both the sample and background, otherwise a sloped baseline will result.

A number of considerations must be made in selecting the proper aperture size for the sample being analysed, a primary one being the diffraction limit of the microscope. A spectral range of 4000–1000 cm^{-1} is possible with a 10 μm sample, but switching to a 5 μm sample limits the range to 4000–2000 cm^{-1}.

Since FTIR microscopes are often employed to analyse very small samples, they require a highly sensitive detector. An MCT detector is usually employed, and as noted above it must be kept cool with liquid nitrogen and it has a reduced range compared to the DTGS

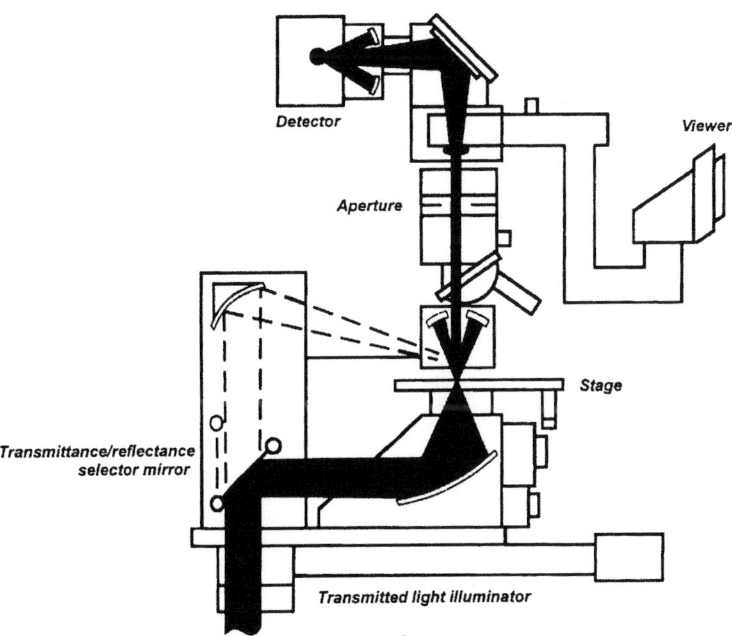

Figure 10.4 Optical diagram of an FTIR microscope

detectors used in FTIR instruments without the microscope accessory, which with CsI optics are transparent to 200 cm^{-1}. The wider range is significant in paint analysis since many inorganic pigments have significant characteristic absorptions in the far-IR [1, 3].

Sample preparation is relatively easy, and is discussed in the following section. A cross-section of a multilayered paint chip cannot only have each layer analysed sequentially in transmission mode, but also may be examined simultaneously for physical attributes such as pigment dispersion.

IR microscopy has been well reviewed by Bartick and Tungol [4] and its specific application to paint analysis has been detailed thoroughly by Ryland [5].

The FTIR microscope is normally used in transmission mode. However, reflection spectra can also be obtained. These are useful in analysis of trace smears of paint which are very difficult to remove and manipulate. However, reflectance spectra are not necessarily 'superimposable' on transmission spectra of the same paint due to reflectance spectral distortion and mathematical corrections. Despite this restriction, reflectance spectroscopy has been described as the method of choice in analysis of multilayered paint chips in a publication from a major automobile manufacturer [6].

Figures 10.5 and 10.6 show FTIR transmission spectra of the same sample of white non-metallic acrylic melamine NBS standard paint KN78A0121 analysed (in different laboratories) with and without a microscope accessory. There are no significant differences between the spectra. The conditions of analysis are given in Appendix 1.

10.4 Preparation of samples for analysis

There are a variety of sampling techniques available to the spectroscopist, including transmission, external reflection, internal reflection, diffuse reflection, IR emission, and various microsampling methods. Each technique has its advantages and disadvantages. The technique selected depends on such factors as the information required, the nature of the material

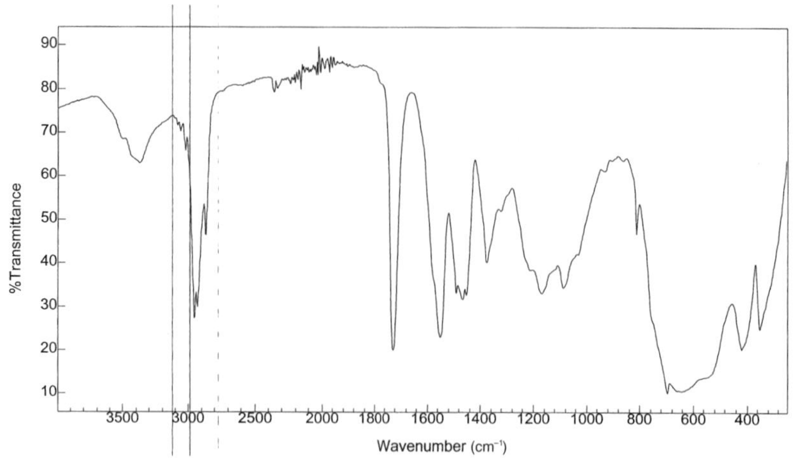

Figure 10.5 FTIR spectrum of a white non-metallic acrylic melamine NBS standard paint KN78A0121 run in a high-pressure diamond cell

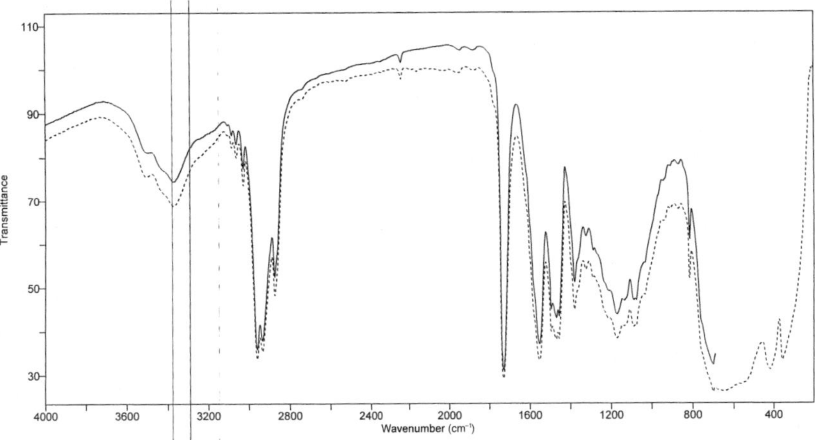

Figure 10.6 FTIR spectrum of NBS standard paint KN78A0121 run on a single-anvil low-pressure diamond cell (dotted line) and as a film (slice sampled without substrate) using an FTIR microscope accessory

being analysed, the size of the sample, the availability of reference libraries developed with that technique and the time involved for sample preparation and analysis. Within a forensic science laboratory, the requirement to analyse extremely small samples, often *in situ*, and the need to preserve exhibits for court, further restrict the infrared technique chosen.

Traditional methods of solid sample analysis include the preparation of mulls and macro alkali halide pellets, but these have almost entirely been replaced by microsampling techniques such as diamond cell transmission, micro-KBr pellets, microreflectance and infrared microscopy. The most commonly used sampling methods for paint are described first for an IR instrument without a microscope accessory, and then for an FTIR instrument with one.

10.4.1 Infrared transmission spectra (no microscope accessory)

The following aspects of paint sampling are discussed:

- layer separation
- KBr micropellets for transmission spectra
- diamond cell for transmission spectra
- other methods used occasionally.

10.4.1.1 Layer separation

An IR spectrum of unseparated multiple layers of paint is of limited forensic value because chemical components cannot be ascribed to specific layers. Therefore, if it is feasible to do so, paint layers should be separated in order to determine the binder and pigment components of each layer. The primary approach to separating individual layers for IR analysis is physical separation by hand or by microtome.

10.4.1.1.1 Separation by hand

Paint layers may be separated with a scalpel while observing the sample through a stereo-binocular microscope at a magnification of 10 to 40 ×. In one commonly used approach, a paint chip is held with a dissecting needle on the surface of a Petri dish and layers sequentially scraped off using a small, curved, no. 15 scalpel blade. Samples of each layer are recovered and placed directly on a diamond surface for analysis (see below). Alternatively, the layers may be cut longitudinally with a no. 15 blade or a small, straight, no. 11 blade. The blades should be changed frequently as they dull quickly with use.

10.4.1.1.2 Separation using a microtome

Instead of a hand-held scalpel, a microtome may be used to separate individual layers [7].

10.4.1.2 Sample preparation for transmission spectra

Most IR spectra of paint are generated by passing IR light of all wavelengths through a sample and recording the wavenumbers of the radiation absorbed or transmitted. The two most common methods of sample preparation are pressing the sample into a film using a diamond cell or mixing the sample with an alkali halide, most commonly potassium bromide (KBr), and applying pressure to form a translucent disc (typically a microdisc with a 1.5-mm diameter).

10.4.1.2.1 Diamond cell

A diamond cell provides an excellent means to sample paint [8]. The diamond anvil cell consists of two diamond windows, a holder for the diamonds, and a mechanism for applying sufficient pressure to squeeze the sample into an appropriate thickness. In one design the diamonds are cemented into steel pistons that fit into a cylindrical holder that keeps them in parallel alignment and prevents their rotation. A rubber O-ring can be used to keep the pistons apart and prevent the diamonds from knocking the sample off during assembly. Preparation involves simply placing a sample of paint on a diamond face, placing a second diamond on top and applying only sufficient pressure to form a film (Figure 10.7). This process is most efficiently accomplished by placing the assembled holder on a microscope stage and viewing the sample as it is pressed to ensure uniform coverage of the small diamond surface (one diamond surface is always smaller than the other). Pressure-related wavelength shifts are minimised by backing off the pressure once the sample is squeezed thin enough to analyse. The cell is then placed in the infrared beam in a reflecting beam condenser mounted in the sample compartment of the instrument. Not only can micro samples be analysed 'as is', but because of the diamond's inertness and the simplicity of sample preparation, the analysis is non-destructive. Samples are neither diluted nor contaminated in the process and may be recovered using a scalpel.

Two types of cell are commonly used – a low-pressure cell (Figure 10.8) and a high-pressure cell (Figure 10.9). Both are effective and any specimen which can be manipulated onto a diamond face can be analysed (for technical details see Appendix 1).

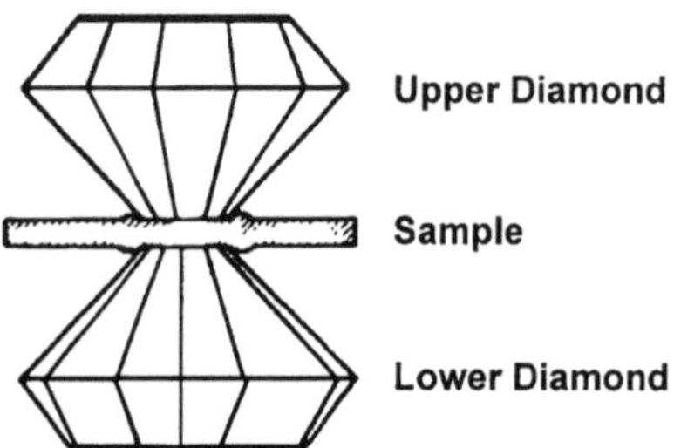

Figure 10.7 Diamonds with sample

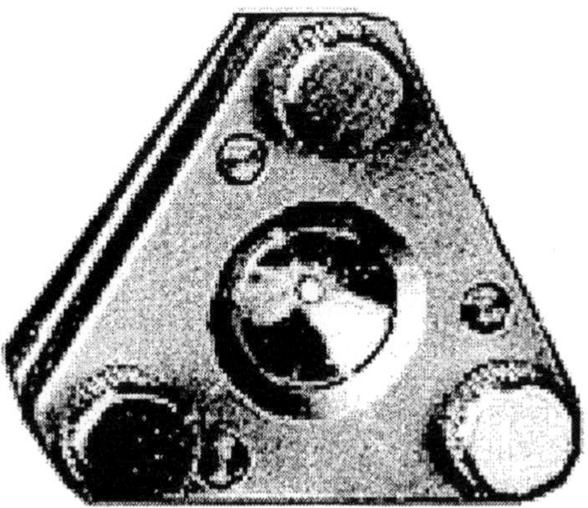

Figure 10.8 Low-pressure diamond cell

Within the RCMP laboratories the diamond anvil cell is the sampling technique of choice. Diamond cells, made from Type II diamonds, are transparent in the region 1800–200 cm^{-1} and are relatively opaque from 4000 to 1800 cm^{-1}. A strong absorption due to the diamond lattice is centred about 2000 cm^{-1}, where few paint peaks are found to occur. They offer the advantages of being chemically inert, hard, durable, and may be loaded with sample relatively quickly. They are costly, however, and have a relatively low-energy throughput (i.e. 5 to 10 per cent of maximum energy) which means that instruments with which they are used must have high throughput and stability.

For certain paint types, especially those which are heavily metallised, mounting paint samples on a diamond anvil may be difficult due to static electricity. In such instances, commercially available anti-static devices are useful.

10.4.1.2.2 KBr pellets

A small sample of paint is ground in an agate mortar and pestle with spectroscopic grade KBr. The homogenising process may be facilitated by adding a drop of methylene chloride. The dry powder is transferred to a die maker and a disc generated using a press. For casework samples, a micropellet with a diameter of 1.5 mm is normally used. Dies are

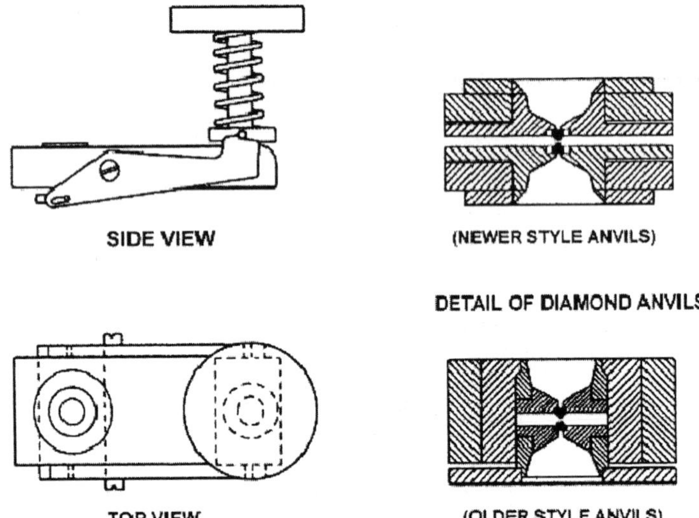

Figure 10.9 High-pressure diamond cell

available both in metal and paper. KBr is transparent to IR radiation between *c.* 4000 and 450 cm^{-1}. KBr is deliquescent, so KBr discs should be stored in an oven or desiccator as should the bulk powder and die components.

Over the past two decades, KBr discs have fallen in popularity in the face of more convenient diamond cell techniques. However, KBr micropellets are cheap and relatively easy to make, and require only a simple beam condenser for analysis. CsI may also be used to make discs.

10.4.1.2.3 Thin film

An effective method of mounting paint samples for IR analysis is to place a thin section across an aperture and pass light directly through it. This method is used more frequently with a microscope accessory but can be used with a transmission instrument.

10.4.2 Infrared reflectance spectra (no microscope accessory)

Reflectance techniques may provide an alternative to transmission methods for poorly transmitting opaque samples or samples requiring *in situ* analysis. Although most conventional attenuated total reflectance (ATR) techniques require samples much larger than those encountered in forensic casework, one accessory, the Harrick SplitPea™, makes use of a semispherical internal reflection crystal to condense the beam onto a spot-sized sampling area. The reflectance crystal is mounted horizontally in the upper surface of a purgable box that fits into the spectrometer's sample compartment (Figure 10.10). A series of mirrors inside the accessory focus the infrared light onto the sampling spot. The light which is not absorbed by the sample is collected and directed onto the detector. The sample

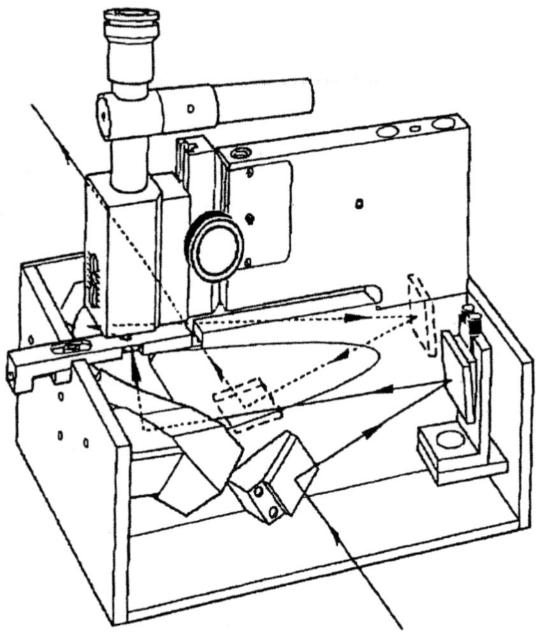

Figure 10.10 SplitPea™ microreflectance accessory – instrumental optical path

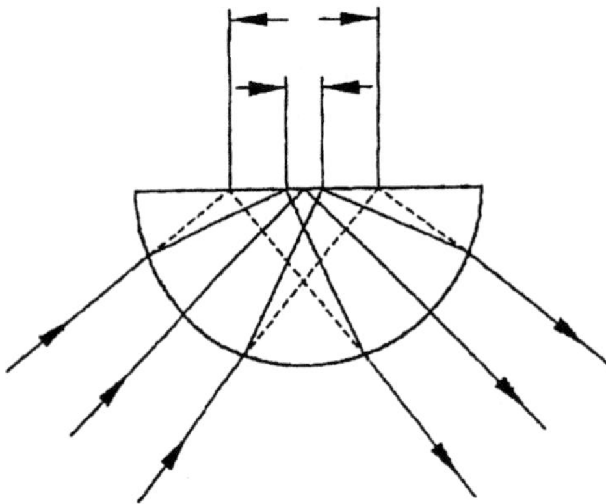

Figure 10.11 Optical path at the sample interface in the SplitPea™ microreflectance accessory

or area of interest, e.g. ink or toner on paper, or paint smears, is inverted and placed in contact with the centre of the crystal. A small adjustable piston is used to apply sufficient pressure to ensure good contact (Figure 10.11). The accessory is also available with a 50 × viewing microscope to locate samples such as fibres, and either a silicon or diamond reflection crystal or an external reflection sample holder.

The SplitPea™ is non-destructive and requires little or no sample preparation. It is well suited to the analysis of documents that may be altered or disturbed by sampling, or smears of paint that cannot be removed from underlying surfaces for fear of loss or contamination. Although contributions from underlying layers or substrate material can be eliminated or reduced, the SplitPea™ is a microsurface analysis technique and is subject to surface effects and heterogeneity of the sample. The analyst must also bear in mind that the effective depth of penetration (sample thickness) varies with the wavelength of light (and index of refraction of the sample and crystal, and angle of incidence) and that the peak heights will be affected.

10.4.3 Spectra of the same sample prepared in different ways

Figures 10.12–10.15 illustrate spectra of the same blue metallic acrylic lacquer automotive topcoat analysed in a high-pressure diamond cell, a CsI pellet, a thin film over an aperture and using a SplitPea™ reflectance accessory respectively. The same information is present in each spectrum.

10.4.4 FTIR spectroscopy using a microscope attachment

Use of a microscope attachment permits FTIR analysis of small multilayered samples of paint without physical layer separation. The attachment can also be used to obtain reflectance spectra.

10.4.4.1 Sample preparation for transmission spectra

When using a microscope accessory, there is no need to separate layers. A thin cross-

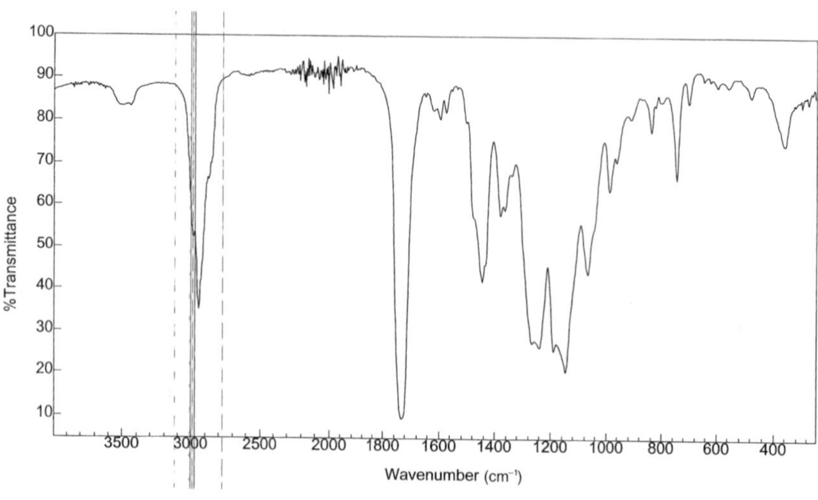

Figure 10.12 IR spectrum of blue metallic paint sample run using a high-pressure diamond cell

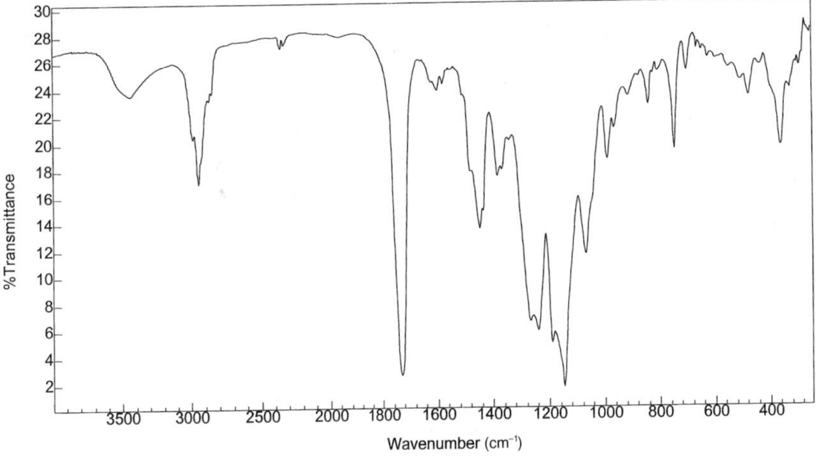

Figure 10.13 IR spectrum of blue metallic paint sample run as a CsI 1-mm micropellet

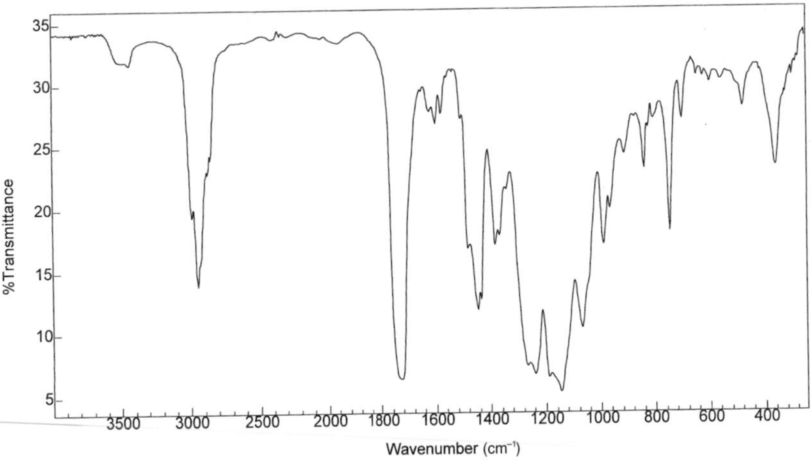

Figure 10.14 IR spectrum of blue metallic paint sample run as a thin film on a 1-mm aperture

section of a multilayered chip which contains all of the layers is cut, typically with a microtome, and placed across or in the aperture of a sample holder on the microscope stage. A single-layered sample may be flattened to a film using either a roller or a diamond cell. The film is then removed and mounted as noted above. Individual layers of multilayered chips may be analysed sequentially by appropriate selection of aperture and magnification.

10.4.4.2 Sample preparation for reflectance spectra

Reflectance spectra may be obtained from samples *in situ* if they can be manipulated onto the sample stage for analysis. An alternative approach for the analysis of multilayered paint

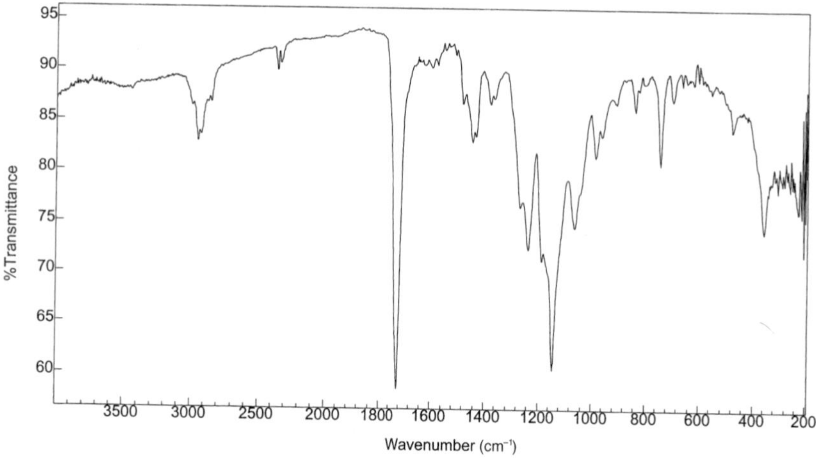

Figure 10.15 IR spectrum of blue metallic paint sample run using a SplitPea™ reflectance accessory

chips has been described by MacEwan and Cheever [6] who sectioned paint chips at a 5° angle to the top surface thereby enlarging the dimensions of the layers tenfold. They did note, however, that the reflectance spectra were not as well defined as transmission spectra and their sensitivity was lower. Also, reflectance spectra had to be adjusted to allow for refractive index changes and they had to be calibrated for band shifts relative to transmission spectra. However, since the authors were not limited by sample size they found the method effective and convenient. The authors illustrated reflectance spectra from 5° and 90° sections and transmission spectra for five typical layers of automotive paint: clearcoat, basecoat, primer surface, conductive primer, moulded coating.

10.5 Infrared analysis of paint

Paint is received at a forensic laboratory as dried fragments which are often composed of several layers. Because paint is such a commonly used product, it can provide valuable physical evidence in the context of many crimes against persons and property. Some examples are traces of paint:

- left at the scene of a hit-and-run motor vehicle accident, e.g. in the hair or on the clothing of a pedestrian victim
- left at the scene of a crime by a 'getaway' car
- left at a scene from a tool or on a tool
- picked up on clothing from the scene of a break-and-enter.

In all such instances investigators ask two typical questions:

- What is the probable origin of the paint?
- Could paint samples have originated from the same source?

The previous sections have described what IR spectroscopy is, and how to generate an IR spectrum. This section deals with the interpretation necessary to answer the noted questions. What sets forensic science apart from some other applications of analytical chemistry is that the scientist must communicate not only whether paint samples have the same composition but also what is *the significance* of the analysis. Thus, a third typical question asked more by scientists than investigators is:

- Could two paint samples having the same physical characteristics and chemical composition have originated from different sources?

Such an opinion must be based on a strong empirical foundation pertaining to the nature of the material, the nature of the analytical data and the conclusions which can be drawn from them. The essence of forensic comparison is to seek significant differences between samples. If such differences are found, an opinion may be given that they did not have a common origin. Infrared spectroscopy is an appropriate method for this purpose since many paints with the same physical characteristics may be discriminated by their chemical composition. If samples cannot be discriminated physically and chemically, then the issue becomes how many other paint samples are likely to have the same physical characteristics and chemical composition? The answer to this question lies in databases (see below).

10.5.1 Paint and coatings

'Paint' consists of pigment(s) dispersed in a resinous binder, which is reduced to a desired viscosity with a solvent. On application, the paint dries to form a hard film adhering to a surface. 'Coating' is a broader term and includes non-pigmented finishes such as varnishes and powders which form films when heated. Additives are incorporated into paint formulations to modify or control physical properties and/or chemical behaviour. Paints are applied to a variety of materials, including wood, metal, rigid and flexible plastics, and composites. The composition of the paint must be tightly controlled and tested to ensure that it meets all of the application, appearance and performance characteristics desired. A detailed classification of paints and their manufacture is given in Chapter 8 of this volume but forensic chemists, influenced by the volume of automobile-related cases and the fact that the IR spectra of automotive paints can often be readily distinguished from those of non-automotive paint, conveniently divide paint samples into only two categories, non-automotive and automotive. This is the classification used in this chapter.

10.5.2 Structural formulae and Infrared spectra of some commonly encountered paint resins and pigments

Paint, in the forensic context of IR analysis, can usefully be regarded as a mixture of two primary components, an organic polymeric resin, which binds the paint to a surface, and organic or inorganic pigments, which give the paint film its colour and hiding power. The pigments are dispersed in the paint resins. Paint films also typically contain many additives in low percentages, but these are rarely in sufficient concentration to be observed in an IR spectrum.

There are many paint resins and pigments on the market, and the IR spectra of most of these can be found in the *Atlas* as noted previously. However, paints received in a forensic laboratory tend to be composed of a fairly small group of resins and pigments.

Paint resins commonly encountered in forensic laboratories include alkyd, amino, acrylic, polyurethane, epoxy, nitrocellulose and vinyl acetate polymers, while the most commonly encountered pigments are titanium dioxide, talc, kaolin, barium sulphate, calcium carbonate, silica and iron oxide. The forensic context and types of paint in which these components may be found are discussed in later sections.

10.5.2.1 Alkyd resins

Alkyd resins are oil-modified polyesters formed by the condensation of polybasic acids or anhydrides and polyols chemically combined with monobasic fatty acids. Drying alkyds use fatty acids derived from unsaturated vegetable oils (soya, linseed, etc.) which allow the resin to cure through oxidation of the double bonds by atmospheric oxygen. Non-drying alkyds use saturated fatty acids (e.g. from coconut oil) to form thermoplastic resins that will not crosslink by air oxidation. Alkyds are often modified with amino resins, urethanes or silicones to achieve desired properties. The most commonly used anhydride is *ortho*-phthalic, but *meta*-phthalic (isophthalic) is also used. An alkyd resin has the structural formula shown in Figure 10.16 in which C_6H_4 represents a substituted benzene ring.

The unsaturated R group, typically containing two or three double bonds, gives alkyds their air-drying property by allowing the film to cure through air oxidation. The IR spectrum of an alkyd would be predicted to have aromatic absorptions, C—H and C—C aliphatic absorptions, a carbonyl absorption, and C—O stretching absorptions. All of these are seen in the characteristic spectrum of a melamine-modified alkyd resin illustrated in Figure 10.17.

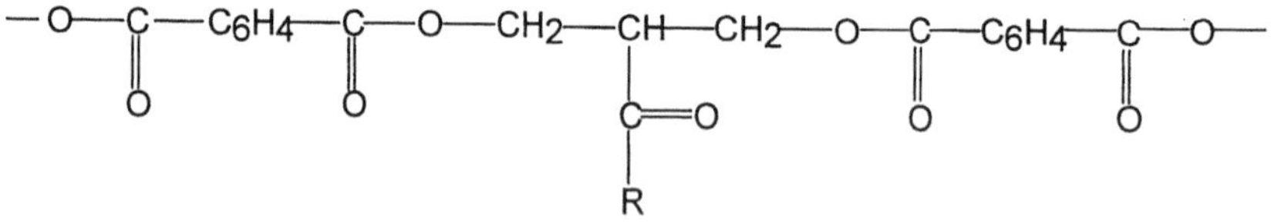

—O—C(=O)—C6H4—C(=O)—O—CH2—CH(—C(=O)—R)—CH2—O—C(=O)—C6H4—C(=O)—O—

Figure 10.16 Structure of an alkyd resin

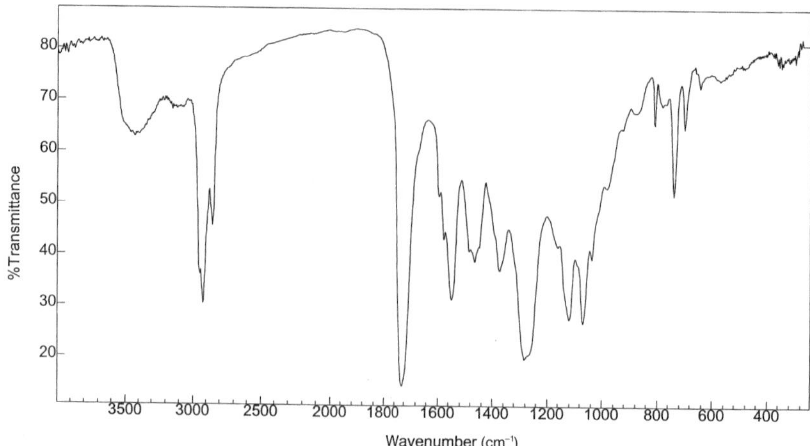

Figure 10.17 IR spectrum of a melamine formaldehyde-modified alkyd resin

The two peaks around 2900 cm^{-1} are due to aliphatic C—H stretching primarily from the drying oils. There is a strong carbonyl C=O absorption at 1730 cm^{-1} and C—O stretches at 1285 and 1122 cm^{-1} indicative of an ester. The two peaks at 1467 and 1376 cm^{-1} are due to methylene and methyl stretching. The peak at 743 cm^{-1} is due to out-of-plane bending of four adjacent hydrogens on the aromatic ring. The small peak at 706 cm^{-1} is due to aromatic ring bending.

10.5.2.2 Amino resins

Amino resins are reaction products of melamine (1,3,5-triaminotriazine) or urea with formaldehyde. They are widely used as crosslinking agents with resins containing hydroxyl groups (e.g. polyesters, epoxys and acrylics). Melamine formaldehyde is widely used for adhesives in furniture and plywood, and as a crosslinking agent for baking enamels in kitchen appliances and automobile finishes. Its structural formula is illustrated in Figure 10.18.

Its IR spectrum is characterised by peaks associated with the triazine ring (C_3N_3) and is illustrated in Figure 10.17. The peak at 1550 cm^{-1} and a small non-diagnostic shoulder at 1450 cm^{-1} are due to in-plane deformation of the triazine ring. A small sharp peak at 815 cm^{-1} is due to out-of-plane triazine ring vibration.

10.5.2.3 Acrylic resin

Acrylic resins are widely used in domestic, industrial, automobile, and anticorrosion and wood protection applications. The major monomers used in the manufacture of acrylic resins are methyl, ethyl, butyl and 2-ethyl hexyl esters of acrylic or methacrylic acids. By altering the proportions of these esters in the formulations or by modification with other monomers such as styrene, vinyl acetate or acrylonitrile, acrylic resins with a wide range of properties can be produced.

Acrylic polymers may be thermoplastic or thermosetting. The former, also called acrylic lacquers, depend on the evaporation of the solvent to form the film while the latter contain reactive sites for further crosslinking after application. Both thermoplastic and thermosetting acrylics have been widely used in the automotive industry. Acrylic lacquers were used extensively by General Motors as topcoats up to the early 1980s. Today, the majority of auto makers in the world are using thermosetting acrylic paint for their clearcoat/basecoat systems.

Acrylic resins have the structural formula shown in Figure 10.19.

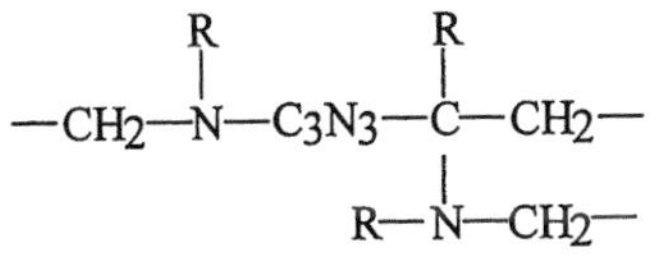

```
          R           R
          |           |
—CH2—N—C3N3—C—CH2—
                      |
                  R—N—CH2—
```

Figure 10.18 Illustration of melamine formaldehyde crosslinking

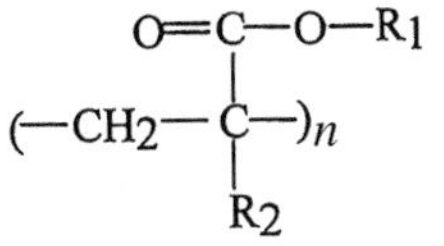

```
          O=C—O—R1
            |
(—CH2—C—)n
            |
            R2
```

Figure 10.19 General structure of an acrylic resin

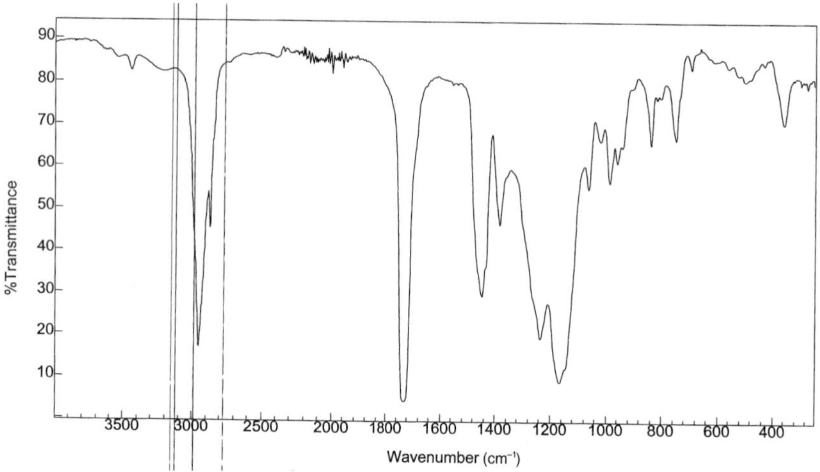

Figure 10.20 IR spectrum of an acrylic resin

Thus, the IR spectra of acrylic resins have characteristic absorptions of carbonyl, C—O, C—C and other absorptions depending on the functional groups in R_1 and R_2.

Figure 10.20 is a spectrum of an acrylic resin. The peaks at 2958 and 2875 cm^{-1} are from aliphatic C—H stretching. There is the familiar 1732 cm^{-1} carbonyl peak, the methylene and methyl C—H bending at 1451 and 1387 cm^{-1} respectively. The broad peaks at 1293 and 1169 cm^{-1} are due to C—O—C stretching of the aliphatic ester.

10.5.2.4 Epoxy resins

Epoxy resins are polyethers made from the condensation of polyols and epichorohydrin. The most commonly used polyol is 'bisphenol A' which is 2,2-*bis*-phenylol propane. Epichlorhydrin has a three-membered ring consisting of two carbons and an oxygen in the form:

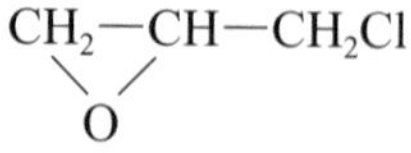

$$\underset{\diagdown\;\;\;O\;\;\;\diagup}{CH_2—CH}—CH_2Cl$$

Epoxys are widely used in automotive thermosetting primers and for corrosion protection. The structural formula for a bisphenol A/epichlorhydrin epoxy polymer is shown in Figure 10.21.

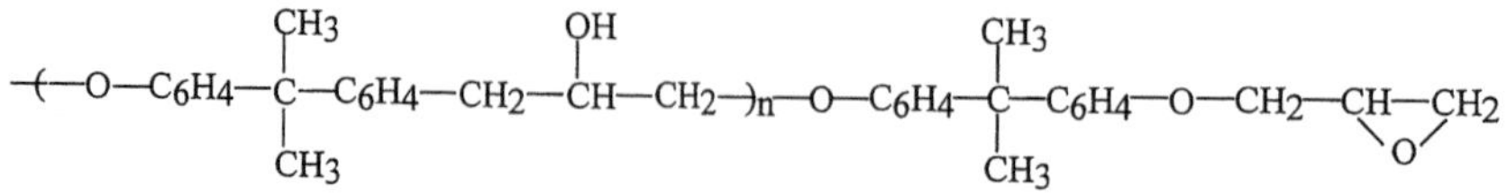

$$—(—O—C_6H_4—\overset{CH_3}{\underset{CH_3}{C}}—C_6H_4—CH_2—\overset{OH}{CH}—CH_2—)_n—O—C_6H_4—\overset{CH_3}{\underset{CH_3}{C}}—C_6H_4—O—CH_2—\underset{\diagdown O \diagup}{CH—CH_2}$$

Figure 10.21 Structure of an epoxy polymer

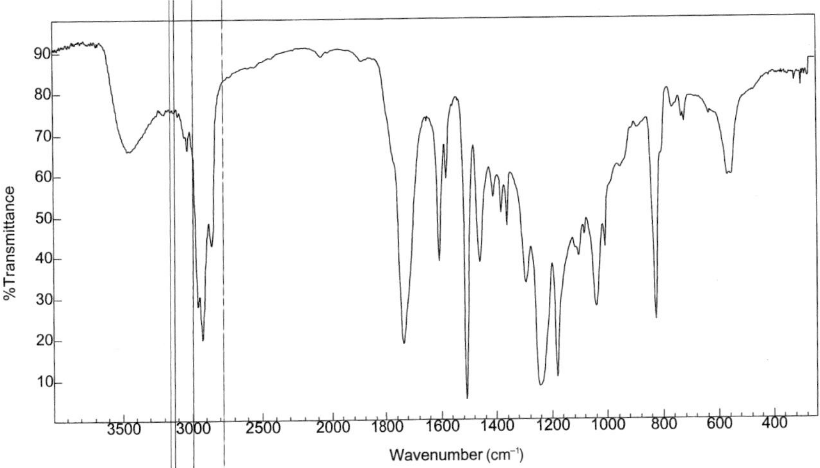

Figure 10.22 IR spectrum of an epoxy resin

The IR spectrum of such an epoxy contains absorptions characteristic of aromatics, ethers and C—O bonds. These are illustrated in Figure 10.22 which shows the IR spectrum of a drying oil modified epoxy resin.

The C—H stretching absorptions between 2800 and 3000 cm^{-1} are contributed to by both the oil and the epoxy. The strong carbonyl peak at 1736 cm^{-1} is from the fatty acid. The sharp peaks at 1607, 1582 and 1510 cm^{-1} are due to C=C stretching of the aromatic bisphenol ring. The broad peak at 1245 cm^{-1} is due to C_6H_4—O stretching. The 1042 cm^{-1} peak is from aliphatic C—O stretching. The peak at 829 cm^{-1} is due to out-of-plane bending of two adjacent hydrogens on the *para*-substituted aromatic ring. The broad absorption around 560 cm^{-1} is due to C—O—C stretching and bending vibrations.

10.5.2.5 Polyurethane resins

Polyurethane resins are condensation polymers formed by the reaction of isocyanates (R—N=C=O) with compounds containing an active hydrogen atom such as hydroxyl groups in polyols, polyesters, epoxies, alkyds and acrylics. The formulations are quite flexible and the resins can be marketed as one-component, two-component or water-cured systems. Polyurethanes offer flexibility, good abrasion and chemical resistance and thus are used extensively as wood floor protectors and in aircraft and automobile coatings. A generic structural formula is shown in Figure 10.23.

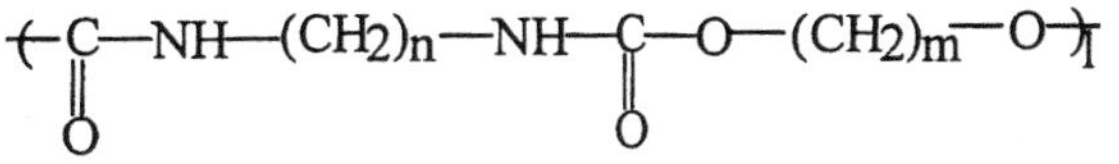

$$\left[\underset{\underset{O}{\|}}{C} - NH - (CH_2)_n - NH - \underset{\underset{O}{\|}}{C} - O - (CH_2)_m - O \right]$$

Figure 10.23 A generic formula for a urethane resin

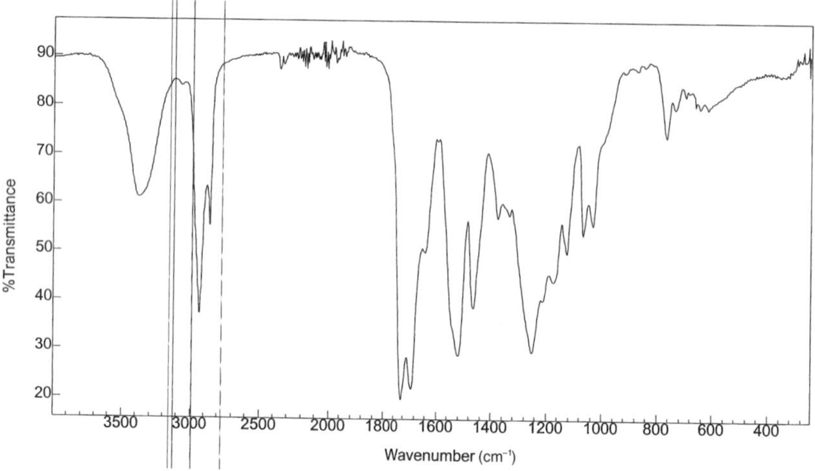

Figure 10.24 IR spectrum of a polyurethane resin

The IR spectrum is characterised by absorptions for carbonyl, carbon chains and amide linkages. Figure 10.24 illustrates an IR spectrum of a polyurethane resin.

The broad band centred at *c.* 3380 cm^{-1} is due to the N—H stretch. The peaks at 2936 and 2861 cm^{-1} are due to aliphatic C—H stretching. The doublet at 1729 and 1691 cm^{-1} is due to carbonyl stretches. The peaks at 1522 and 1468 cm^{-1} are respectively due to N—H and methylene bending. It is interesting to note the relatively weak methyl bending absorption at 1380 cm^{-1}. The broad absorption at 1254 cm^{-1} is due to C—N—H vibration and the C—O stretching of the carboxyl group.

10.5.2.6 Nitrocellulose

Nitrocellulose, also known as cellulose nitrate, is produced from the reaction of cellulose, nitric acid and sulphuric acid. It is marketed for applications which depend on its degree of nitration. The 'low' nitrogen formulation consisting of *c.* 10 per cent nitrogen by mass has the general formula $C_{12}H_{17}O_7(ONO_2)_3$ and is used as a lacquer coating for tools and in some older automotive repaint body fillers and undercoats.

The structural formula of nitrocellulose is based on a cellulose molecule bearing the nitrate ester group $O—NO_2$. Thus the IR spectrum would be predicted to consist of strong C—O and C—H stretches plus the characteristic $O—NO_2$ absorptions at *c.* 1650, 1280 and 830 cm^{-1} plus a broad O—H stretch at *c.* 3500 cm^{-1}. The spectrum illustrated in Figure 10.25 shows the predicted C—H stretches in the 2900 cm^{-1} region, the $O—NO_2$ absorptions at 1651, 1280 and 832 cm^{-1}, the cellulose C—O stretches as broad absorptions centred on 1065 cm^{-1} and a broad O—H absorption at *c.* 3500 cm^{-1}.

10.5.2.7 Vinyl resins

Vinyl resins are made by polymerisation of vinyl monomers such as vinyl chloride and vinyl acetate resulting in polyvinyl chloride (PVC) and polyvinyl acetate (PVA) with the structures shown in Figure 10.26.

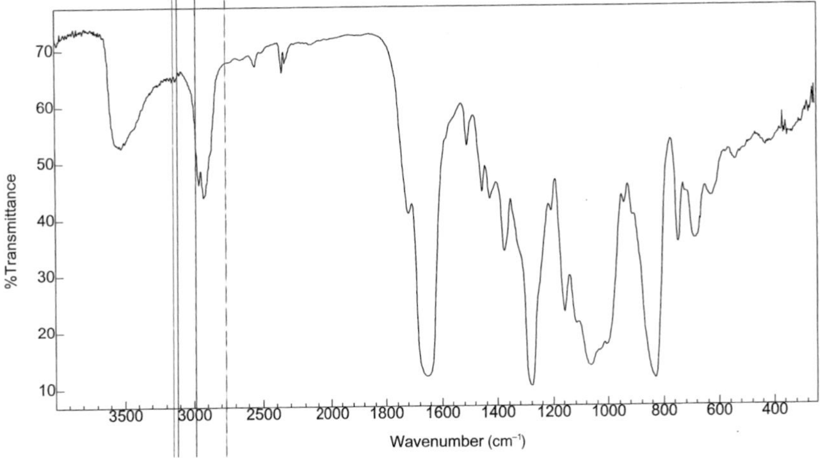

Figure 10.25 IR spectrum of nitrocellulose resin

$-CH_2-CH(Cl)-$

PVC

$-CH_2-CH(C(=O)-OCH_3)-$

PVA

Figure 10.26 Structures of polyvinyl chloride and polyvinyl acetate

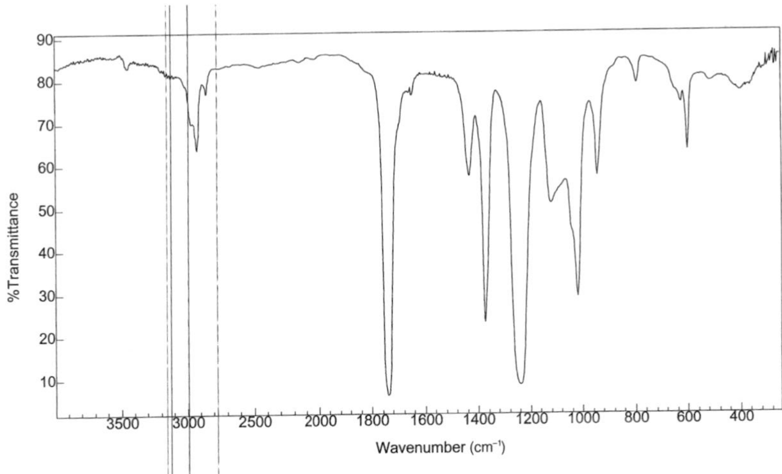

Figure 10.27 IR spectrum of polyvinyl acetate

Vinyl resins are used for interior and exterior house paints due to their low cost. Figure 10.27 illustrates the IR spectrum of polyvinyl acetate. The major absorptions are due to aliphatic C—H stretching in the 2900 cm^{-1} region, a carbonyl at 1738 cm^{-1}, a small methylene bending peak at 1434 cm^{-1} and an enhanced methyl bending peak at 1373 cm^{-1}, C—O stretching from the O=C—O single bond at 1240 cm^{-1} and from the O—CH_3 bond at 1021 cm^{-1}. A small peak at 605 cm^{-1} is a bending mode of the acetate group. There are also characteristic peaks at 1122 and 946 cm^{-1}.

10.5.2.8 Inorganic pigments

Inorganic pigments absorb strongly in the IR. Their structural formulae are less important than the ions which they contain. The IR spectra of the most commonly used pigments are illustrated in Figures 10.28 to 10.35.

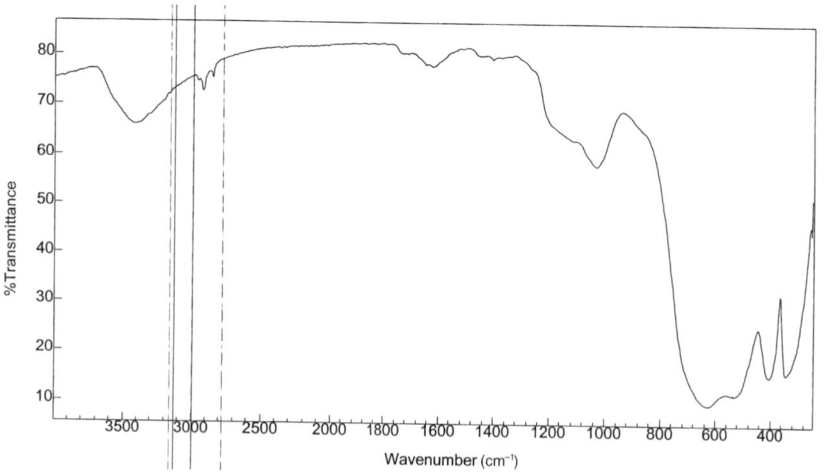

Figure 10.28 IR spectrum of titanium dioxide (TiO_2, rutile)

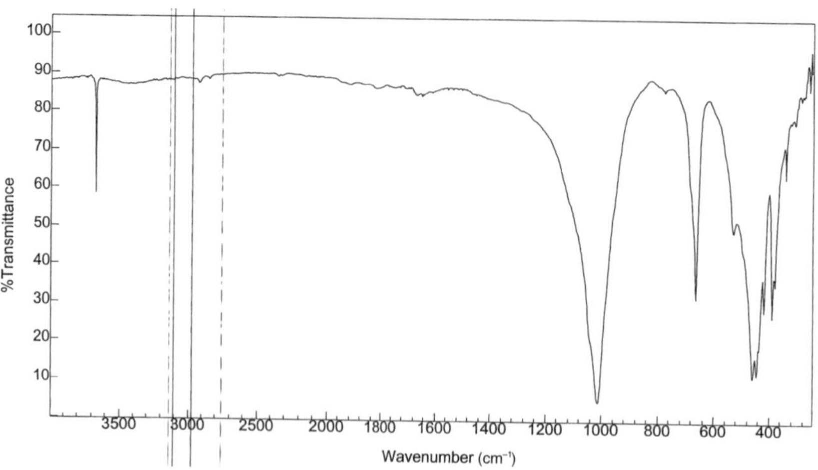

Figure 10.29 IR spectrum of talc ($3MgO{\cdot}4SiO_2{\cdot}H_2O$)

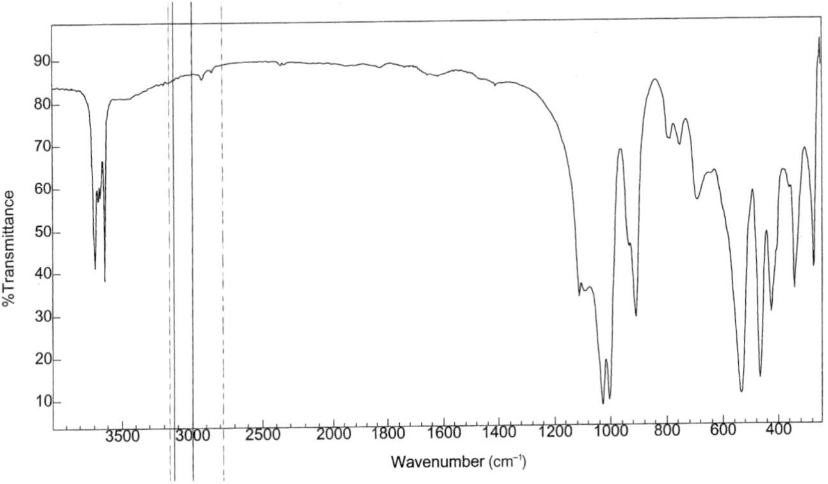

Figure 10.30 IR spectrum of kaolin ($Al_2O_3 \cdot 2SiO_2 \cdot 2H_2O$)

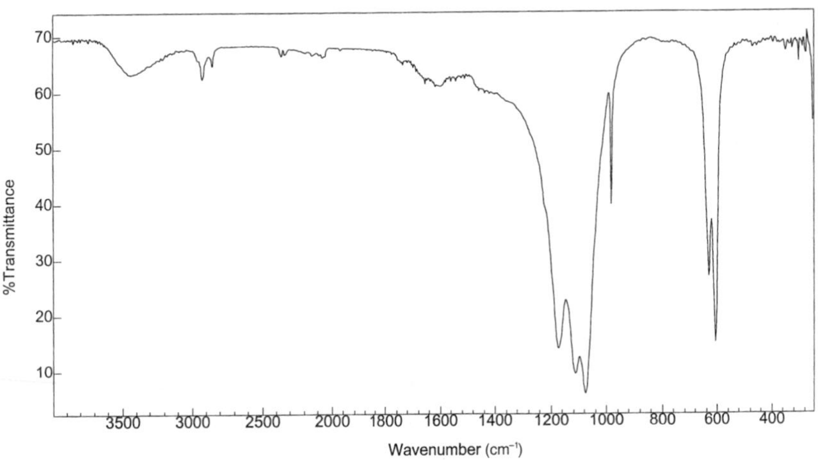

Figure 10.31 IR spectrum of barium sulphate ($BaSO_4$)

10.6 Interpretation of IR spectra of non-automotive (domestic) paints

The majority of domestic paint samples encountered in a forensic laboratory are in the form of paint chips or paint smears from doors, windows or walls left on house-breaking tools and vice versa. Occasionally, law-enforcement agencies submit industrial paint smears from concrete barriers or lamp standards left on car bumpers in suspected fraudulent insurance claims.

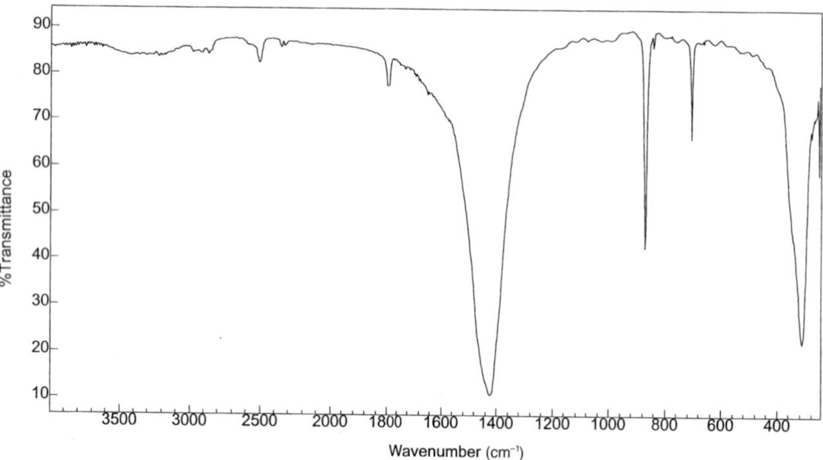

Figure 10.32 IR spectrum of calcium carbonate ($CaCO_3$)

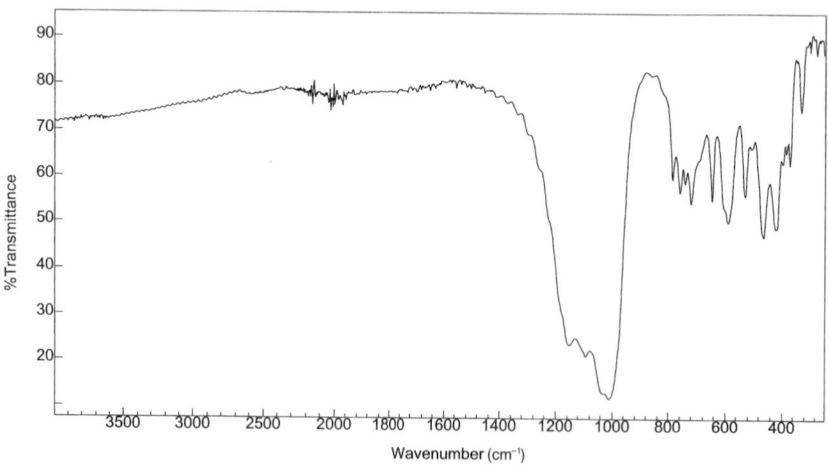

Figure 10.33 IR spectrum of a sodium potassium aluminosilicate pigment ([Na,K]$AlSiO_4$)

In general, domestic paints contain a high proportion of inorganic pigment and extenders such as calcium carbonate, titanium dioxide and clays to give good hiding power and a flat appearance. IR spectra of such paints are therefore useful for identifying both the resin and pigment.

Figures 10.36, 10.37 and 10.38 are spectra of three domestic paints. The question is: what are they? That is, what information can be obtained from the IR spectra about the resin and pigment components of each?

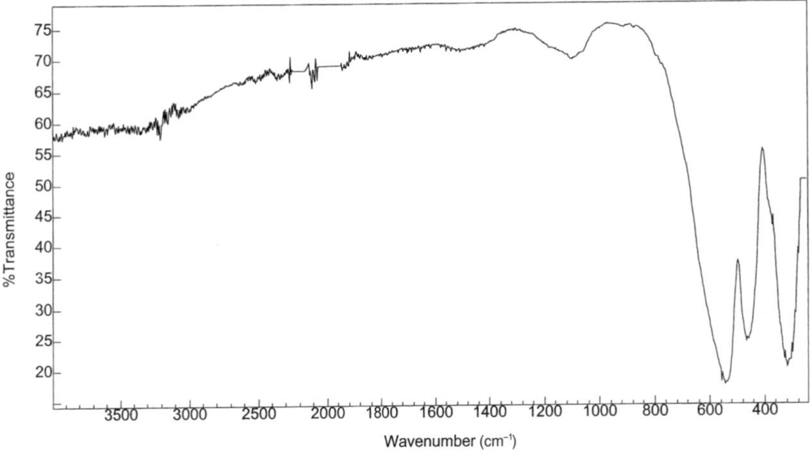

Figure 10.34 IR spectrum of iron oxide (Fe_2O_3)

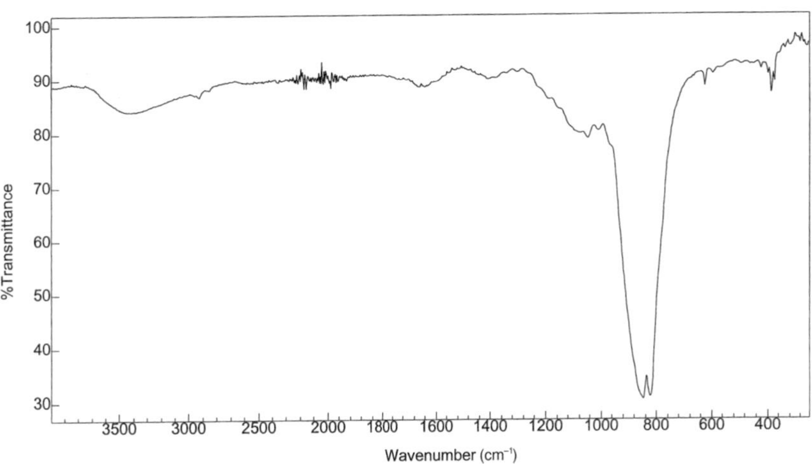

Figure 10.35 IR spectrum of chrome yellow pigment ($PbCrO_4$)

10.6.1 Unknown paint sample #1

A quick scan of the spectrum indicates that there is a carbonyl peak at 1734 cm^{-1} arising from the resin and that there are broad and very strong peaks at 1423 and *c.* 713–318 cm^{-1} probably due to pigment. Comparing the pigment absorptions to pigment standard spectra (Figures 10.28 to 10.35) indicates that the peaks at 1423, 876 and 318 cm^{-1} correspond to the primary peaks in the IR spectrum of calcium carbonate (Figure 10.32), and the broad, very strong peaks in the far-IR region from 700 to 350 cm^{-1} correspond to titanium dioxide (Figure 10.28). This identification can be made with high confidence, but if there are

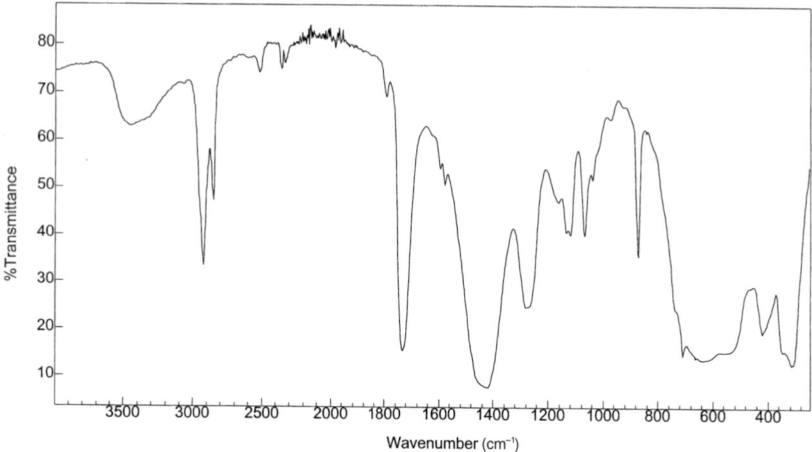

Figure 10.36 IR spectrum of unknown paint sample #1

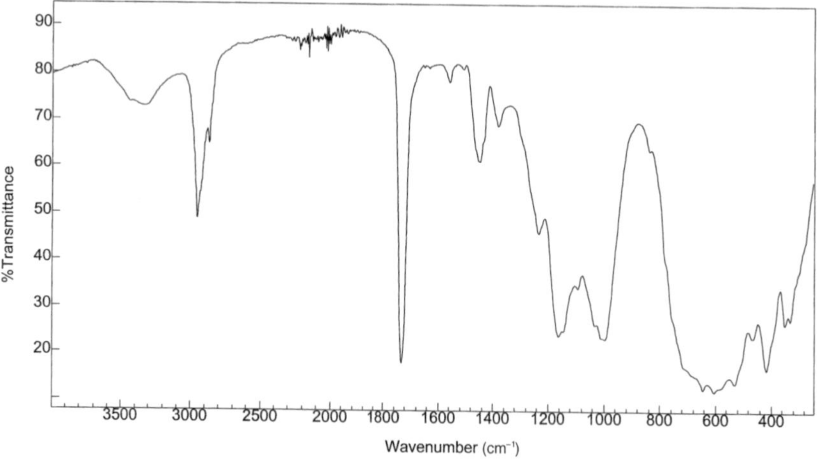

Figure 10.37 IR spectrum of unknown paint sample #2

any doubts, these can be set to rest by elemental analysis using energy-dispersive X-ray analysis, usually in the form of scanning electron microscopy (SEM/EDX), which would show titanium and calcium as the principal elements present.

The resin can be determined from the peaks at 1734, 1283, 1124 and 1072 cm^{-1} to be an alkyd. The most practical way to do this is by visual comparison with the IR spectra of known standards (see Figures 10.17, 10.20, 10.22, 10.24, 10.25, 10.27). The IR spectrum in Figure 10.17, a modified *ortho*-phthalic alkyd, corresponds to many major unassigned peaks in the spectrum of the unknown resin (but the melamine formaldehyde absorptions are not present in Figure 10.36).

A more systematic approach is to use a wavenumber flow chart as shown in Figure 10.39. There are several approaches to interpretation, but whichever approach is used, the

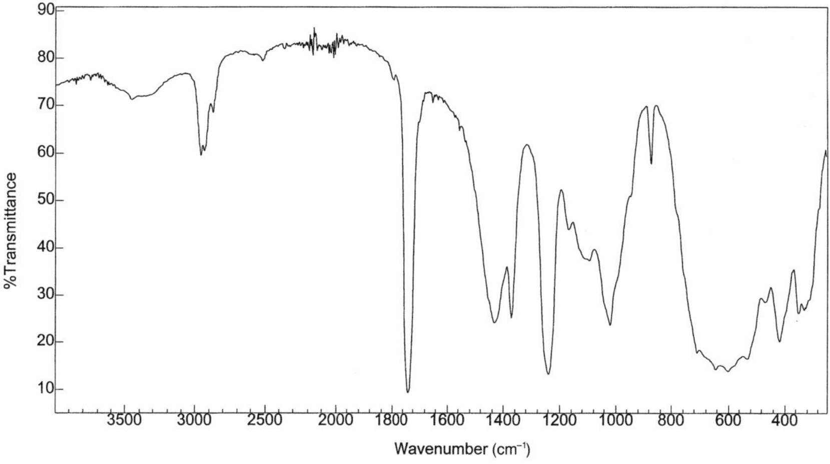

Figure 10.38 IR spectrum of unknown paint sample #3

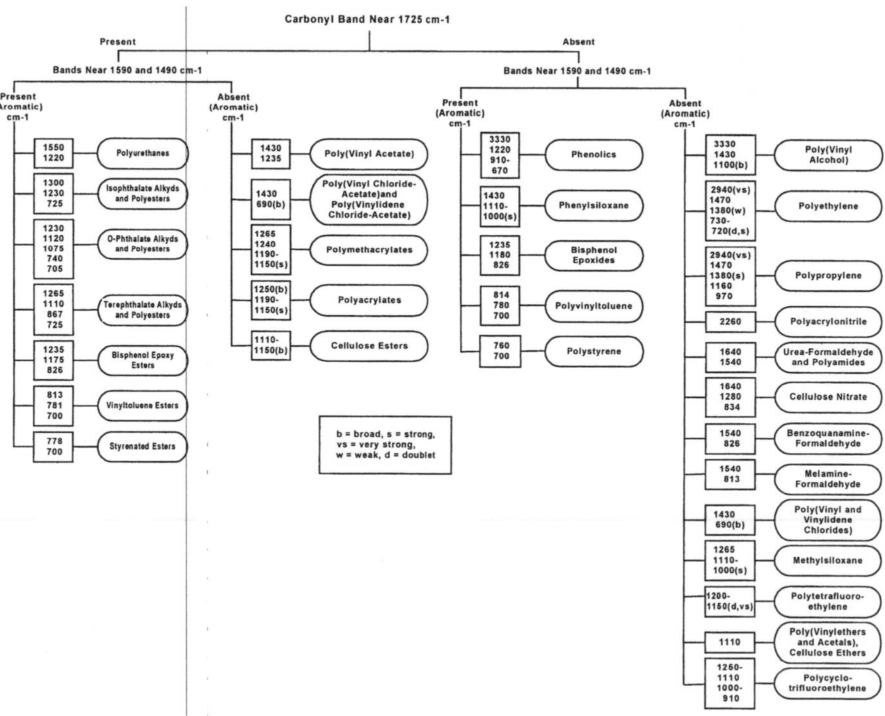

Figure 10.39 IR peak identification flow chart from *An Infrared Spectroscopy Atlas for the Coatings Industry,* reproduced with permission

answer should be that Figure 10.36 is the IR spectrum of an unmodified *ortho*-phthalic alkyd paint pigmented with titanium dioxide and calcium carbonate. Figure 10.40 illustrates IR spectrum of the pure resin used to make the paint, and shows how overlapping pigment peaks have the potential to impair identification of the resin.

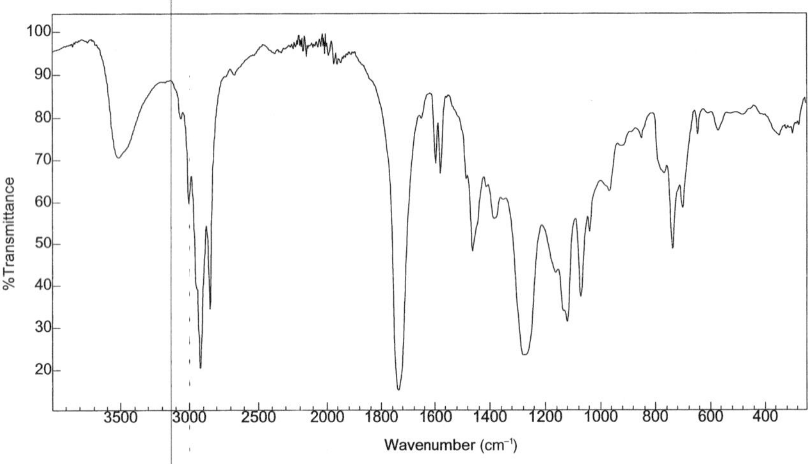

Figure 10.40 IR spectrum of the alkyd resin component of unknown paint #1 (Figure 10.36)

10.6.2 Unknown paint sample #2

Figure 10.37 illustrates a relatively simple spectrum. There is a strong carbonyl absorption at 1734 cm^{-1} from the resin, and the characteristic absorptions of titanium dioxide are present in the 700–350 cm^{-1} region (Figure 10.28). Thus the issue becomes what is causing the major peaks at 1150 and 1000 cm^{-1}. Comparison with Figures 10.17, 10.20, 10.22, 10.24, 10.25, 10.27 and 10.28 indicates that the spectrum of an acrylic resin in Figure 10.24 is similar to that in Figure 10.37 especially with respect to the strong C—O stretching vibration at 1164 cm^{-1}. The remaining broad, strong peak at *c.* 1000 cm^{-1} is characteristic of the Si—O stretch as illustrated in Figure 10.33 and can therefore be described as a silicious pigment. This assignment could be confirmed by SEM/EDX analysis which would identify the principal elements as titanium and silicon. Thus, comparison with standards or working through a flow chart such as Figure 10.39 would identify unknown paint #2 as an acrylic resin pigmented with titanium dioxide and a silicious pigment.

The particular pigment used in this paint is a clay called 'nepheline syenite' – a sodium potassium aluminium silicate – and it is sold under the trade name of Minex 7. The IR spectrum of the acrylic resin used to formulate the paint is shown in Figure 10.41. In comparison to Figure 10.37, it can be seen that several weak- to medium-intensity peaks from both the Minex 7 and the resin are hidden by the strong absorptions of titanium dioxide.

10.6.3 Unknown paint sample #3

Figure 10.38 shows some familiar features from the previous two 'unknowns'. There appear to be three pigments in this sample. The familiar titanium dioxide peaks are present between 750 and 353 cm^{-1}, carbonate peaks are present at 1437 and 876 cm^{-1}, and a silicious pigment is indicated at *c.* 1000 cm^{-1}. These interpretations would be supported by SEM/EDX analysis showing the principal elements to be Ca, Ti and Si. The issue then

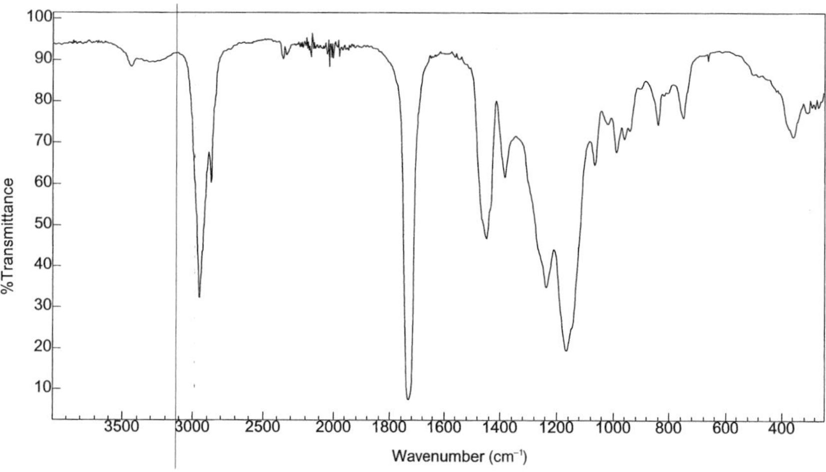

Figure 10.41 IR spectrum of an acrylic resin, Rhoplex AC-388 (General Paint Co.)

becomes how to identify the resin. There is a carbonyl at 1739 cm^{-1}, and a strong C—O stretch at 1243 cm^{-1}. There also appears to be an unusually strong absorption at 1374 cm^{-1}. Comparing the spectrum in Figure 10.38 with Figures 10.17, 10.20, 10.22, 10.24, 10.25 and 10.27 shows similarities to some of the major peaks in the vinyl acetate spectrum in Figure 10.27.

Alternatively, use of the flowchart in Figure 10.39 narrows the search to a vinyl acetate resin. The process should take no more than about five minutes. The sample was thus identified as a vinyl acetate polymer with titanium dioxide, carbonate and silicious pigments.

Figure 10.42 illustrates the IR spectrum of the unpigmented resin. It is noteworthy that

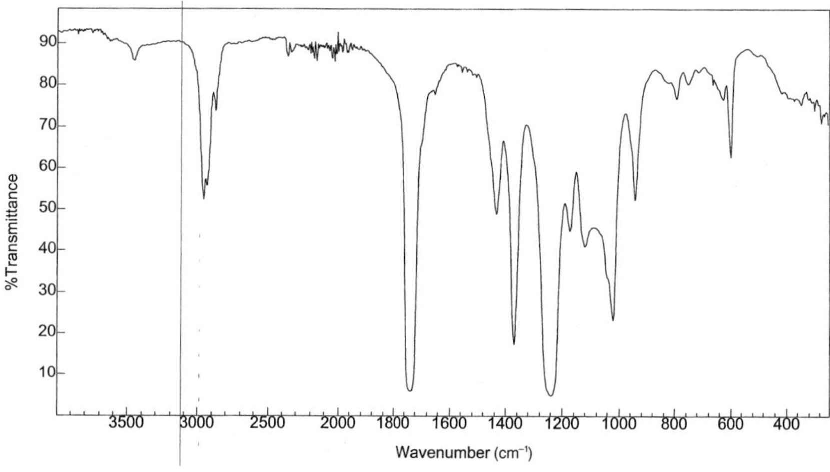

Figure 10.42 IR spectrum of a vinyl acetate/acrylic copolymer

Table 10.1 Manufacturer's data on unknown paints #1, #2 and #3

Unknown sample #1	Unknown sample #2	Unknown sample #3
22-010 Premium interior semi-gloss alkyd	70-010 Breeze exterior flat latex	55-010 Breeze interior eggshell latex
Alkyd resin (Beckosol AA203)	Acrylic polymer resin (Rhoplex AC2388)	Vinyl acetate/acrylic copolymer resin (Rovace GP-92)
Titanium dioxide (Titanox 2101)	Titanium dioxide (Titanox 2102) Clay (Minex 7: 'nepheline syenite', a clay-like mineral)	Titanium dioxide (Titanox 2102) Clay (Minex)
Calcium carbonate (Imasco 6HX)		Calcium carbonate (Imasco 6HX)

the resin has a strong absorption at 1024 cm^{-1} which can be easily confused with an Si—O absorption. Without evidence of elemental analysis showing a significant quantity of silicon, assigning the peak to Si—O could be incorrect. That is, IR alone may not be sufficient to fully characterise the components of a paint. This caution is further borne out by comparison with the components identified in the manufacturer's data (Table 10.1) which indicate that the identification is incomplete since the resin is described as a vinyl acetate/acrylic copolymer. Re-examination of the spectra in Figures 10.38 and 10.42 shows a peak at *c.* 1180 cm^{-1} which could be attributed to an acrylic polymer, but the *Atlas* also contains spectra of non-acrylic vinyl acetate copolymers with a similar absorption, so complete reliance on IR to identify the resin system in this paint is not justified. In this instance, pyrolysis gas chromatography might reveal the presence of the acrylic component in the form of acrylate pyrolysates.

10.6.4 Manufacturer's data on the three unknown domestic paints

Table 10.1 lists the contents of the three unknown paints dealt with above. Each is a product retailed by the General Paint Company.

10.6.5 Reflective traffic paint

Occasionally, smears of reflective paint or traffic paint are encountered in forensic casework. The term traffic paint refers to the material applied to bituminous (asphaltic) or concrete pavement surfaces to identify traffic lanes and excludes reflective adhesive-backed decals and tape inlays. As with other paints, it consists of an organic binder and pigments brought up to a certain viscosity with a solvent. Glass beads or titanium oxide are incorporated into the paint in a two-stream application process to provide the paint layer with its reflective properties. The yellow or white pigments which colour the paint also provide a reflective surface on the back of the glass beads to reflect light back to the source (retroreflectivity). Pigments may include mixtures of titanium dioxide, calcium carbonate, sodium potassium aluminosilicate (nepheline syenite), talc, clay and chrome yellow. Reflective paints are available as both solvent-borne and water-borne formulations. Acrylic latex emulsions are seen as binders in water-borne paints (Figure 10.43), whereas short-oil alkyds and medium-oil alkyds are common binders in solvent-borne paints (Figure 10.44). The choice between a short-oil or medium-oil alkyd depends on the drying time required by the application (highway versus parking lot). Short-oil alkyds dry within 1 to 1.5 minutes, as opposed to 20 minutes for medium-oil alkyds.

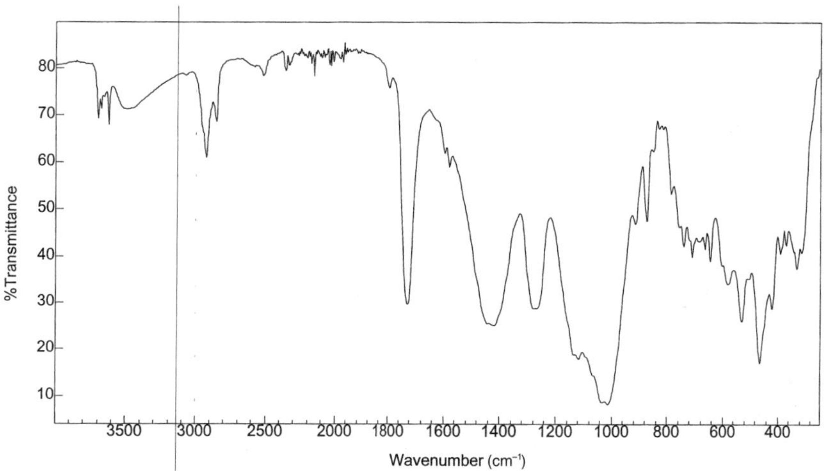

Figure 10.43 IR spectrum of an acrylic latex reflective traffic paint

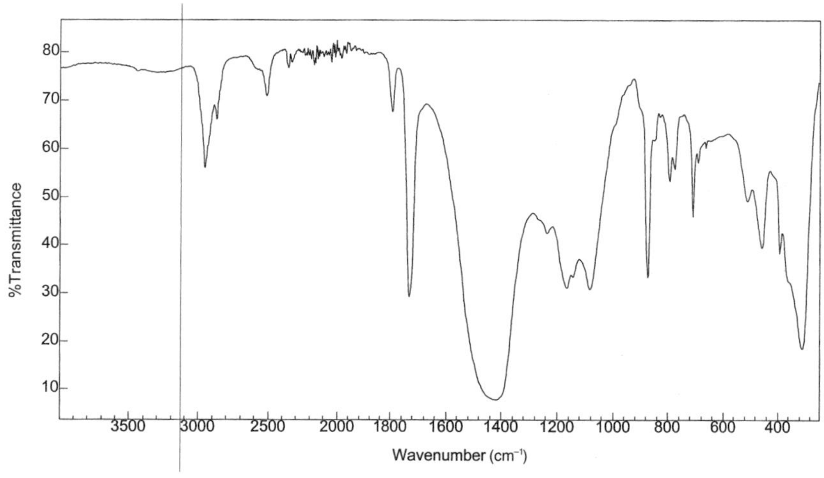

Figure 10.44 IR spectrum of an alkyd reflective traffic paint

Stringent environmental laws have caused a swing away from solvent-based paints toward the newer technology employing lead-free, high-solids, water-borne paints. More expensive organic yellow pigments are replacing the lead chromate pigments. The composition, physical and performance requirements of traffic paints are carefully controlled to meet the specification standards set out by government regulatory bodies.

10.6.6 Examples of infrared spectra of paints found on tools

Tools, such as prybars and axes, are often used in the commission of crimes ranging from break-ins to murders. Paint transfer and toolmarks from these tools are therefore of interest to forensic scientists.

Paint types applied on tools are predominantly alkyd, alkyd melamine and nitrocellulose, although other types such as polyester, acrylic, acrylic styrene and epoxy are occasionally used. Sometimes more than one type of paint is present on the same tool if it was assembled from components made in two or more factories. Figure 10.45 shows the IR spectrum of a yellow paint from a hacksaw blade. The paint components may be identified as a nitrocellulose (see Figure 10.25) modified alkyd (see Figure 10.40) pigmented with chromate (see Figure 10.35).

10.6.7 Heavily pigmented paint

When inorganic pigments or metallic particles are present in such high concentrations that they essentially mask the spectrum of the paint resin, then improvised sampling techniques are sometimes helpful. Examples of non-metallic, heavily pigmented paints are marine antifouling paint and paint from safes; heavily metallicised paints are sometimes encountered on tools or amateur refinishes, e.g. a refinished car bumper.

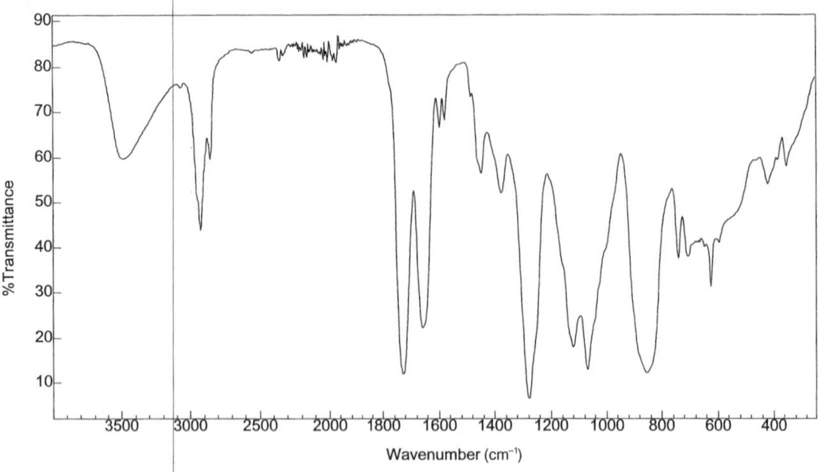

Figure 10.45 IR spectrum of yellow paint from a hacksaw blade

10.6.7.1 Marine antifouling paint

Figure 10.46 illustrates the spectrum of a marine antifouling paint which is primarily that of the antifouling agent cuprous oxide. To separate the binder from the interfering pigment, the paint is crushed (e.g. in an agate mortar and pestle) and extracted with a small volume of methylene chloride or a similar solvent. The suspension is centrifuged to yield a clear supernatant solution which is pipetted and dried on a glass slide. Figure 10.47 shows the IR spectrum of the clear film obtained by chloroform extraction of the paint illustrated in Figure 10.46. The binder is an alkyd.

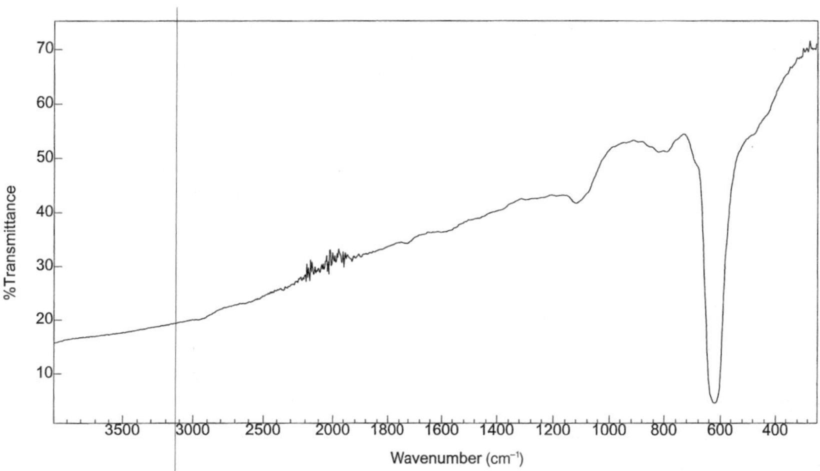

Figure 10.46 IR spectrum of marine antifouling paint

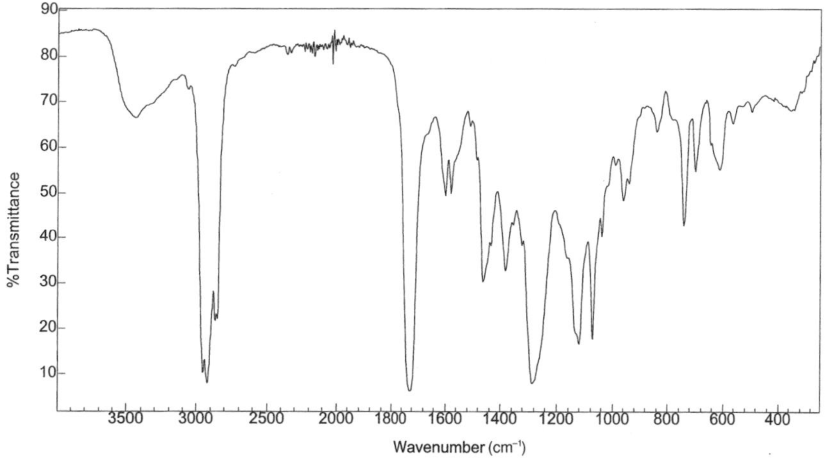

Figure 10.47 IR spectrum of a chloroform extract of marine antifouling paint (see Figure 10.46)

10.6.7.2 Heavily metallised paint

Heavily metallised paints can be treated with concentrated hydrochloric acid (HCl) for a few hours to dissolve enough of the aluminium flakes to allow the binders to be analysed relatively free from interference.

10.7 Identification of automotive paints by interpretation of their infrared spectra

Automotive paint systems consist of multiple layers of topcoats (most often a clearcoat and coloured basecoat) over one or more primer layers. An original factory finish has a typical layer sequence of: substrate, primer, surfacer, topcoat. Common substrate materials include steel (galvanised and non-galvanised), aluminium and aluminium alloys, plastics (rigid and flexible) and composites. A number of substrate types can be found on each vehicle. The paint applied to each must be specifically formulated (e.g. flexible paints for flexible parts) and must be compatible with other layers in the paint system. The pigments used to achieve a particular colour of topcoat are the same whether they are applied to metal or plastic. This is to minimise colour differences under different lighting conditions/sources (metamerism). Thus, differences between similarly coloured paints tend to be in their resins rather than in their pigments.

Each layer in the paint system has a specific purpose. Primers are applied to cover and protect the substrate, hide defects and provide adhesion for subsequent layers. Almost all metal primers are electrocoats (E-coats) that are applied by cathodic electrodeposition. Plastic parts are dipped in conductive primers if they are to receive electrostatically applied topcoats. Many of the parts coming into an assembly plant are pre-primed, and may receive common final coats.

Topcoats are applied for their aesthetic value and also serve to protect the layers below. Topcoats are available in non-metallic or metallic (aluminium flake) colours and, increasingly, as pearlescent (mica) colours. Clearcoats, first used on European and Japanese vehicles, are now applied in North America to metallic coats and to most non-metallic finishes. UV additives are added to clearcoats to protect against degradation caused by sunlight.

An intermediate surfacer, also called a ‘filler’ or ‘middle’ coat, may be used between the primer and topcoat layers. Many manufacturers apply an antichip layer to the lower sections of side panels, behind wheels and along the leading edges of fenders (bumpers) to protect against stone chipping. The surfacer may also be a colour-coordinated (keyed) primer which is matched to the topcoat to provide greater colour depth and make stone chipping less obvious [9, 10].

Automotive paint layers, of course, may also be repaint finishes, and paint fragments from repainted vehicles may contain both original finish and repaint layers. Samples received at a forensic laboratory range from discrete chips to thin, contaminated smears. The latter may provide only limited information about the physical and chemical properties of the paint sample because of physical damage and chemical contamination. Repaint chips, on the other hand, may contain layers from multiple repaint operations in addition to any original paint layers present. Such multilayered repaint fragments provide very significant evidence in paint comparison cases, but are of minimal value in the identification of make, model and year if no factory original layers are present.

If factory original layers are present, however, the paint chips can often be sourced to a specific make and model of motor vehicle through the use of databases of physical attributes and chemical compositions.

10.7.1 Methodology for paint source identification

The standard analytical procedure in our laboratories for identification of the source of a paint sample follows several steps:

- microscopic examination
- IR
- primer colour coding
- paint database query (PDQ)
- topcoat colour coding.

10.7.1.1 Microscopic examination

An initial microscopic examination is conducted using a stereo microscope to determine the number, order, colour, thickness and texture of each layer in a paint system. The layers of a paint chip are revealed by making a diagonal cut with a scalpel along the edge of a horizontally held chip. The chip may also be viewed along its edge or as thin cross-sections under polarised light should the number of layers be unclear. Unique or accidental features such as overspray, irregularities or peculiar characteristics in, on or between each layer are noted. As layers are sampled from the horizontal chip, they are cut or scraped away to reveal the layers below. Samples for infrared analysis are obtained mid-layer to avoid any chances of surface contamination and weathering effects or to limit the possible contributions of component migration from substrate materials or adjoining layers, especially in 'wet-on-wet' applications.

10.7.1.2 Infrared analysis

Instrumental analysis is normally conducted on each topcoat and primer layer. Because of the sensitivity, speed, and discriminating ability offered by extended range mid-infrared spectroscopy, FTIR is the primary method of analysis. Data are collected and plotted as transmission spectra, at $4\,cm^{-1}$ (wavenumber) resolution, over a spectral range of 4000 to $250\,cm^{-1}$. The extended spectral range is achieved through the use of CsI optics which allows for the simultaneous determination of both the organic and inorganic components in paint.

Stability and sensitivity requirements are key factors in equipment selection because of casework sample size limitations and the low throughput of the accessories. Digilab model FTIR models FTS-40/60, 40A and 575 have all been found by the Royal Canadian Mounted Police (RCMP) laboratories to be suitable for use with either the high-pressure diamond cell or mini cell and a Harrick 6X beam condenser (see Appendix 1).

The next stage in the process is to interpret the infrared spectrum generated for each layer. Key components are recognised by their characteristic 'fingerprint' patterns and the diagnostic peaks empirically determined from analysis of unpigmented resins/binders, pigments and extenders obtained as raw ingredients from the paint industry, or by reference to spectra in the *Atlas*.

In cases involving automotive paint, the identification process involves combining IR data with colour and layer sequence data and searching a database. The database used in RCMP laboratories is called 'Paint Data Query' or PDQ [11] (see Sections 10.7.1.5 and 10.7.3.2).

10.7.1.3 Colour coding

The PDQ database requires that the colour of primer layers be assigned using the hue, value, and chroma designations in the *Munsell Book of Color* (1976) [12, 13]. Topcoat colours are assigned according to a list of standard colours in the database. The absence or presence of metallic flakes and/or mica particles is also noted. This is to prevent missing a 'hit', should a sample with the exact topcoat colour not be in the database (Section 10.7.1.5).

10.7.1.4 Paint data query (PDQ) database

The next step is a search of the automotive paint database using the colour and chemical information generated from the complete paint layer system. The database unknown spectra are compared to the hard-copy spectra of each hit to ensure that the components are common to each associated layer and present in the same relative amounts. Spectral searches have met with some success, but since automotive paint libraries are composed of large numbers of very similar spectra, the search algorithms have not proven to be sufficiently sensitive at distinguishing subtle but significant features such as shoulders, unique shapes and patterns, and minor peaks.

The operation of the PDQ database is described in detail in Section 10.7.3.2.

10.7.1.5 Topcoat colour confirmation

The final step is to confirm that the colour of the topcoat on a paint identified as originating from a particular make and model of a motor vehicle was in fact used by the possible manufacturer in the year listed in the query results. Reference is made to topcoat colour collections such as the US National Bureau of Standards, topcoat colour standards and trade refinish books (e.g. Dupont). Often the list of possible 'hits' can be further narrowed by eliminating the manufacturers and years in which a specific topcoat colour is not listed as having been used.

Where only a repaint topcoat is left at a scene, only the colour of the vehicle can be given to an investigator.

10.7.2 Characteristic peaks in infrared spectra of automotive paint

Paint systems of the 1970s and early 1980s generally consisted of low-solid solvent-borne coatings which were quite distinctive for different manufacturers. Since the mid-1980s, however, global competition, consumer expectation and environmental regulations have combined to force plant mergers and technology changes. From these have emerged new formulations consisting of high-solid paints.

Three paint manufacturers, Du Pont Automotive, PPG Industries Inc. and BASF, now dominate the global automotive paint market to such an extent that paint systems used by manufacturers around the world have become much more uniform [14]. However, differences in the colour, layer sequence and chemistry of the topcoat and primer layers in a paint system can still be used to differentiate the paint applied to different vehicles.

The essence of interpretation of IR spectra is to determine the resin and pigment components of each layer. Table 10.2 displays the diagnostic peaks of common binders and resins used in automotive paint, and Table 10.3 likewise displays common pigments and extenders. By coding the composition data and combining it with other physical data, the make, model and year of vehicle from which the paint originated may be deduced using databases. Tables 10.2 and 10.3 include the codes used in the RCMP PDQ system.

10.7.2.1 Coding of infrared spectra of a four-layer paint chip from a 1995 Ford vehicle

Figures 10.48–10.51 show the IR spectra of the layers of a four-layer paint chip which is an original factory finish from a 1995 Ford vehicle manufactured in Oakville, Ontario.

The layer sequence (top to bottom) is: clear/green/grey/grey. The technical description and coding for these layers are illustrated below using the abbreviations 'OT' for the original topcoat and 'OU' for the original undercoat. The numbers increase 'up' and 'down' from the topcoat/undercoat interface as shown:

clearcoat [OT2]/basecoat [OT1]/primer surfacer [OU1]/electrocoat [OU2]

Table 10.2 Diagnostic peaks of common binders/resins used in automotive paints

Binder/Resin	Coding	Key peaks (cm^{-1})
Acrylic	ACR	1450 1380 1260 1170 1150
Ortho-phthalic alkyd	ALK OPH	1450 1380 1270* 1130* 1070* 740 700
Isophthalic alkyd	ALK IPH	1475 1373 1305 1237* 1135 1074 730*
Terephthalic alkyd	ALK TER	1270 1250 1120 1105 1020 730
Benzoguanamine	BZG	1590 1540 825 789 710
Cyano	CYA	
Acrylonitrile N≡C	CYA NIT	2238
Isocyanate residue N=C=O	CYA ICN	2272
cf. ferrocyanide, $Fe(CN)_6$		(2092)
Epoxy	EPY	1610 1510* 1240 1180 830*
Melamine	MEL	1550 815
Nitrocellulose	NCL	1650 280 840
Polybutadiene	PBD	970 915
Polyurethane	PUR	1690 1530 1470 1250 1070
single peak		1690
modified epoxy		1730 1510 (non-asymmetric broadening)
water based		1690 770
Styrene	STY	1490 1450 760 700
Urea	REA	1655

The layers were separated and analysed by IR spectroscopy (Appendix) and the undercoats were coded in the Munsell colour system. Table 10.4 shows the components identified and the coding.

Table 10.3 Diagnostic peaks of common pigments and extenders used in automotive paints

Pigment and extender	Coding	Key peaks (cm^{-1})
Calcium carbonate	CAR CAC	
Aragonite	CAR CAC ARA	1445 870 857 712 317
Calcite	CAR CAC CAL	1445 870 712 317
Chromate	CHR	
Barium chromate	CHR BCH	935 896 860
Potassium zinc chromate	CHR KZC	950 880 805
Strontium chromate	CHR SCH	920 885 860 840
Oxide	OXI	
Iron oxide	OXI FEO RED	560–530 480–440 350–310
Iron oxide	OXI FEO YEL	899 797 606 405 278
Lead oxide	OXI PBO	530 450
Zinc oxide	OXI ZNO	420–500
Silicon dioxide	OXI SIO	
Cristobalite	OXI SIO CRI	1090 795 621 485 387 300
Opal, diatomaceous silica	OXI SIO OPA	1099 795 475
Quartz	OXI SIO QUA	1081 798 779 512 460 397 373
Titanium dioxide	OXI TIO	
Rutile	OXI TIO RUT	600 (broad suppression) 410 340
Anatase	OXI TIO ANA	600 (broad suppression) 340
Zinc phosphate	PHO ZNP	1120 1080 1020 950 630
Silicate	SIL	
Magnesium (talc)	SIL MGS TAL	1015 670 465 450 420 390 345
Aluminium (kaolinite)	SIL ALS KAO	1035 1005 940 910 540 470 430 350 280
Barium sulphate	SUL BAS	980 630 610

Table 10.4 Composition of paint layer system on 1995 Ford from Oakville, Ontario

Figure	Layer	Components	Coding
10.48	OT2	Acrylic, melamine, styrene	ACR, MEL, STY
10.49	OT1	Acrylic, melamine, styrene	ACR, MEL, STY
10.50	OU1	Isophthalic alkyd, melamine, barium sulphate, kaolin, strontium chromate	ALK, IPH, MEL, SUL, BAS, SIL, ALS, KAO, CHR, SCH
10.51	OU2	Epoxy, polyurethane, kaolin, titanium dioxide (rutile)	EPY, PUR, SIL, ALS, KAO, OXI, TIO, RUT

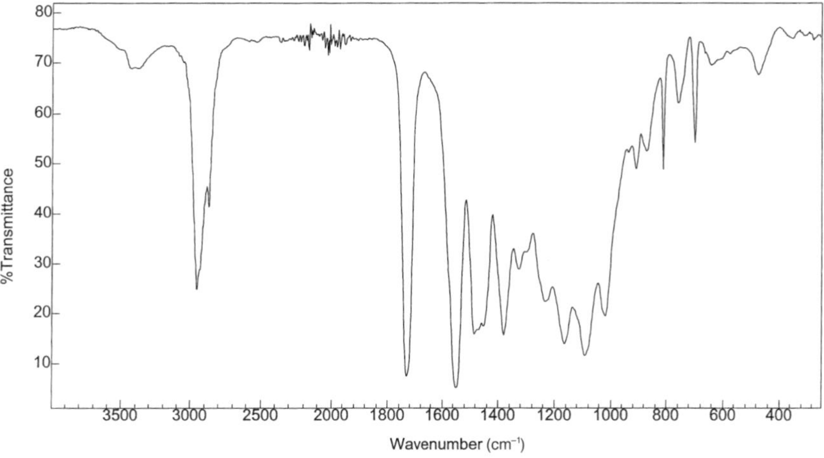

Figure 10.48 IR spectrum of clearcoat [OT2] Ford, Oakville, paint chip

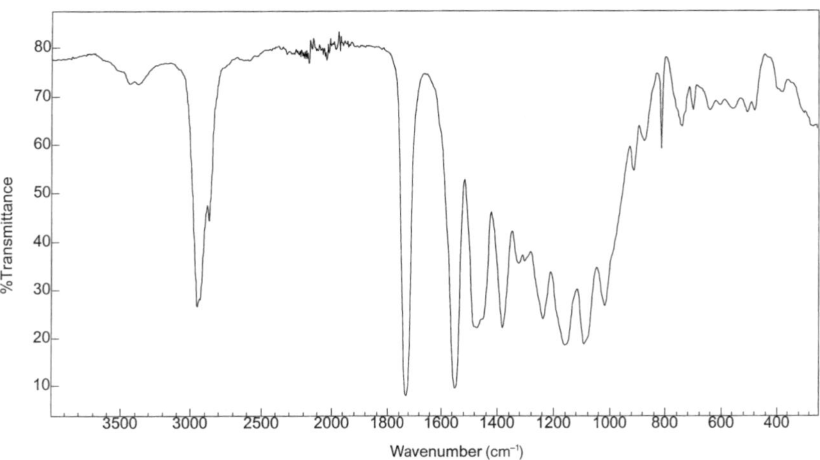

Figure 10.49 IR spectrum of basecoat, green [OT1] Ford, Oakville, paint chip

10.7.3 Databases used to identify the origin of automobile paint

Identification of the probable origin of paint chips is achieved through databases. Two databases are described below, an in-house visual database and a comprehensive computer-based system (PDQ).

10.7.3.1 A visual database

In the 1960s and 1970s, major car manufacturers such as General Motors, Ford and Chrysler developed paint systems for their production lines which were quite specific to

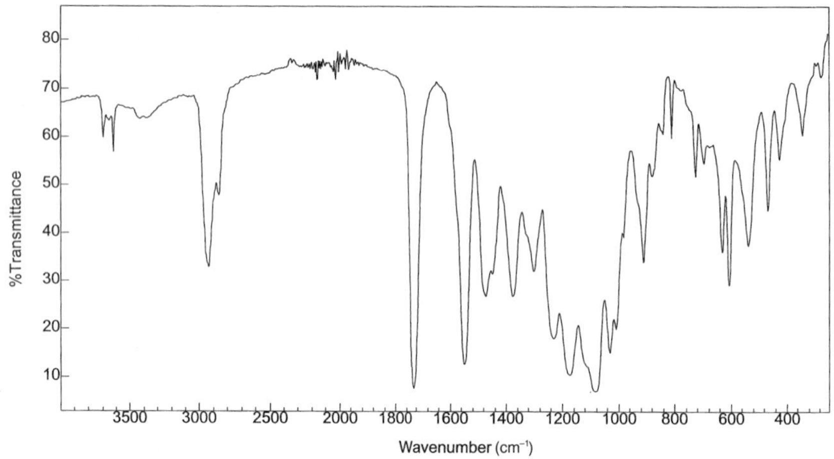

Figure 10.50 IR spectrum of primer surfacer [OU1] Ford, Oakville, paint chip

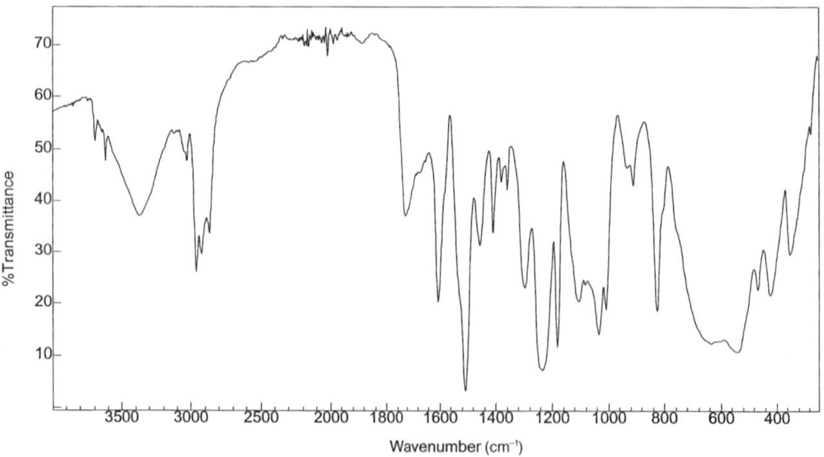

Figure 10.51 IR spectrum of electrocoat [OU2] Ford, Oakville, paint chip

certain makes and models. It was therefore useful, at that time, for individual laboratories to set up automotive paint databases by archiving authenticated street and casework samples. Paint chips used in the collection were cut obliquely with a scalpel so that all undercoat layers were evenly exposed. These paint chips were then arranged logically in a number of ways which permitted undercoat layer sequence and colour to be compared. Once matches were found, the IR spectra were compared.

One arrangement was to mount paint chips on glass slides using a transparent medium and sorting according to assembly plant and year of manufacture. For example, a slide for Chrysler might have Detroit, St Louis and Belvedere (assembly plant locations) across the top and years, e.g. 1986, 1987, 1988 and 1989, down the side. The number of plants and the number of years on a particular slide depended on the number of paint chips to be mounted and estimated spaces for future addition. An alternative was to mount the paint chips from a specific automobile manufacturer on cardboard cards according to microscopic similarities of the undercoat systems. For example, a card for General Motors had a row of single-layer black undercoats (mainly trucks) and another row of grey/black undercoats from passenger vehicles. Similarly, a card for Hyundai had rows of single-layer grey and light grey/dark grey for trucks and cars respectively.

Besides physical displays, such databases should include:

- a catalogue to identify each paint chip in the collection, containing the VIN (vehicle identification number), information obtained from it (make, model, year and assembly plant) and layer sequence.
- a collection of the actual paint samples from which the paint chips on the visual displays were prepared. A convenient process was to store the samples in numbered glass vials so that they were readily available for IR analysis if the layer sequences and colours matched an unknown sample.
- albums of hard-copy IR spectra from all analysed layers to provide chemical information and for comparison with IR spectra from unknown paint samples.

Such in-house paint databases were relatively comprehensive and proved to be fast and accurate identification tools for most automotive paint samples from the 1960s to the early 1980s. However, as paint usage by different car makers became more uniform in the last decade, it has become more difficult to differentiate paint chips from different cars using such a system and reliance on a more sophisticated computer database became necessary.

10.7.3.2 Computerised database: automotive paint database and search software – the paint data query (PDQ) system

The paint data query (PDQ) system is composed of two components:

- the colour and chemical information of complete paint-layer systems (topcoats and primers) applied to known vehicles
- the search and retrieval software used to query the database for vehicles having similar paint systems to the unknown being searched.

Originally the data and program resided on a centralised mainframe where it was known as PSAS© for 'Paint Sample Analysis System' and was only accessible through terminals within the RCMP network. Each laboratory was supplied with hard-copy infrared spectra of the samples in the database. In 1993 the database was off-lined to the PC environment and the software was rewritten in-house as a Paradox® for Windows™ run-time application. The infrared spectra were compiled into spectral libraries linked to the text database for ease of retrieval and comparison in order to make the system widely accessible.

The automotive paint database was designed as a general text-based search and retrieval system to eliminate the dependency on any one instrument or software package. The data used to generate a query can originate from a variety of instrumental sources and contain as much detail as is desired, depending on the analyst's instrumental methods and degree of confidence. The idea behind a PDQ search is to narrow down the list of possibilities to a reasonable number of suspects, not to identify a single vehicle. Allowances must be made in drawing forensic conclusions to accommodate the samples not contained in the database ('holes') and manufacturers' use of trial runs or alternative paint suppliers, and paint application anomalies, and so on.

The database itself contains information on the complete topcoat and undercoat systems applied to most of the domestic and foreign vehicles marketed or imported into North America since the mid-1970s. Currently there are over 7500 paint systems (over 24,500 layers) represented in the database with a minimum of 1000 samples added per year. Approximately 33 per cent of the samples are factory panels that are received directly from the automotive manufacturers and paint suppliers. Factory panels represent the paint applied on all of the models produced on each assembly line in each year. The remaining 67 per cent are actual street samples which are collected to validate the manufacturers' information. The information generated for each sample is compiled into a record. Each record is assigned a sequential alphanumeric catalogue (PDQ) number which identifies the country, province/state and laboratory providing the sample.

The automotive paint database is a relational database that comprises linked individual databases containing fields of information describing:

- the source of the sample; and
- the colour, chemical composition, and layer sequence of each layer in the standard paint systems.

Data entry

Information entered into these fields is restricted to single- or three-letter code words that are selected from drop-down lists and controlled (validated) by look-up tables. The only exception to this format is the comments field which is open and unrestricted.

An example of a record is illustrated in Figure 10.52. The record illustrates data about a sample bearing the PDQ number UNCR00022 (Figure 10.52, top left). The number

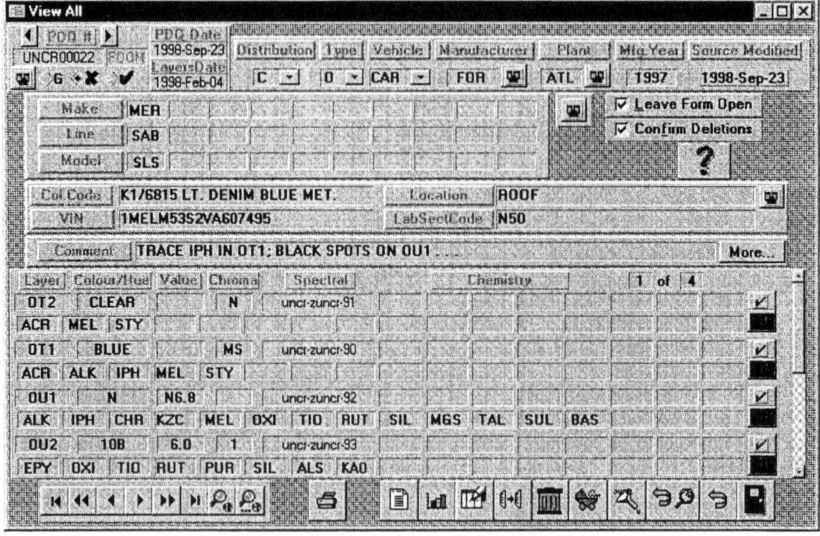

Figure 10.52 A PDQ record

indicates that the sample was provided by a USA laboratory (U) in North Carolina (NC) in the city of Raleigh (R) and that it has a sequential catalogue number of 00022.

The record contents can be divided into two halves, the information about the source of the sample on the top half and information about the layers on the bottom half. Each half bears a stamp showing the date on which data were entered. The certainty of the information in each record is indicated by a validation flag. For example, Figure 10.52 shows, near the top left, a tick (✓) which indicates that the contents have been validated, a 'G' which indicates that the record has been grand-fathered, that is copied unchanged from the older PSAS mainframe database, and an 'X' indicates that the record contains current information which has not been fully validated.

The sourcing information of the sample in the top half of each form or record describes the vehicle type (car or truck, etc.), manufacturer, plant and year of assembly, make, line, model, topcoat colour code and name, location on the vehicle (hood, fender, door, etc.), vehicle identification number (VIN), comments and the contributor's internal sample number.

Figure 10.52 describes a car (CAR) manufactured by Ford (FOR) at the Atlanta assembly plant (ATL) during the 1997 production year (1997). The make is Mercury (MER), the line is Sable (SAB) and the model is LS (SLS). The manufacturer's topcoat colour code is K1/6815, the colour name is 'light denim blue metallic' and the Dupont refinish colour code is B9743. The sample came from the ROOF of the vehicle. The VIN is 1MELM53S2VA607495 and the internal catalogue number of the laboratory which provided the sample was N50. General comments arising from the observation and analysis of the paint sample indicate that a trace of isophthalic alkyd was present in the basecoat (OT1), black spots were observed in the primer surfacer (OU1), and that trace phosphate (PHO) was present in the primer (OU2).

The bottom half of each form contains the colour, chemical and sequence information of each layer in the paint system. The topcoat colours are given a general description using the CTS/NBS designations:

white, grey, black, red/pink, maroon/burgundy, orange, brown/bronze/copper, tan/beige, yellow, green and blue/purple.

The topcoat appearance is described as non-metallic (N), metallic (M), pearlescent (P), or combination (C) of metallic and pearlescent. Queries are normally entered with only the topcoat appearance and chemistry in order to identify possible sources with the same topcoat resin system but different pigmentation.

As noted above, the layer sequence is described by:

- type of layer, topcoat (T) or undercoat (U)
- whether it is original (O), repaint (R) or combination (C), and
- numerical location of the layer (1, 2, 3, etc.) as numbered from the topcoat undercoat interface.

An example of a layer sequence for an entire paint system would be:

RT2/RT1/RU1/OT2/OT1/OU1/OU2/SUB

The paint system shown in Figure 10.52 consists of four original layers – two topcoats and two primers. The clearcoat (OT2) is colourless non-metallic and the basecoat (OT1) is blue metallic silver (MS). The original grey primer surfacer (OU1) was assigned a Munsell colour designation of N6.8 (i.e. hue of Neutral, value of 6.8, no chroma), and the original blue/grey primer (OU2) was assigned a Munsell colour designation of 10B6/1 (i.e. hue of 10Blue, value of 6 and chroma of 1).

The chemical composition of each organic and inorganic component in a layer is described using a hierarchical system composed of three-letter code words (Tables 10.2 and 10.3). Two levels are currently being used to describe the binder or resin content, for example alkyd (ALK) might be used as the main category, then *ortho*-phthalic (OPH), isophthalic (IPH) or terephthalic (TER) as the subtype. Three levels are used to describe the inorganic components by their anionic group, cation and crystalline form. As an example the calcium carbonates would be described as CAR for carbonate, CAC for calcium carbonate and CAL or ARA for the calcite and aragonite crystalline forms, respectively. The nature of the substrate (SUB) is described, e.g. PLA for plastic, MET for metal, GAL for galvanised layer. Depending on the quality and nature of the analytical data, and the certainty of the examiner, the data can be entered and queried with any level of confidence. Table 10.5 shows the composition and coding of the paint layers in the example shown in Figure 10.52.

The 'chemical and colour properties screen' used to enter the information for each layer of a record is shown in Figure 10.53. The same screen is used to enter layer information of the unknown being searched. The screen contains hierarchical check boxes with headers that can be switched to display either the three-letter code words or the chemical names which they represent (e.g. ALK or alkyd). Drop-down boxes are available to input a layer sequence number, the hue, value and chroma of primer layers, and the colour and appearance of topcoats.

Data query

A query containing the information from all layers in the unknown paint system is compiled and entered into the PDQ database (Figure 10.54). The unknown paint fragment illustrated is a two-layer paint chip: one topcoat (OT1) and one primer (OU1). The query is run and printed and the results, indicating the possible suspect vehicles, are presented accord-

Table 10.5 Composition and coding of paint layers illustrated in Figure 10.52

Layer	Components	Coding
OT2	Acrylic, melamine, styrene	ACR, MEL, STY
OT1	Acrylic, isophthalic alkyd, melamine, styrene	ACR, ALK, IPH, MEL, STY
OU1	Isophthalic alkyd, melamine, potassium zinc chromate, titanium dioxide (rutile), magnesium silicate (talc), barium sulphate	ALK, IPH, MEL, CHR, KZC, OXI, TIO, RUT, SIL, MGS, TAL, SUL, BAS
OU2	Epoxy, urethane, titanium dioxide, zinc phosphate, aluminium silicate (kaolinite)	EPY, PUR, OXI, TIO, RUT, PHO, ZNP, SIL, ALS, KAO

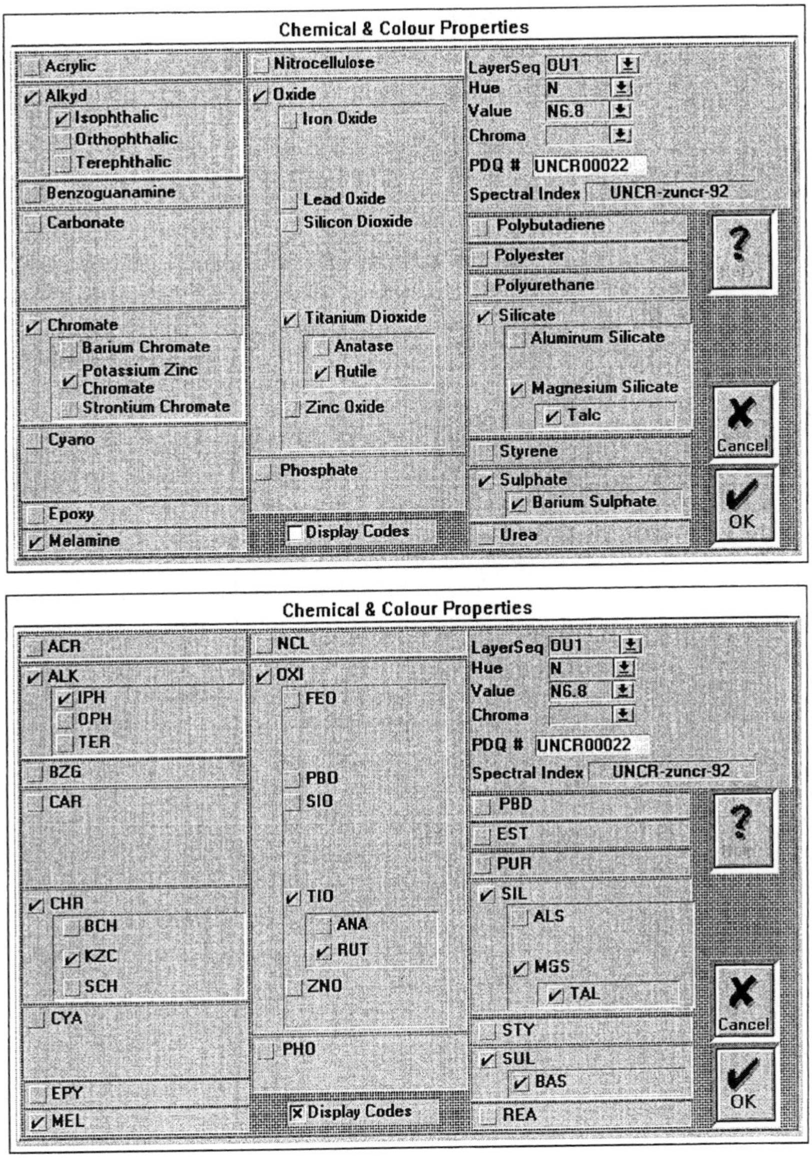

Figure 10.53 PDQ screen for chemical and colour properties

ing to the manufacturer, plant and year (Figure 10.55), and by the vehicle lines (Figure 10.56) in each of those plants. In the example illustrated, 63 'hits' were obtained and all were Fords, namely 'F' series pickup trucks, Mustangs and Rangers from 1983 to 1993.

As mentioned in Section 10.7.1.4, after the PDQ search has been conducted, the spectrum of each 'hit' is compared with that of the unknown to check both for uncoded components and for differences in relative amounts of components. In this case all the Ford Mustangs were eliminated because of differences in the infrared spectra of the topcoats.

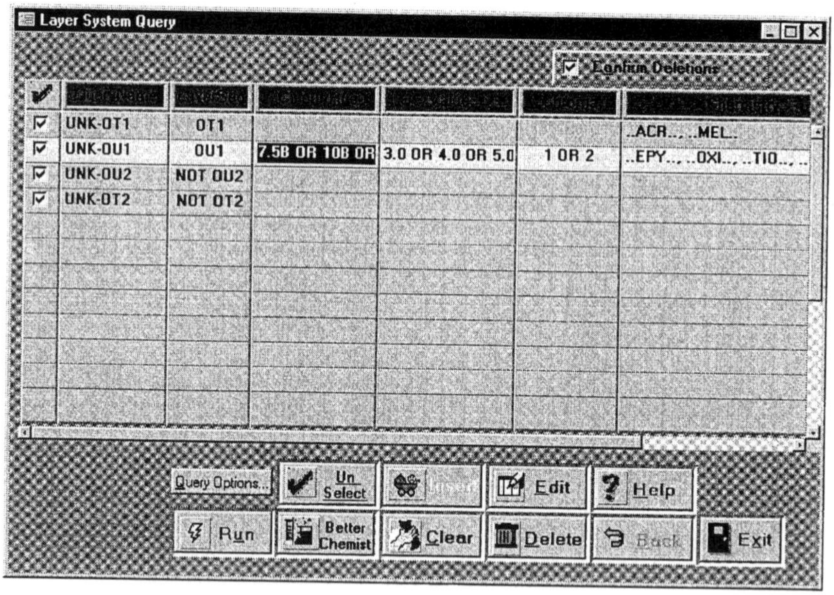

Figure 10.54 A PDQ system query

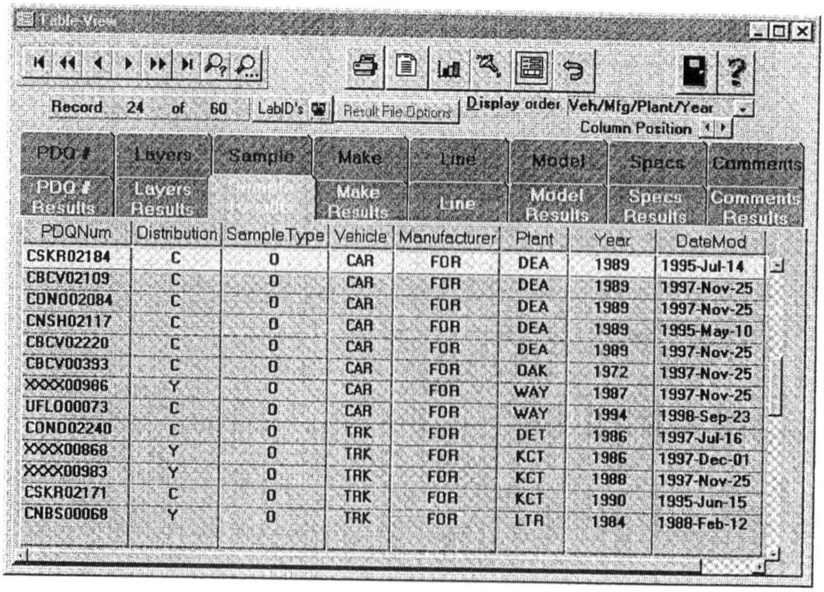

Figure 10.55 PDQ results by manufacturer, plant and year

The final stage is to confirm possible 'hits' by topcoat colour comparison. Further narrowing is possible by ruling out the years in which the manufacturer was known not to have used the specific topcoat colour.

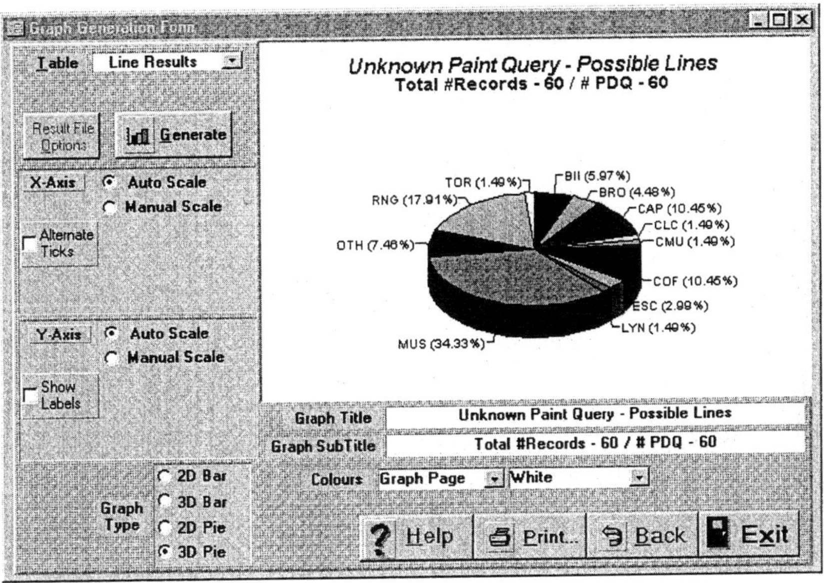

Figure 10.56 PDQ results by vehicle lines

10.8 Application of infrared spectroscopy in comparison of paint fragments

In the comparison of paint fragments, IR spectroscopy is usually the first analytical method to be applied after microscopic examination. By determining the resin and pigment components of each layer of paint in a multilayered fragment, a firm foundation is established for a forensic opinion on the significance of a comparison. Indeed, if there are several repaint layers in paint chips with corresponding layer sequences, the comparison of physical properties may make instrumental analysis unnecessary.

With single-layered fragments, however, analysis is crucial. In such instances, IR spectroscopy, while very useful, is best used in conjunction with other techniques, particularly pyrolysis gas chromatography (PGC). The reasons are that single layers of paint can rarely be individualised by physical characteristics and that visually indistinguishable layers from different sources may have the same IR spectrum and different pyrograms or vice versa. This is demonstrated in Figures 10.57–10.62.

Figures 10.57 and 10.58 show the IR spectra and pyrolysis gas chromatograms (pyrograms) for two white acrylic melamine original automotive paints. The IR spectra are similar but the pyrograms are different and permit the samples to be differentiated analytically. Likewise, Figures 10.59 and 10.60 illustrate IR spectra and pyrograms for two red acrylic lacquer automotive refinish paints in which the two samples have similar IR spectra but different pyrograms. By contrast, Figures 10.61 and 10.62 illustrate the IR spectra and pyrograms of two white alkyd automobile refinish paints, in which the pyrograms are very similar but the IR spectra are clearly different (primarily due to the presence of urate absorptions at 1680 and 1520 cm^{-1} in Figure 10.61).

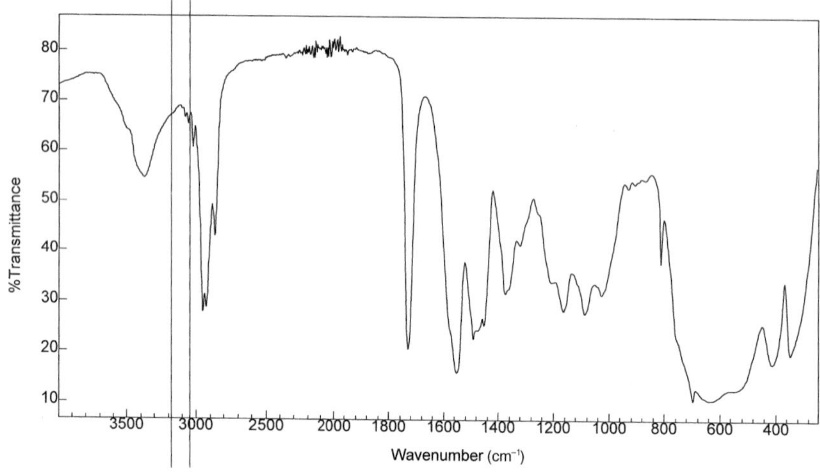

(a)

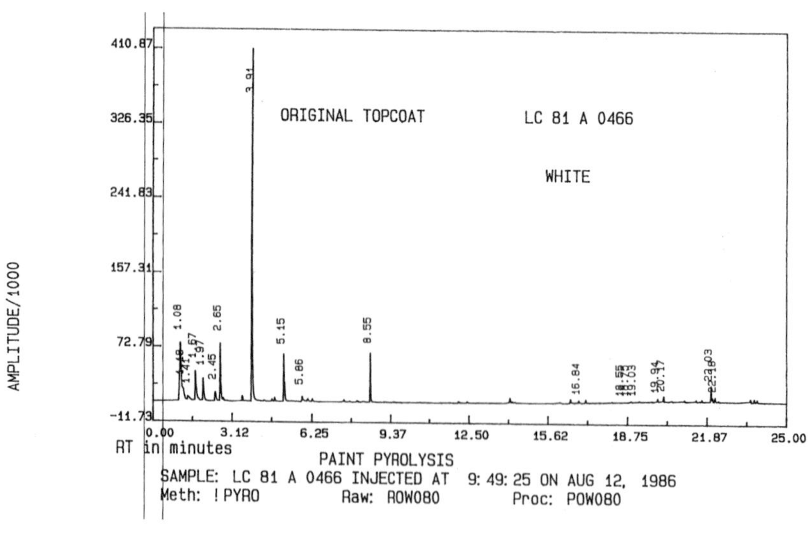

(b)

Figure 10.57 IR spectrum (a) and pyrolysis gas chromatogram (b) of a white acrylic melamine original automotive paint LC81AO466 (NBS)

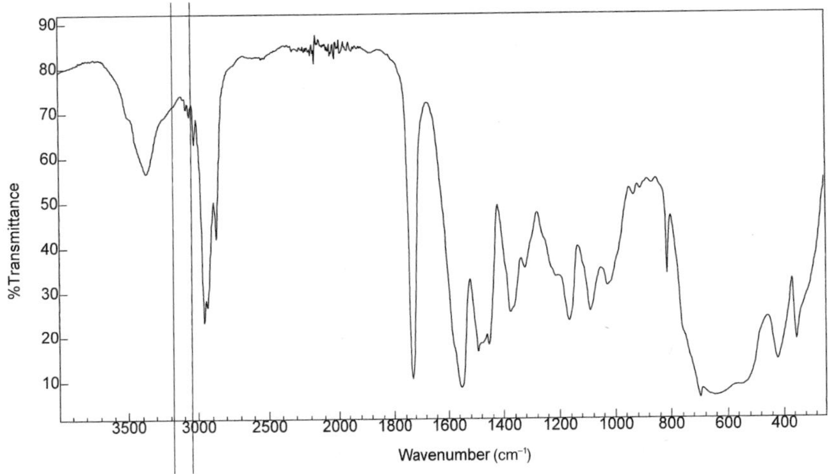

(a)

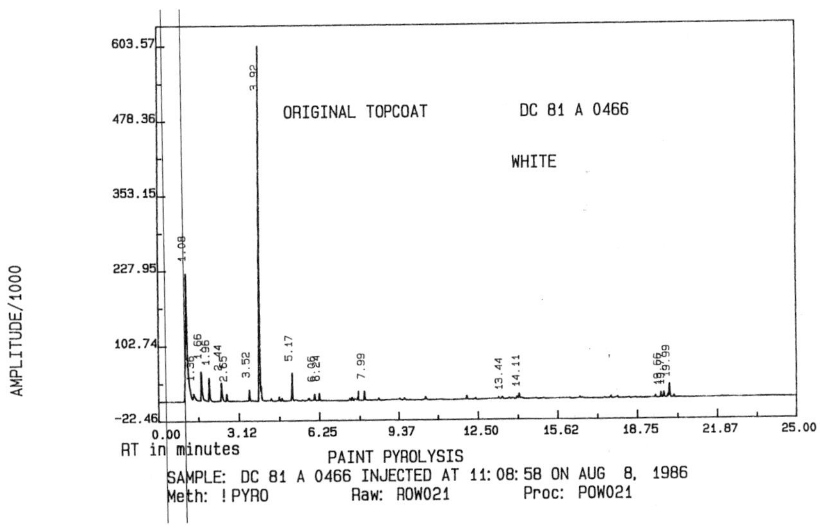

(b)

Figure 10.58 IR spectrum (a) and pyrolysis gas chromatogram (b) of a white acrylic melamine original automotive paint DC81AO466 (NBS)

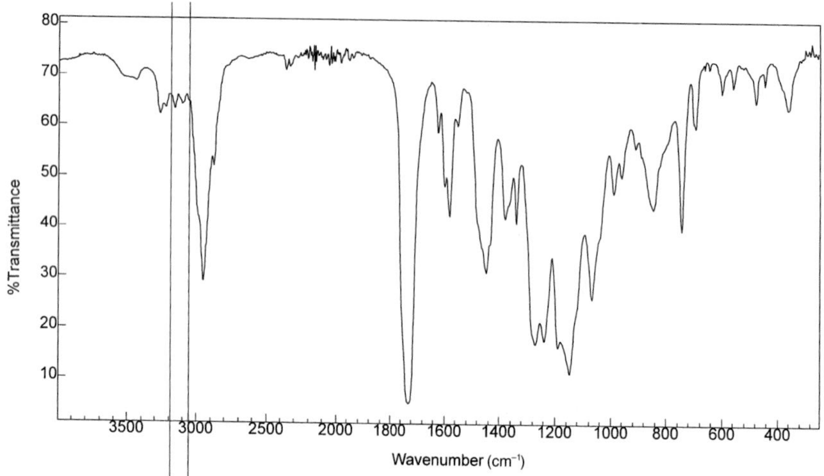

(a)

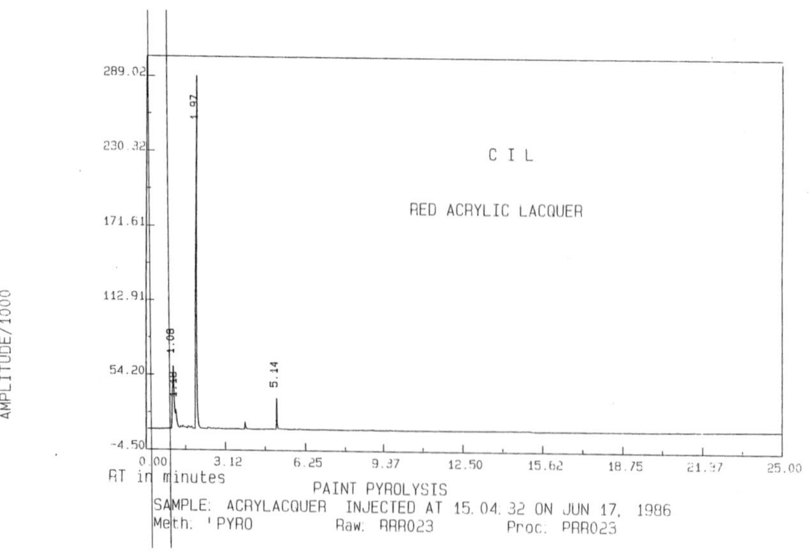

(b)

Figure 10.59 IR spectrum (a) and pyrolysis gas chromatogram (b) of a red acrylic lacquer automotive refinish paint (Canadian Industries Ltd.)

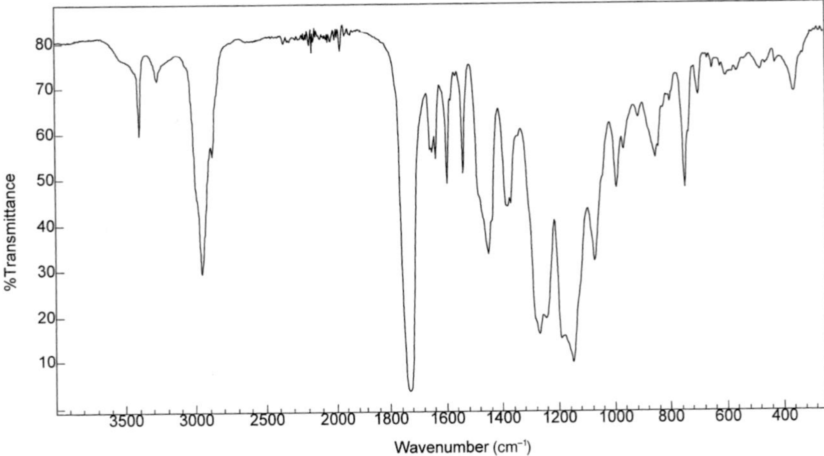

(a)

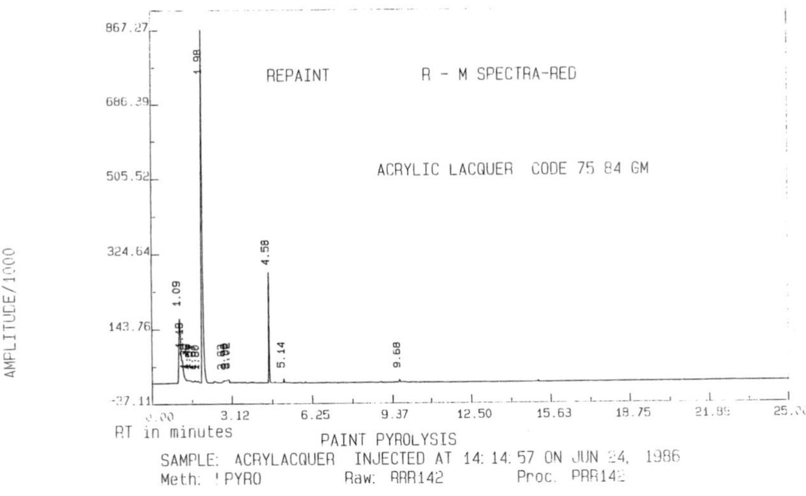

(b)

Figure 10.60 IR spectrum (a) and pyrolysis gas chromatogram (b) of a red acrylic lacquer automotive refinish paint (Rinshed-Mason, Inmont, Canada)

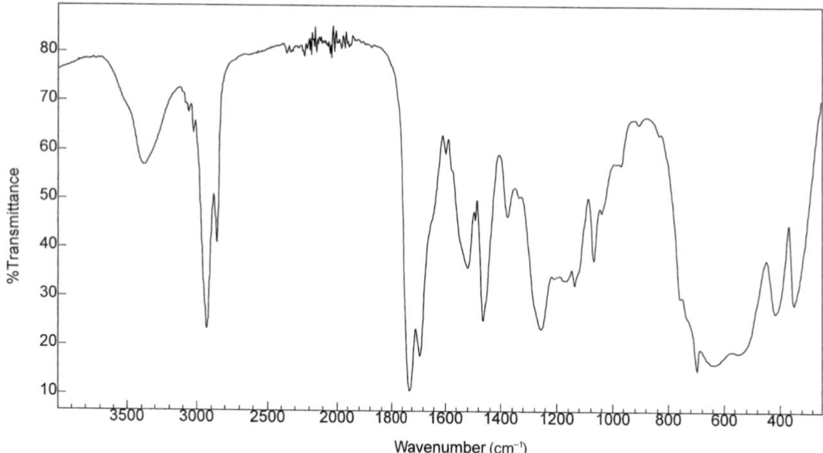

(a)

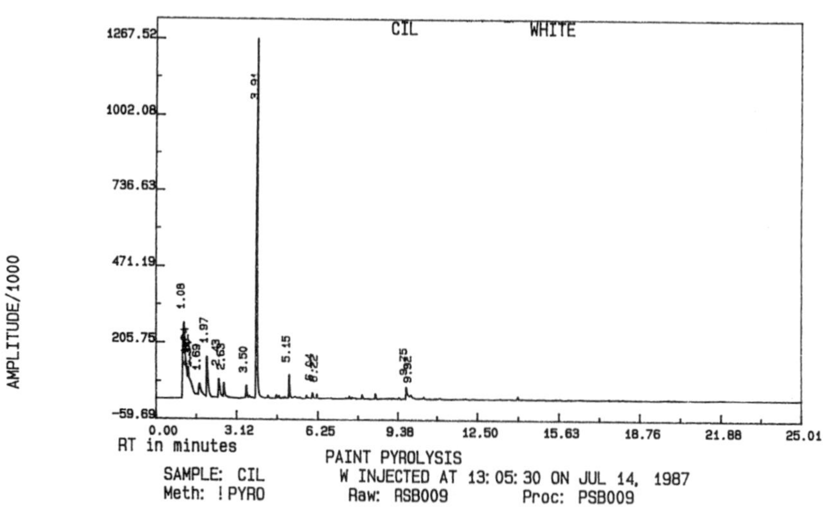

(b)

Figure 10.61 IR spectrum (a) and pyrolysis gas chromatogram (b) of a white alkyd automotive refinish paint (Canadian Industries Ltd.)

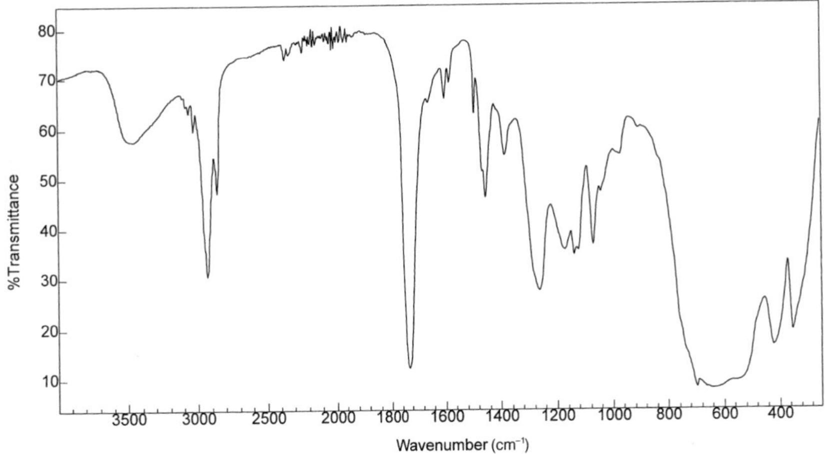

(a)

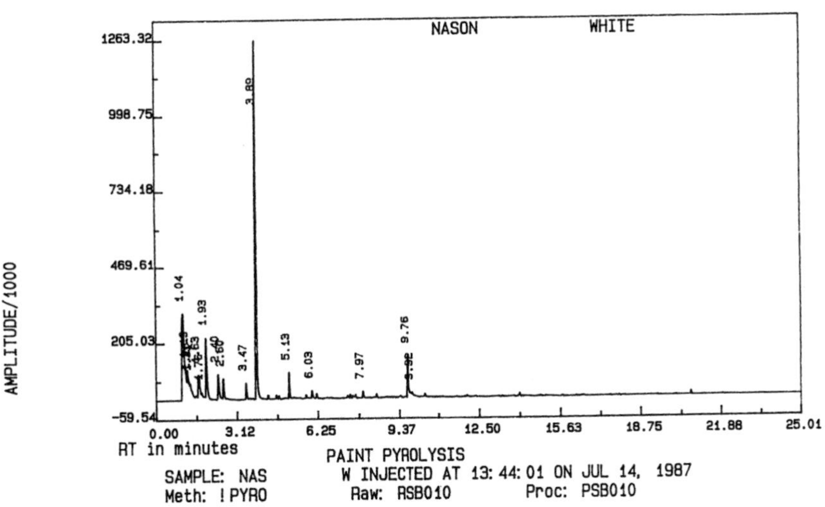

(b)

Figure 10.62 IR spectrum (a) and pyrolysis gas chromatogram (b) of a white alkyd automotive refinish paint (Nason)

These paints were taken from a study of panels of original and refinish automotive paint in five colours: red, black, white, blue metallic and silver metallic. Within each colour group, the samples were (nominally) microscopically indistinguishable. The purpose was to determine which analytical technique or combination of techniques was most discriminating. The US National Bureau of Standards (NBS) authentic samples of original finishes were used and refinish samples obtained from various manufacturers in 1986 were blended to the same colour specifications. The refinish samples were prepared authentically on metal panels in an automobile refinish shop. Analysis was by IR spectroscopy, PGC, solvent tests and SEM/EDX (elemental analysis by energy dispersive X-ray analysis using a scanning electron microscope). In a typical set of results, 30 microscopically indistinguishable white paint samples were subdivided into 13 distinguishable groups by IR, into 16 by PGC and into 20 by IR and PGC combined. Elemental analysis and solvent testing (acetone, conc. sulphuric acid) increased the number of distinguishable groups to 21 [15]. The results demonstrated that the combination of IR spectroscopy and PGC is more discriminating than either technique alone and that when comparing the chemical composition of paint fragments which have only one layer, it is best to use complementary techniques.

10.9 References

1. Federation of Societies for Coatings Technology. (1991). *An Infra-red Spectroscopy Atlas for the Coating Industry.*
2. Noble D. (1995). FT-IR spectroscopy; it's all done with mirrors. *Analytical Chemistry* **June**:381A–385A.
3. Rodgers P., Cameron R. C., Cartwright N. S., Clark W. H., Deak J. S., Norman E. W. W. (1976). The classification of automotive paint by diamond window infrared spectrophotometry. Part I: binders and pigments. *Canadian Society of Forensic Science Journal* **9**:1–14.
4. Bartick E. G., Tungol M. W. (1993). Infrared microscopy and its forensic applications. In *Forensic Science Handbook* (ed. R. Saferstein), Vol. III, pp. 196–592. Reagents/Prentice Hall, Englewood Cliffs, New Jersey.
5. Ryland S. G. (1995). Infrared microspectroscopy of forensic paint evidence. In *Practical Guide to Infrared Microspectroscopy* (ed. H. J. Humecki), pp. 163–243. Marcel Dekker, New York.
6. MacEwen D. J., Cheever G. D. (1993). Infrared microscopic analysis of multiple layers of automotive paints. *Journal of Coatings Technology* **65**:35–41.
7. Cartwright L. J., Cartwright N. S., Rodgers P. D. (1977). A microtome technique for sectioning multilayer paint samples for microanalysis. *Canadian Society of Forensic Science Journal* **10**:7–12.
8. Tweed F. T., Cameron R., Deak J. S., Rodgers P. G. (1974). The forensic microanalysis of paints, plastics, and other materials by an infrared diamond cell technique. *Forensic Science* **4**:211–218.
9. Fettis G. (ed.) (1995). *Automotive Paints and Coatings*. VCH Publishers, New York.
10. Brezinski D. (1995). *Introduction to Coatings Technology*. Handout. Consolidated Research Inc., Kingsford, MI, USA.
11. Buckle J. L., MacDougall D. A., Grant R. R. (1997). PDQ – Paint data queries: the history and technology behind the development of the Royal Canadian Mounted Police Forensic Science Laboratory Services automotive paint database. *Canadian Society of Forensic Sciences Journal* **30**:199–212.
12. *Munsell Book of Color* (1976). Munsell Color, Neighboring Hues Edition, Matte Finish Collection, Macbeth Division of Kollmorgen Corporation, Baltimore, Maryland 21218.
13. Cartwright L. J., Cartwright N. S., Norman E. W. W., Cameron R., MacDougall D. A., Clark W. H. (1984). The classification of automotive primers using the Munsell Color Coordinate System – a collaborative study. *Canadian Society of Forensic Science Journal* **17**:14–18.

14. Weiss K. D. (1997). Paint and coatings: a mature industry in transition. *Progress in Polymer Science* **22**:203–245.
15. Beveridge A. D. (1987). Analysis of automotive repaints by infrared spectroscopy, pyrolysis gas chromatography, solvent tests and elemental analysis. Presentation to the 6th International Association of Forensic Sciences, Vancouver, BC.

Appendix 1: Instrumental conditions

Infrared spectroscopy

Digilab FTS-40A FTIR

Source:	high-intensity ceramic element, water cooled
Beamsplitter:	cesium iodide, germanium coated
Interferometer:	60° Michelson, single air bearing
Detector:	deuterated triglycine sulphate (DTGS)
Spectral range:	4000–250 cm^{-1}
Purge gas:	0.5 cfm dry air, atmospheric H_2O and CO_2 removed
Aperture:	open
Sensitivity:	8
Number of scans:	100
Scan speed:	5 kHz
Apodisation type:	triangular
Gain range radius:	40
Undersampling ratio:	2
Zero filling factor:	2
Beam condenser:	6× transmission beam condenser, Harrick Scientific
Diamond cell:	high-pressure diamond cell, anvil diameters 0.6 mm and 0.8 mm, high pressure optics

Pyrolysis gas chromatography

Chemical data systems model 190 pyroprobe

Pyrolysis temperature:	980°C
Duration:	10 seconds
Sample size	c. 10 μg

Perkin Elmer model 8420 gas chromatograph

Column:	30 m capillary J & W Durabond-WAX 320 μm o.d. 25 μm film thickness
Carrier gas:	helium @ 2 ml/min
Make-up gas:	nitrogen @ 28 ml/min
Injector:	split/splitless horizontal for CDS Pyroprobe; split ratio 70:1 240°C
Detector:	flame ionisation; hydrogen @ 30 ml/min; air @ 200 ml/min; 300°C attenuation 1; range 5×10^{-12} amps/mvolt
Oven temperature programme:	start 60°C; hold 1 minute; ramp at 10°C/min final 200°C; hold 11 minutes

11

Scanning electron microscopy and energy dispersive X-ray spectrometry (SEM/EDS) for the forensic examination of paints and coatings

M. LYNN HENSON and TAMMY A. JERGOVICH

11.1 Introduction

Paints and coatings have been recognised as an important type of forensic evidence for several decades. During this time, the materials themselves, the analytical methods used to examine them and the legal environment surrounding the use of scientific evidence in courts have changed. This chapter provides an overview of scanning electron microscopy/energy dispersive X-ray spectrometry (SEM/EDS) and its current application to the forensic examination of paints and coatings.

11.1.1 Forensic considerations

Historically, a paint has been described as a liquid composition composed of both a pigment and a vehicle portion which dries or cures to form a protective and/or decorative coating. The vehicle contains the binder, the solvent and the additives, while the pigment portion may consist of a combination of organic and inorganic colouring pigments as well as extenders and decorative (effect) pigments. Because today's coatings are not always pigmented (e.g. automotive clearcoats) or even liquid (e.g. powder coatings), the terms paint and coating will be used interchangeably throughout this chapter to refer to both pigmented and unpigmented formulations.

Forensic paint samples are typically those that have been previously applied, cured and exposed to potentially harsh environments for a period of time. They are usually small (e.g. less than 1 square millimetre); the solvent is no longer present and they are often composed of multiple layers. In many instances they are found as trace smears which are mixed or contaminated with adjoining layers and/or the underlying substrate material (e.g. fabric, metal or other paint). The analytical challenges presented by forensic samples can be formidable. These challenges are further complicated by the desire to retain a portion of the sample for examination by other scientific experts who may be involved in the legal process.

While considering these limitations, analytical techniques should be chosen to provide a comprehensive examination protocol. Because questioned samples are often well below a square millimetre in size, it is desirable to use non-destructive tests or tests that will consume the smallest amount of sample that is practical. In order to maximise the ability to discriminate between similar paint formulations, techniques must be selected that examine all major components remaining in the cured coating.

A comprehensive forensic examination should employ tests to examine both the binder (organic) and the pigment (inorganic and organic) portions of the coating. The organic portion of the paint is commonly analysed by either infrared (IR) spectroscopy or pyrolysis gas chromatography (PGC). Some laboratories have opted for the use of pyrolysis gas chromatography combined with mass spectrometry (PGC-MS) in addition to, or in place of, PGC.

11.1.2 Elemental analysis of paints and coatings

Forensic laboratories employ a variety of instrumental methods of elemental and inorganic analysis. However, the most commonly used method for the elemental analysis of paints and coatings is scanning electron microscopy/energy dispersive X-ray spectrometry (SEM/EDS). The summary information distributed by the Collaborative Testing Services Inc. for their 1999 paint proficiency test (Test No. 99-545) indicates that the method most commonly used by the 162 test participants was SEM/EDX (Table 11.1). It should be noted that SEM/EDS is alternatively referred to as scanning electron microscopy/energy dispersive X-ray spectroscopy (SEM/EDS or SEM/EDX), scanning electron microscopy/energy dispersive X-ray analysis (SEM/EDX or SEM/EDXA) and scanning electron microscopy/energy dispersive X-ray microanalysis (SEM/EDX).

Scanning electron microscopy/energy dispersive X-ray spectrometry (SEM/EDS) is an ideal tool for performing non-destructive analyses on small as well as large single or multi-layered paint fragments. SEM/EDS has the ability to simultaneously detect a wide range of elements on the periodic chart. When used properly, the technique will enhance discrimination potential by revealing the identity of various elements in the coating. Elements can be detected in quantities as low as 0.1 weight per cent. In addition to using this elemental profile for direct comparison, the identities of the elements can be used when trying to decipher which extenders, inorganic pigments and even organic pigments are present.

Table 11.1 Instrumental usage for coating analyses

Method of elemental analysis	Percentage of participants
Scanning electron microscopy/energy dispersive X-ray analysis (SEM/EDX)	56.0
X-ray fluorescence spectroscopy (XRS/XRF)	14.0
X-ray diffraction spectroscopy (XRD)	8.0
Emission spectroscopy (ES)	1.8
Inductively coupled plasma–atomic emission spectroscopy (ICP-AES)	0.6
None reported	19.6

Summary information from the CTS 1999 paint proficiency test (Test No. 99-545).

Although inferences can be made as to which specific extenders or pigments exist in the specimen, SEM/EDS alone does not provide definitive pigment identification, as it is not a structural elucidation technique.

11.2 Pigments and extenders

The pigment portion of a coating includes both the colouring pigments and extenders. Coatings may contain the same binder and differ only in their pigments. This phenomenon is quite common in the automotive industry where a single vehicle model might be offered in eight or more colours. The choice of extender pigments in primers and architectural paints will be determined by the manufacturer's desire to adjust cost and physical properties of the paint such as gloss, weatherability and viscosity. Extenders are also used to prevent the settlement of other heavy pigments. Tables 11.2, 11.3 and 11.4 list some commonly used extenders, colouring pigments and their elemental indicators.

Table 11.2 Common extenders and their identifying elements

Extender	Elemental indicator(s)
Barium sulphates ($BaSO_4$)	
Barytes (natural mineral)	Ba, S
Blanc fixe (precipitated form)	Ba, S
Calcium carbonates ($CaCO_3$)	
Whiting (natural mineral)	Ca
Precipitated calcium carbonate	Ca
Calcite (crystalline calcium carbonate)	Ca
Calcium magnesium carbonate	
Dolomite ($CaCO_3/MgCO_3$)	Ca, Mg
Calcium sulphate	
Gypsum ($CaSO_4{\cdot}2H_2O$) crystalline calcium sulphate	Ca, S
Calcium silicate	
Wollastonite ($CaSiO_3$)	Ca, Si
Calcium magnesium silicate	
Asbestine	Ca, Mg, Si
Aluminium silicates	
Kaolin ($Al_2O_3{\cdot}2SiO_2{\cdot}2H_2O$) china clay	Al, Si
Bentonite (hydrated aluminium silicate)	Al, Si, possible Mg or Na
Potassium aluminium silicate (hydrated)	
Mica ($K_2O{\cdot}2Al_2O_3{\cdot}6SiO_2{\cdot}2H_2O$)	K, Al, Si
Magnesium carbonate (precipitated)	
$MgCO_3{\cdot}3Mg(OH)_2{\cdot}11H_2O$	Mg
Magnesium silicate	
Talc ($3MgO{\cdot}4SiO_2{\cdot}H_2O$)	Mg, Si
Silicas	
Quartz (crystalline silicon dioxide)	Si
Diatomaceous earth (amorphous silicon dioxide)	Si
Synthetic silica (amorphous silicon dioxide)	Si

Table 11.3 Coloured inorganic pigments and their identifying elements

Coloured inorganic pigments	Elemental indicator(s)
Chrome pigments (yellow, red, orange)	
Lead chromes (yellow, red)	
Primrose (lead chromate, lead sulphate and alumina)	Pb, Cr, S, Al
Lemon (lead chromate and lead sulphate)	Pb, Cr, S
Middle (lead chromate only)	Pb, Cr
Orange (lead chromate and lead hydroxide)	Pb, Cr
Scarlet (lead chromate, lead molybdate and lead sulphate)	Pb, Cr, Mo, S
Basic lead silicochromate (lead chromate, lead silicate)	Pb, Cr, Si
Zinc chromes (yellow)	
Zinc potassium chromate	Zn, K, Cr
Zinc tetroxychromate	Zn, Cr
Barium chromate (yellow)	Ba, Cr
Strontium chromate (yellow)	Sr, Cr
Cadmium colours (yellow, red)	
Cadmium sulphide	Cd, S
Cadmium sulphide/cadmium selenide	Cd, S, Se
Titanium nickel yellow	Ti, Ni, Sb
Iron oxides (yellow, red, brown, black)	
Ochres (iron oxide and mineral silicates)	Fe, Si
Siennas (iron oxide and about 1% manganese dioxide)	Fe, Mn
Umbers (iron oxide and about 15% manganese dioxide)	Fe, Mn
Synthetic red and yellow oxides	Fe
Magnetite (natural black oxide of iron)	Fe
Red lead (orange–red)	Pb
Lead cyanamide (yellow)	Pb
Inorganic blue pigments (blue)	
Iron blues (Prussian blues)	Fe
Potash or non-bronze blue	Fe, K
Bronze blue	Fe
Cobalt blue (CoO and Al_2O_3)	Co, Al
Cerulean blue (cobalt oxide, tin oxide and silica)	Co, Sn, Si
Inorganic green pigments (green)	
Chrome greens (lead chromes and Prussian blue)	Pb, Cr, Fe
Chromium oxide (Cr_2O_3)	Cr
Viridian ($Cr_2O_3 \cdot 2H_2O$)	Cr

Information summarised from [1].

The most common white pigment in use today is titanium dioxide (TiO_2). The rutile form of TiO_2 is unrivalled in its opacity and durability. It is also non-toxic, unlike some of its lead-based predecessors. The anatase form of TiO_2 has a lower opacity and has a tendency to chalk. It is used in some applications where good whiteness is important, but weather resistance is not (e.g. hospital equipment). The anatase form is also used in applications where it is desirable for chalking to occur, such as in an exterior architectural paint designed to refresh itself by sloughing off its dirtied surface. The use of zinc oxide as a

Table 11.4 Coloured organic pigments and their identifying elements

Coloured organic pigments	Elemental indicator(s)
Nitroso group	
Pigment Green B	Fe
Nitro group	
Lithol Fast Yellow GG	Cl
Azo group (yellow, red, orange)	
Monoarylide Yellows (several in the class)	Cl
Benzidine Yellow	Cl
Lithol Red (calcium or barium salts)	S, Ca, Ba
Nickel Azo Yellow (olive yellow)	Ni, Cl
Metallised Azo Reds	S, (Ba, Ca or Mn)
Non-metallised Azo Reds	Cl
Phthalocyanine pigments (deep blue to yellow–green)	
Copper Phthalocyanine Blue	Cu, possible Cl
Copper Phthalocyanine Green	Cu, Cl or Br
Vat pigments (pink, red, violet)	
Thioindigo derivatives	S, Cl
Anthraquinone pigments	
Synthetic Madder Lake (calcium salt of alizarin)	Ca
Brominated Pyranthrone Red	Br
Pyranthrone Red	Br
Quinacridone pigments (red, violet, gold)	Cl
Dioxazine pigments (violet)	Cl
Lake pigments (pigment precipitated onto a white base)	
Bases for lakes	
Alumina $Al_2(OH)_6$	Al
Blanc fixe	Ba, S
China clay	Al, Si

Information summarised from [1, 13].

white pigment has declined drastically and it is used only in certain specific roles such as a fungistat in emulsion paints or to assist with gloss retention in some decorative alkyd finishes. Antimony oxide use has also dwindled. Its primary use today is in fire-retardant paints [1]. Lead-based pigments are almost extinct, as the use of lead in coatings has been tightly controlled because of its toxic nature.

Most black paint pigments come in the form of carbon which has been generated from a variety of sources. Carbon blacks are derived from petroleum gases and oils and are classified as channel or furnace blacks according to their method of production. Channel blacks are produced by burning hydrocarbon gases in air. Furnace blacks are produced by combustion of gases in a limited supply of air. Graphite and iron oxide are also used as black pigments.

Decorative pigments, also known as effect pigments, are added to coatings to achieve a glittery or flamboyant appearance. Aluminium metal flake has traditionally been used to put the metallic appearance in an automotive finish coat, but its use is declining with the increased use of water-borne coatings. The use of titanium dioxide coated micas (pearlescent pigments) has increased because they do not have any adverse reaction to water. Pearlescent pigments are also known as interference pigments because they derive their effects from the production of interference colours through constructive and destructive interference of light. The thickness of the TiO_2 layer determines the colour of the pigment. Additional colours can be formed by covering the mica particles with thin coats of heavy metal oxides (e.g. iron oxide).

Another type of interference pigment was introduced to the automotive market in 1996. The 1996 Ford Mustang Cobra coupe sported a paint colour called 'Mystic'. The Mystic colour automotive paint was formulated by BASF Corporation and contains colour-shifting interference pigments. These ChromaFlair® interference pigments, developed and manufactured by Flex Products, Inc., show dramatic shifts in colour as the viewing angle changes. The colour is produced by the constructive and destructive interference of light as it is reflected. The pigment is made from alternating layers of material. The outermost layers consist of transparent metal. The inner layers are made up of dielectric materials and a metal reflector [2, 3]. Dupont has also produced its own line of paints (i.e. ChromaLusion™) [4] using this new type of pigment.

Another major breakthrough in pigment technology came in 1997 in the form of optical pigments. In July of 1997, PPG Industries introduced a new paint system called Prizmatique™ on a Lotus PPG Cup race car at the Molson Indy. The pigments that formed the basis for the system were called Geometric Pigments™ and were created by Spectratek Technologies Inc. The pigment is referred to as a holographic pigment by Spectratek. Holographic pigments are created by embossing a specific holographic light recording on a 0.004-inch particle cut from a polyester film. The colour is created by physical grooves in the surface of the particle. Each particle is sealed with an aluminium coating and a clearcoat [5].

11.3 History of scanning electron microscopy/energy dispersive X-ray spectrometry (SEM/EDS)

SEM/EDS has been used in the forensic arena to analyse pigments and extenders since the mid-1970s. The scanning electron microscope (usually abbreviated to SEM) allows the examiner to image the sample and also serves as an excitation source. The interaction of the SEM beam with the sample creates X-rays that are detected and identified by the energy dispersive X-ray spectrometer.

The first transmission electron microscope was constructed by Ruska and Knoll in 1932 [6]. In 1935, Knoll [7] demonstrated the theory of scanning electron microscopy but it was not until 1938 that von Ardenne [7] actually built the first scanning transmission electron microscope. The first 'modern' scanning electron microscope with electromagnetic lenses and an Everhart–Thornley detector was built in 1960 [7]. The first commercially available scanning electron microscopes were produced by Cambridge Scientific Instrument Co. in 1965 [7].

In 1913, Moseley [7] reported that the frequency of an emitted X-ray is a function of the atomic number of the emitting source. This concept was the basis for the construction of the first electron microprobe by Castaing and Guinier in 1949 [6]. This microprobe utilised wavelength-dispersive technology to sort and identify X-rays. The advent of semiconductor technology opened the door for the development of the energy dispersive X-ray detector that sorts and identifies X-rays based on their energy rather than their wavelength. The energy dispersive X-ray spectrometer utilising a lithium-drifted silicon Si(Li) detector was developed in 1968 by Fitzgerald *et al.* [7].

11.4 Theoretical principles of scanning electron microscopy

Understanding the operation of the scanning electron microscope allows an examiner to optimise the instrument not only for imaging purposes but also for elemental examination techniques. Understanding the interaction between the electron beam and the sample is the key to understanding the results of the elemental analysis. An in-depth discussion of the SEM can by found in several books [7, 8]; what follows is a basic overview.

11.4.1 Basic operation of a scanning electron microscope

The scanning electron microscope operates by generating a beam of electrons in a vacuum. The beam is focused by electromagnetic lenses within a column, and directed downwards toward the sample. This beam of electrons is scanned (rastered) back and forth over the selected area of the sample. The interaction of the electron beam with the sample causes electrons to be dislodged from the atoms within the sample. The electrons generated by this sample–beam interaction are detected, amplified and displayed on a screen. The image on the screen can be viewed, captured on photographic film or, in the case of newer instruments, digitised, stored and printed. High magnifications are made possible by scanning smaller areas of the sample and displaying that smaller area on the fixed-size display screen. The smaller the area scanned, the greater the magnification.

The most common method of generating an electron beam uses a hairpin-shaped tungsten filament, a Wehnelt cylinder and an anode. The filament and Wehnelt cylinder have a large voltage applied to them, usually variable in the range of 1000 to 40,000 volts (1–40 kV). The filament is heated by applying a current to the level where thermionic emission occurs and electrons are released from the tungsten filament. The anode is at earth and the electrons are accelerated toward the anode. A portion of the electron beam passes through a hole in the anode and continues down the column toward the sample. This portion is called the beam current. The remaining portion that does not pass through the anode is referred to as the emission current. A resistor is used to control the difference between the voltage of the filament and the voltage of the Wehnelt cylinder. By maintaining the Wehnelt cylinder at a slightly higher negative voltage than the filament, electrons leaving the gun are forced into a crossover, the first focusing point for the electron beam (Figure 11.1).

As the electrons continue down the column, electromagnetic lenses are used to focus the beam. The number of lenses varies with the microscope manufacturer. When current is applied to the lenses, a magnetic field is generated that acts upon the beam. These lenses cause the electrons to move toward the centre of the column into focused crossover points. The repeated focusing of the beam results in crossovers beginning at the gun and ending at the sample. There are typically two sets of electromagnetic lenses in the microscope column. The first set is referred to as the condenser lenses. The stronger the magnetic field

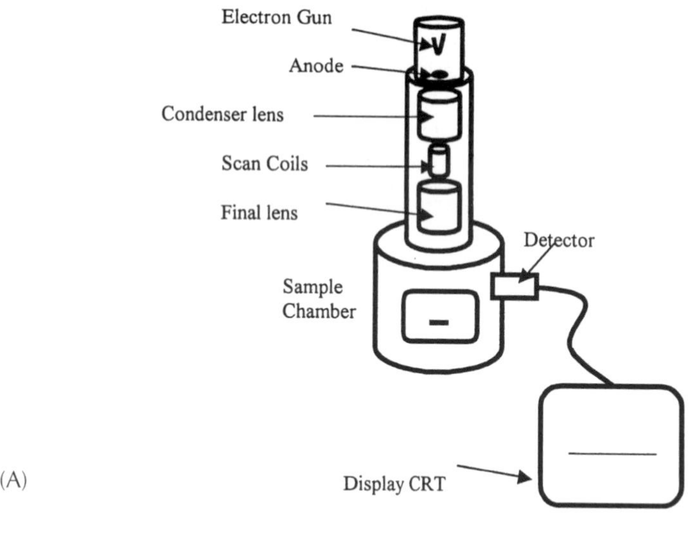

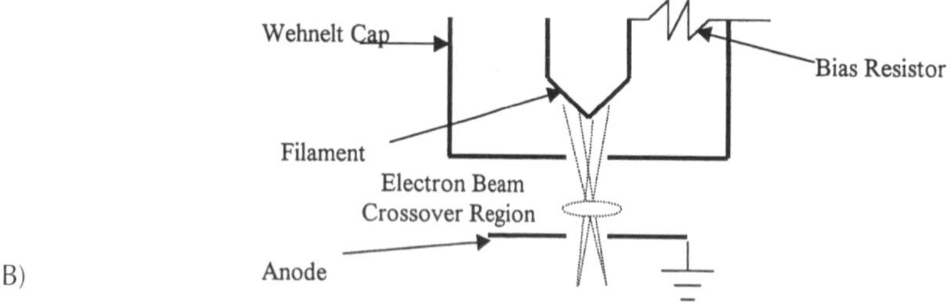

Figure 11.1 (A) Block diagram of a scanning electron microscope. (B) Electron gun assembly; the bias between the Wehnelt cap and the filament creates the first focused spot of the electron beam

applied to the beam, the smaller the focused spot that will result. However, as the size of the focused spot (beam spot size) is decreased, more and more beam current is lost before the beam reaches the sample.

With each focus event, there are widespread electrons that become grounded through contact with the walls of the column. This 'loss' of electrons results in a significantly reduced beam current. The beam current is reduced by several orders of magnitude as it passes from the filament to the sample.

The second set of electromagnetic lenses is referred to as the final focus lens. This set of lenses moves the focused beam spot in a plane perpendicular to the sample in order to locate the focused beam at the surface of the sample. An alternate method of focusing the sample is to physically move the sample to the focused spot of the beam by adjusting the sample height.

Scan coils deflect the beam so that the focused spot is scanned back and forth across the sample surface. The scan coils are synchronised with a cathode-ray tube (CRT) display so that, as the beam scans the sample, the response registered by the detector is seen on the CRT (essentially a television screen).

Some SEMs use apertures between the lenses to limit and further collimate the electron beam. Variable apertures are generally placed at the bottom of the focusing portion of the column. These apertures help to increase the depth of field of the SEM. Smaller apertures allow a greater thickness of a sample to be in focus at one time. Because the smaller apertures limit the beam and create a finer column of focused electrons, the sample does not see as much signal. The larger apertures allow more of the beam to pass to the sample at the expense of depth of field.

At the base of the column is the pole piece. This is a metal plate constructed of a non-ferrous material with a small hole through which the beam passes into the sample chamber. The pole piece blocks electrical and magnetic fields generated by the lenses and scan coils from entering the chamber. The pole piece forms the top of the sample chamber.

The sample chamber of the SEM contains a stage on which the sample can be mounted as well as ports for access to the chamber by a variety of detectors or accessories.

11.4.2 Electron beam–sample interactions

Understanding beam and sample interaction is crucial to proper SEM/EDS analysis. When a primary electron beam strikes a sample in the SEM, a wide variety of signals are produced and may be detected for a number of different purposes. Among these signals are backscattered electrons, secondary electrons, characteristic X-rays, photons of light and Auger electrons. We are primarily interested in the emission of electrons to be detected by the SEM and of X-rays to be detected by the EDS system.

As the beam strikes the sample, some electrons come in close proximity to sample atoms and are elastically scattered without significant loss of energy (i.e. they retain just slightly less energy than the beam). These high-energy electrons leave the sample and are known as backscattered electrons. Other electrons penetrate the sample and through multiple inelastic collisions with atoms in the sample, dissipate their original energy. During these collisions, the beam electrons may cause the sample atom to ionise. The lower energy electrons (i.e. less than 50 eV) emitted as part of this ionisation are known as secondary electrons. Before the high-energy backscattered electrons leave the sample, they too may engage in these types of collisions and also generate secondary electrons. Electrons from outer energy shells of an atom fill holes left by the emission of secondary electrons from inner energy shells. When electrons drop down into the lower energy shells, the energy difference may be released in the form of a characteristic X-ray. It is these X-rays that allow us to perform EDS analysis and identify the sample elements.

When a high-energy beam of electrons strikes the surface of our sample, it penetrates and what we see is the response from a three-dimensional excited volume on and within the sample. It is from this excited volume and not just the area of beam penetration that characteristic X-rays arise. The size and shape of the excited volume is related to the energy of the SEM beam, the size of the beam spot, the topography of the sample, the average atomic weight of the sample and the secondary fluorescence that occurs within the sample. Generally speaking, the excited volume created when the electron beam strikes a sample is teardrop shaped for low average atomic number matrices and is shaped like a hemisphere for high average atomic number matrices. A graphical representation, known as a Monte Carlo plot, can be generated using computer software to describe the path an electron of a given potential can be expected to take in a given matrix (Figure 11.2). A Monte Carlo plot demonstrates the area of beam penetration but not the entire excited volume of the sample. The higher the accelerating voltage of the beam, the deeper the penetration into the sample. Deeper beam penetration and larger spot sizes will result in larger excited volumes.

The resolution of the SEM is directly related to the excited volume. When the beam scans a surface, the beam spot has a defined area but the excited volume associated with that beam spot must also be considered. Secondary electrons used for imaging will also arise from within the excited volume. The beam spot size is controlled by the bias on the electron gun and the focusing of the electromagnetic lenses as discussed previously.

11.4.3 Image formation

Image formation in the SEM is achieved by the detection of the backscattered and secondary electrons that arise from the sample. The most common SEM detector is the Everhart–Thornley detector. This detector can be used for the detection of both secondary and backscattered electrons. The Everhart–Thornley detector face is composed of a scintillator material. As electrons strike the scintillator material, they produce photons that then pass through a total internal reflecting light pipe to a photomultiplier tube. The photomultiplier tube converts the light photons back to electrons and amplifies the signal. The signal display is synchronised with the scan coils in the SEM. As the beam is rastered across the sample, each point on the sample is assigned to a corresponding XY position on the CRT display.

A cage on the front of the detector can be biased with a positive voltage (approximately +300 V) to attract the low-energy secondary electrons toward the detector. The scintillator is held at a much higher voltage, typically 10 to 12 kV, to attract electrons to its surface. A low voltage can be applied to the cage to repel the lower-energy secondary electrons and thereby enhance the ratio of backscattered electrons detected. The high-energy backscattered electrons are unaffected by these low voltages on the cage. The detection of mainly backscattered electrons gives a visual impression that a light is shining on the sample from one side. If the sample surface has a bump on it, the lower-energy secondary electrons that originate from the side of the bump away from the detector will not be drawn toward the detector. They will be obstructed. Because X-rays act much like backscattered electrons in that they cannot be attracted to the detector, a backscattered image can be very useful for X-ray work.

11.4.3.1 Backscatter detectors

Specialised backscattered electron detectors are also common accessories on SEMs. These detectors are typically mounted beneath the pole piece perpendicular to the electron beam path. This position is the most efficient for collection of backscattered electrons that are essentially deflected back from a flat sample. Backscatter detectors are popular in situations where it is desirable to see topographical contrast and atomic number contrast.

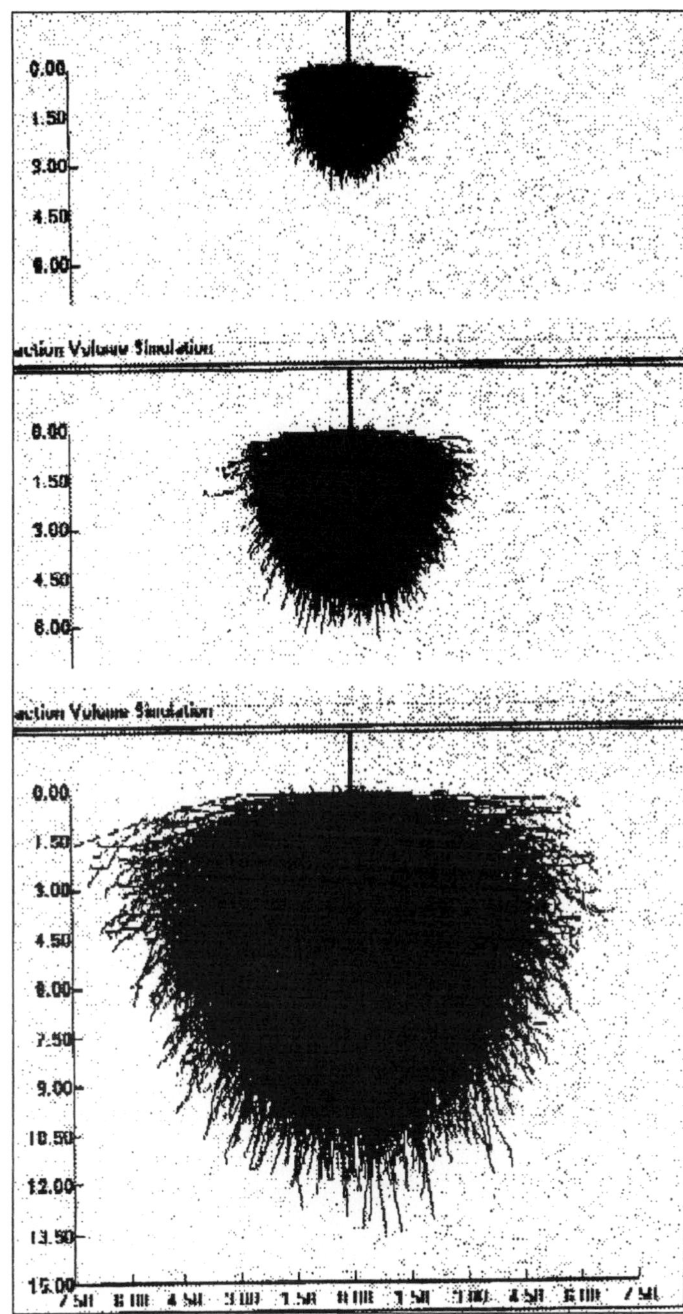

Figure 11.2 Monte Carlo plots of the electron path in a bulk sample of carbon. The top plot is for an electron beam with 10 kV potential, the middle is for a beam with a 15 kV potential and the bottom plot is for a beam at 30 kV potential. The vertical and horizontal scales are in 1.5 μm increments. Note that the horizontal spread of the beam is roughly equal to its penetration depth

Relatively smooth samples are particularly suited for viewing by backscatter detectors, as even slight irregularities in topography will appear distinct due to shadowing (Figure 11.3). Samples containing elements of differing atomic numbers will show interesting effects. Because backscattering increases with atomic number, elements with higher atomic

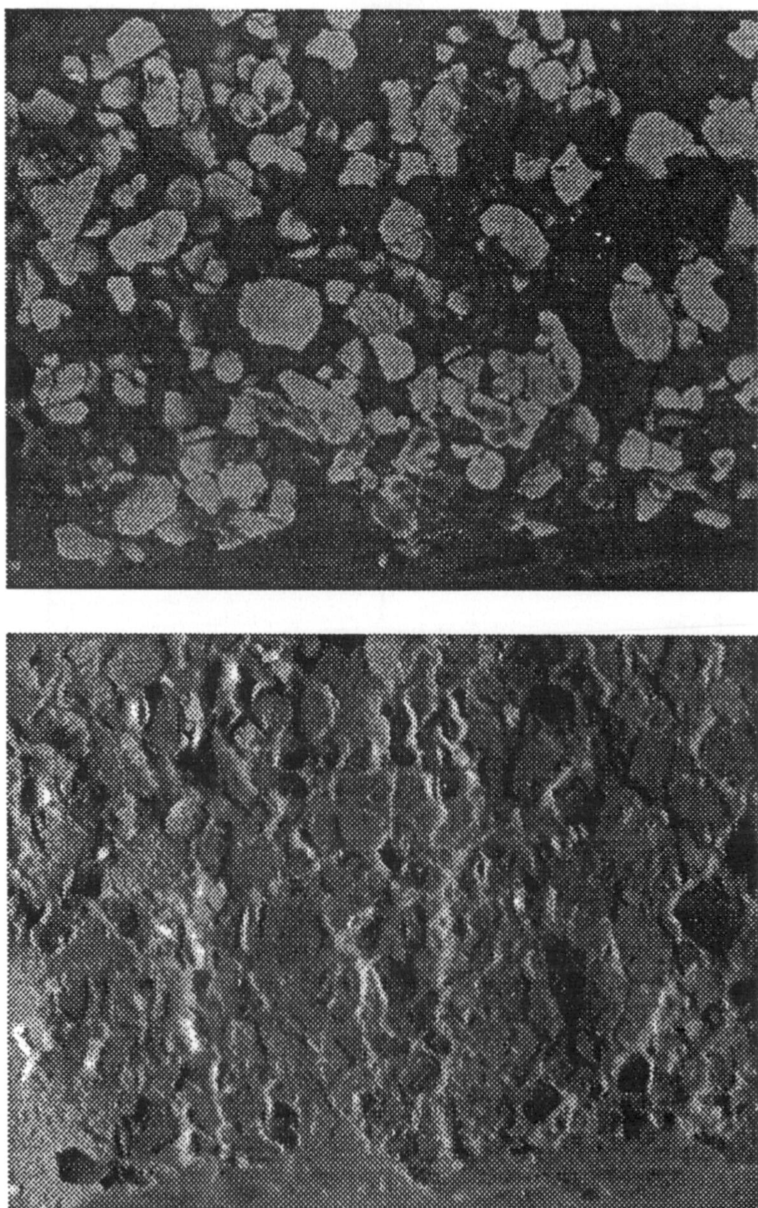

Figure 11.3 Backscattered electron images, 20 kV, 1.9 mBar, rastered area ~166 μm × 125 μm. Top image emphasises atomic number differences. Bottom image emphasises topographical differences

numbers will appear brighter than those with lower atomic numbers. This characteristic can be useful in imaging differences between paint layers.

11.4.3.2 Variable-pressure SEMs

The advent of the economical variable-pressure or environmental SEM has enhanced the ease of producing images of paint. The variable-pressure mode of these instruments allows the SEM to operate with a higher pressure in the sample chamber than previously possible. The system operates by using a limiting aperture at the pole piece so that a higher vacuum can be maintained in the electron gun area (i.e. at least 10^{-4} millibars) while a very low-pressure atmosphere can be maintained in the sample chamber (i.e. 0.1–2.5 millibars). This low-pressure atmosphere allows electrical charges that accumulate on non-conductive samples, such as paint, to be dissipated by ionising the gases in the chamber. Thus, samples can be imaged without coating. Although the variable-pressure SEM is convenient for imaging, it does have some disadvantages when used for EDS analysis. The interaction between the electron beam and the gases in the sample chamber causes three main effects. The beam tends to spread and will degrade resolution. Characteristic and continuum X-rays from the gas molecules will be generated and, finally, the total X-ray count will be reduced [9].

11.5 Theoretical principles of energy dispersive X-ray spectrometry (EDS)

When the electron beam strikes the sample, electrons travel through the sample dissipating their energy as they collide with sample atoms. Some of these electron–sample collisions result in the ejection of electrons from the atoms within the sample. This ionisation creates an atom in an excited state. The atom returns to a stable state by transitions of electrons from one shell or subshell to a lower level shell. Because these outer-shell electrons have a higher potential energy than the inner-shell electrons, the extra energy must be released in some manner. It can be released in the form of an X-ray.

The energy of the X-rays produced is determined by the structure of the atom (Figure 11.4). Electrons revolve around the nucleus of an atom in defined shells. Quantum numbers are used to uniquely identify each electron in an atom. These quantum numbers identify the shell in which an electron resides (i.e. the principal quantum number), the angular momentum of the electron, the spin of the electron and the angular momentum under the influence of a magnetic field. The significance of these quantum numbers to X-ray analysis is that we can predict the energy of an X-ray produced by a particular atom. The energy of the X-ray created is dependent on the amount of energy released in the transition between the shells of the atom and is characteristic of the atom. Energy dispersive X-ray analysis uses this characteristic X-ray emission to identify the elemental content of a sample. Energy tables correlating energy lines to elements are available in a tabular slide-rule format provided by many EDS manufacturers.

Principal quantum shell numbers 1, 2 and 3 correspond to the K, L and M shells of an atom. The innermost shell is the K shell and will, for every element above hydrogen, contain two electrons. When a K-shell electron is ejected from an atom, energy released by electrons dropping to the K shell from either the L shell or the M shell are identified as

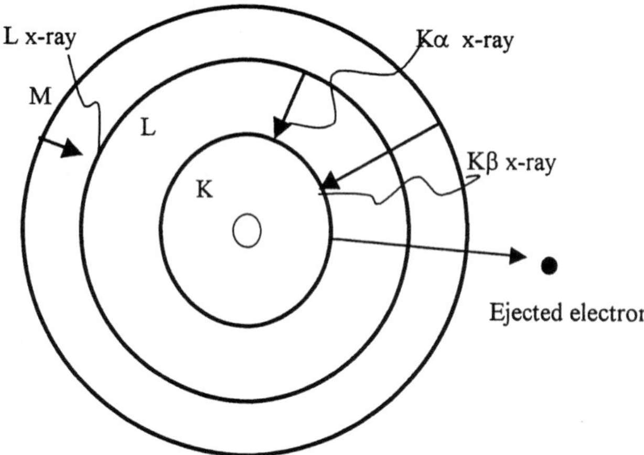

Figure 11.4 When inner-shell electrons are ejected from the atom, outer-shell electrons drop to fill the void and stabilise the atom. X-ray energy released by these transitions is characteristic for each element

K-level X-rays. L-level X-rays are created by electrons releasing energy as they drop to the L shell and M-level X-rays are created by the release of energy from electrons dropping to the M shell. As the atomic number of an atom increases, the atoms become more complex and sets or families of K or L X-ray lines are created. Families of X-ray lines are created by additional transitions between subshells of each major shell level. For example, Kα lines are created by L-shell electrons dropping to the K shell. Kβ lines are created by M-shell electrons dropping to the K shell. Sodium (Na) is the first element with electrons in the M shell and is, therefore, the first element that will have a Kβ line. The L shell can contain up to eight electrons in three distinct subshells. Replacing the electrons ejected from the K shell with electrons from each of these subshells requires slightly different energies to be released. Many of these energies are too close to be resolved as individual lines by the EDS system (e.g. $K\alpha_1$ and $K\alpha_2$ may be seen as one peak).

X-rays emitted from a sample are detected and then analysed through the use of a multichannel analyser. The semiconductor-type detector utilises a crystal comprised of silicon. Holes in the crystal lattice are filled or drifted with lithium (i.e. Si(Li)) to improve performance. As a characteristic X-ray strikes the detector crystal, a signal is generated which is proportional to the energy of the X-ray. These signals are very small and the detector must be at liquid-nitrogen temperatures to reduce electronic noise. In order to maintain a vacuum in the liquid-nitrogen Dewar and to allow X-rays to reach the detector, a window is placed at the end of the detector snout protruding into the SEM chamber. This window can be constructed from a variety of materials. Beryllium has traditionally been the window material of choice, but most modern systems use either a thin polymer window or sometimes no window at all. Windowless detectors are becoming less common because they cannot be used in the newer variable-pressure SEM environment. In a standard high-vacuum SEM, there must be a window in place when the chamber is vented. A turret allows a window to be rotated into place over the snout when the SEM chamber is not under full vacuum and then out when the SEM chamber is under a high vacuum. The composition of the window material dictates which X-rays can pass through the window to reach the X-ray detector. Thin-window or windowless detectors can detect elements down to boron. Beryllium window detectors will only detect elements from sodium and above, because only X-rays from the elements sodium and above have enough energy to pass through the window. A detailed description of the Si(Li) detector can be found in [7].

Sample geometry and SEM chamber geometry also impact on which X-rays will be detected. When the electron beam strikes the sample, X-rays are generated throughout the excited volume. Generally, X-rays that are generated in the top half to two-thirds of the sample are the only X-rays that will escape. Furthermore, X-rays that travel in a direction perpendicular to the surface of the sample will have the highest odds of escaping the sample boundary. X-rays that travel at less than right angles to the sample surface will physically pass through more of the sample and have a greater probability of being absorbed. The term 'take-off angle' is used to describe that angle between the surface of the sample and the line connecting the face of the EDS detector to the surface of the sample. (Figure 11.5). Theoretically, the EDS detector would 'see' more X-rays if it were located directly above the sample, at the base of the pole piece. For obvious reasons (e.g. it would block the electron beam and many backscatter detectors that already occupy this position) the EDS detector cannot be placed at the optimum position. Instead, EDS manufacturers design detectors to be mounted in the SEM chamber port at the highest take-off angle possible. The actual angle at which the detector is placed varies depending on the SEM chamber configuration. The EDS manufacturer supplies this chamber geometry information to the user. In addition to the angle of inclination, the manufacturer will provide information on the optimum working distance for X-ray analysis. The optimum working distance is defined as that position where the beam of electrons intersects a line drawn perpendicular to the face of the EDS detector. This point may or may not correspond to the optimum working distance for imaging purposes. If the sample is not placed at the optimum working distance for EDS analysis, a significant reduction in the count rate will be observed.

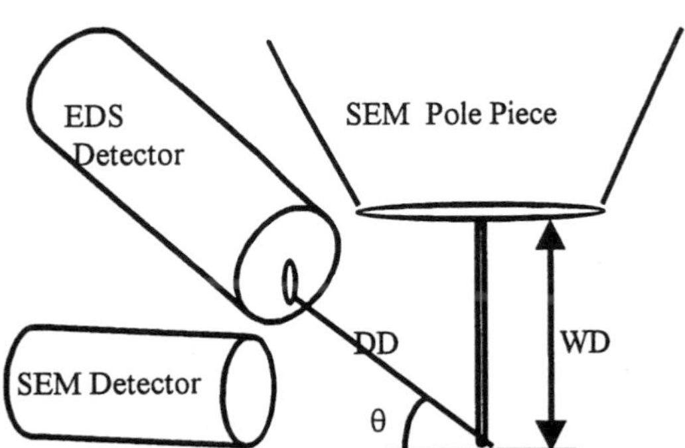

Figure 11.5 SEM chamber configuration. The point at which the EDS detector at its given angle of inclination intersects the beam of electrons from the SEM defines the optimal working distance (WD) for X-ray analysis. The distance from the face of the EDS detector to the intersect point is the detector distance (DD). The angle between the surface of the sample and a line from the sample to the detector face is the take-off angle (θ)

The sample topography also has an effect on the collection of X-rays. Raised features on the surface of the sample may block X-rays from reaching the detector. Pits and crevices may also yield X-rays from deeper within the sample than expected. Ideally, we would like to work with flat polished samples but that option is often not available. Surface imperfections may not be evident in a secondary electron image. However, the backscatter topographical image will show a dark area from which no electrons are being detected. Remember that X-rays behave like backscattered electrons in the sense that they travel in straight lines. The backscatter image can serve as a good indicator of what the X-ray detector is seeing.

11.6 Sample preparation methods and materials

There are several methods used by the forensic community for the preparation of paint samples for analysis by SEM/EDS. The common goal is to expose and present each individual layer of a multilayered chip for analysis. Some examiners prefer to embed intact chips in a resin to form a block. This block can be cut to expose the cross-section of the chip for polishing to yield a smooth surface suitable for analysis. Other examiners forgo the embedding step and simply mount a paint chip on edge on a sample stub to expose the cross-section to the SEM beam. Still others choose to prepare paint chips by exposing portions of each layer in a stair-step fashion using a scalpel or other sharp cutting tool. This results in a chip that resembles a small staircase when mounted flat. It is also common to excise a thin peel of each layer and mount the thin peels on a stub.

When comparing questioned and known samples, it is important that the samples be prepared and analysed in the same manner and under the same conditions. If differences are introduced during sample preparation, differences can be expected in the analytical results. The choice of sample preparation technique will depend on the sample size and condition, the equipment available in the laboratory, the skill level of the examiner in mechanically manipulating small samples, and personal preference. Each of the listed methods has its relative merits and disadvantages, but no matter which one is chosen, the quality of the results will be proportional to the care taken in preparation of the sample.

11.6.1 Embedding and polishing

Embedding and polishing is lauded as the ideal in SEM/EDS sample preparation as it will yield a sample with a flat, smooth surface. A variety of embedding materials for the microtomy of paint are listed by Derrick [10]. However, suitable resins for polishing are usually somewhat harder than those used for microtomy. Samples can be polished using an abrasive of very fine particle size such as a diamond paste. The preparation of flat, smooth surfaces is just one key to obtaining reproducible results. Embedding both known and questioned chips in the same block will also aid in controlling sample geometry. Care must be taken to ensure that both samples are in the same orientation and are mounted perpendicular to the electron beam in the SEM. Tilted samples can cause the electron beam to penetrate into an adjacent layer resulting in the simultaneous analysis of two or more layers. Polishing both embedded samples at the same time should leave them with the same smooth finish. Embedding paint chips also significantly reduces the amount of sample manipulation performed by the examiner and therefore the likelihood that the sample will be lost or destroyed.

The obvious disadvantage of this technique is the difficulty in reclaiming the sample from its resin prison. Once the sample is permanently mounted, it is difficult to retrieve it intact for other examinations. For those who run PGC on individual layers, this is obviously not the appropriate sample preparation technique. However, with a large amount of questioned sample, this will not be an issue. It is also not an issue for those who prepare an embedded sample and then use a microtome to prepare thin sections for microscopy and infrared microspectroscopy. The remaining block can be used for SEM/EDS and for colour analysis by microspectrophotometry.

The greatest disadvantage of this technique is peculiar to multilayered paint samples. Thin layers present a very limited area for analysis. It is not uncommon for layers in an automotive coating system to be less than 10-μm thick. The average thickness of a coloured basecoat has decreased over the past decade as automotive clearcoats have been improved to provide better physical and chemical protection. The advent of colour-coordinated primers has also allowed the automotive industry to reduce the thickness of the expensive coloured basecoat and still achieve a great-looking finish. The sanding of paint during the vehicle refinishing process can also leave behind thin layers of both finish coats and primers.

When performing SEM/EDS analysis, the examiner must take care to ensure that the electron beam does not leave the physical boundaries of the layer being analysed. The examiner must also take into consideration the volume of sample being excited outside of the visible window being targeted. In order to ensure that only a single layer is being analysed and that the excited volume is not spilling over into the adjoining layers, the beam raster window must be kept well away from the layer interfaces. This usually results in long, thin areas of analysis. This in turn can lead to homogeneity problems. Extenders in architectural paints and primers, as well as metal flake or other effect pigments in finish coats, can have large particle sizes. In order to encompass an overall 'average' of the elemental constituents of the paint layer, a large area must be sampled. Gardiner reported that homogeneity problems were seen when raster areas were reduced to squares of 30 micrometres per side in architectural paints [11]. As a general rule of thumb, rectangular areas measuring at least 50–100 micrometres per side will yield reproducible results. This will vary from paint to paint and can be assessed by performing multiple analyses on a layer using successively smaller sample areas until results are no longer reproducible. One possible solution to overcome the problem of the thin layer is to collect multiple spectra along the length of the layer and to mathematically average them using the spectrometer software.

Another potential pitfall to be avoided can occur during the polishing phase. Care must be taken to polish in a direction parallel to the layers. If a back and forth or circular polishing motion that crosses over the layers is used, one layer may be smeared into the next, thereby creating interlayer contamination.

11.6.2 Unembedded cross-sections

For those who attempt to place unembedded questioned and known paint chips on edge on a sample stub, they will encounter the same challenges as described for embedded cross-sections. In addition, the chance of samples being tilted in relation to the electron beam as well as to one another is increased. It is also extremely challenging to hold and polish paint chips that are not encased in a block of embedding medium.

11.6.3 Stair-step preparation

The method referred to as the 'stair-step' requires the greatest skill to perform; however, with a little practise anyone can become adept in its use. This technique requires a stereomicroscope, a sharp cutting implement such as a #11 scalpel blade, a pair of small forceps and a pair of steady hands. While holding the paint chip flat on a glass microscope slide, begin by making an angled cut on one corner starting on top slightly away from the edge of the chip and ending at the bottom edge. This will form a bevelled edge on the chip. From here, the top layer can be peeled back or carved away to expose the underlying layer. The carving is continued until an area of each layer is exposed in a stair-step fashion. The advantage to this technique is that a large flat surface area can be exposed for analysis (Figure 11.6). This avoids the homogeneity concerns experienced using cross-sections. While peels of each layer are being carved away, they can be retained for analysis using other techniques such as IR spectroscopy or PGC.

One disadvantage of this technique again results from thin layers. Layers are stacked on top of one another and the electron beam may penetrate through a very thin layer into the layer below it. Therefore, it is important to prepare the questioned and known samples with an equivalent thickness to their 'steps'. When interpreting data, it is important to remember that there may be contributions from two layers.

11.6.4 Thin peels

The preparation of thin peels of individual layers falls somewhere in-between embedding and stair-stepping on the sample manipulation scale. This technique avoids penetration into underlying paint layers or intrusion of exited volumes into adjacent layers. In addition, sample geometry is controlled by mounting a single layer flat on a platform.

As always, care must be taken to prepare questioned and known samples of the same thickness. The greatest danger associated with thin peels is the failure to detect low-level components. Because the electron beam is interacting with a thin layer of sample rather than an 'infinite' thickness, the number of atoms of trace elements available to interact with the beam is reduced. Long analysis times must be used to ensure that these low-level elements have time to generate enough X-rays to be detected above the sample back-

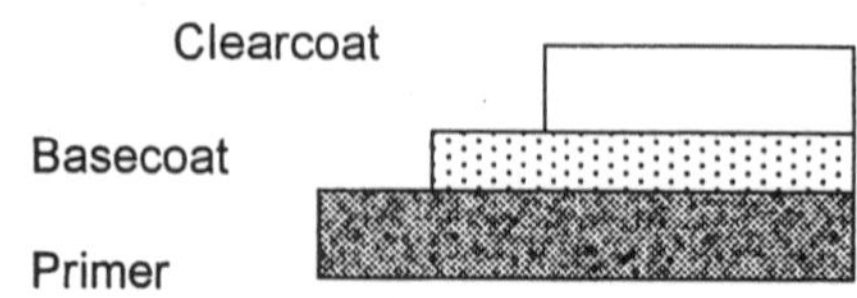

Figure 11.6 Stair-step carving of a three-layer paint chip to expose each layer for EDS analysis

ground. These low-level constituents are important to the ability to discriminate between similar paint formulations as they often arise from additives or tinting pigments. Discrimination power is decreased if only the more common major pigments or extenders are observed and compared.

11.6.5 Mounting materials

Aluminium sample stubs, carbon sample stubs and aluminium sample stubs topped with carbon planchettes are the most common specimen platforms for use in SEM/EDS analysis. If penetration of the electron beam through the sample is suspected, it is better to use a carbon base so that aluminium is not detected. Aluminium may be the element that provides discrimination between paint samples, whereas carbon will be present in all polymeric coatings. Samples can be secured to the stub by using a variety of conductive materials. Carbon- or aluminium-based paints (e.g. Electrodag 502) that contain little organic binder can act as an adhesive to 'glue' samples to the stub. Carbon tape or adhesive discs (e.g. carbon conductive tabs) are more convenient to use as they alleviate some of the manipulation required with the liquid adhesive. For larger items, a plastic conductive carbon cement (Leit-C-Plast™) may be useful. This material has the texture of putty. Any mounting material used should be analysed separately under the same conditions as the paint to determine what, if any, additional elemental information may be coming from the mounting material. It is not uncommon to find trace elements in mounting paints or tapes.

11.6.6 Coating materials

In a standard high-vacuum SEM, paint samples must be coated to make them electrically conductive. If no path to earth is available, an accumulation of electronic charge will build up on the sample surface. This charge interferes with both sample imaging and efficient EDS analysis. The charge will repel or deflect the electron beam rather than allowing it to penetrate the sample. Beam deflection can prove to be detrimental particularly when an EDS examination is performed with the SEM in a spot mode as opposed to a raster mode. When an examiner is operating in a spot mode, a frozen image is typically displayed on the SEM screen. Sample charging may cause the beam to be deflected far from the spot the examiner visualises on the screen.

The choice of materials used to coat samples for SEM imaging may be different from the materials used to coat samples for EDS examinations. In order to achieve good resolution in a high-magnification image, a multilayer system such as carbon and gold–palladium can be used. This will allow a very thin coating which will effectively dissipate the charge on a sample and not obscure fine detail. However, a gold or a gold–palladium alloy coating is not recommended for elemental analysis as the gold and palladium may interfere with the X-ray analysis of a sample. Practically speaking, the forensic examination of paints by SEM/EDS does not require high-magnification images. Owing to homogeneity concerns, much forensic work is performed at magnifications of 1000× or less. Because we are not concerned with obtaining high-quality, high-magnification images, we can tolerate a coating that may obscure fine detail. Coating with carbon alone is acceptable for most paint examinations.

The most common methods of sample coating include sputter coating, thermal evaporation and spray coating. The first two methods require a sample coater that functions by applying a current and heating the coating material. The heated carbon rod or other coating material will begin to evaporate and deposit itself on the cooler surface of the sample. One problem which can arise from thermal evaporation or sputter coating is that the heat required to evaporate the coating material may damage the sample. If no sample coater is available, an alternative method can be used. After the sample is adhered to the stub, it can be coated using a conductive graphite spray (e.g. Aerodag G) to lightly cover the sample. Care should be taken when using a spray coating to ensure it is applied to create a thin, even coating.

Although carbon coating (spray or vacuum) will require a thicker layer to achieve good conductivity, its use will not provide any elemental contributions to the X-ray spectrum such as gold or palladium. It should be noted that phosphorus has been detected in thick coatings applied using Aerodag G. Care should be taken to keep any coating as thin as practical. A thick coating will absorb a portion of the beam and block it from reaching the sample while also absorbing the low-energy X-rays emitted from the sample.

11.7 Practical applications of SEM/EDS

The primary use of SEM/EDS in a forensic paint examination is for comparison and discrimination. A less frequent use is as an aid to the identification of pigments and extenders in the sample. The identity of elements contained in the sample can be used in conjunction with data from other techniques to identify or classify pigments and extenders. This information can be used when trying to assess the significance or uniqueness of a particular paint. The use of EDS for comparison purposes consists of an evaluation of both qualitative results (which elements are present) and quantitative results (what are their relative amounts). To appropriately evaluate these results, one must learn to recognise what is truly a significant difference between samples. A significant difference is sufficient reason to conclude that two samples did not originate from the same source. Repetitive analyses are usually required to demonstrate the range of variation that can be seen within one sample. If an element or the detected amount of an element in the questioned sample does not fall within the range of variation demonstrated by the known sample, a significant difference has been observed. Questioned samples that are less than ideal (e.g. contaminated, insufficient size or poor condition) must be evaluated accordingly. Care must be taken to avoid false inclusions and exclusions.

11.7.1 Collection parameters

The first step in the elemental analysis of a paint sample is the acquisition of a spectrum. In order to evaluate the elemental profile of a paint, the examiner must be assured that all the detectable elements present will be observed. This will happen only if the appropriate accelerating voltage is chosen on the SEM. Higher atomic number elements require more energy to eject the K- or L-shell electrons. However, electrons having too much energy will not efficiently excite the lower atomic number elements. The most efficient excitation occurs when the electrons have approximately 1.5 to 2.5 times the energy of the X-ray line observed. To ensure that all elements present are observed, multiple collections may be taken using both high and low excitation energies. When using a lower accelerating voltage, keep in mind that higher atomic weight elements may be visible because of their lower energy L or M lines. It is recommended that the presence of these elements be confirmed by looking for the higher energy lines using a higher accelerating voltage. For most paint samples, the use of an excitation voltage of 20 kV will prove to be appropriate for an initial analysis.

After selecting an accelerating voltage, the sample position should be optimised along the Z axis. Remember that the sample surface should sit at the intersect point of the electron beam and the line perpendicular to the face of the X-ray detector. The distance between the face of the detector and the sample can be varied by moving the detector along the axis of the elevation angle of the detector (Figure 11.5). Usually, the closer the detector is to the sample, the more efficient the X-ray detection.

Next, the count rate should be adjusted to balance peak resolution with examination time. Higher X-ray count rates result in a loss of spectrometer resolution. Typically, the beam current is adjusted to establish a count rate that does not exceed a dead time of 30 to 40 per cent on the spectrometer. Adjusting the spot size controls the beam current. As the spot size is increased, more X-rays are generated, the spatial resolution of the microscope decreases and the size of the excited volume within the sample increases. A decrease in the spot size reduces the number of X-rays generated and collection time for the X-ray data can become unacceptably long.

The length of the collection time should be chosen so that the spectrum has smooth, well-formed peaks. Some rules of thumb for length of count time include: (a) the spectrum should contain at least 100,000 total counts; (b) the largest peak in the spectrum should contain at least 10,000 counts (exclusive of carbon and oxygen); or (c) enough counts should be collected so that the smallest peak in the spectrum is Gaussian shaped and somewhat smooth. The pitfall to be avoided is using a short collection time that will not allow trace elements to appear before collection is ceased. Another common mistake is the use of the same preselected count time for every type of paint. Extender-rich primer layers may require only a few moments to generate a spectrum with good signal-to-noise ratio. Automotive clearcoats and coloured basecoats usually have a much lower pigment-to-binder ratio, which will produce a spectrum with a higher background (bremsstrahlung). In order to achieve a good signal-to-noise ratio for all elements detected, longer collection times will be necessary.

11.7.2 Peak identification

Examination of a spectrum can begin by first identifying the most prominent peak and all of that element's associated peaks. Most spectrometer software will automatically identify peaks, but manual confirmation is still advisable. A number of charts listing the X-ray energy lines for each element are also available. These charts list the peak energy and the relative intensity of associated peaks. For example, when a copper (Cu) Kα peak at 8.041 keV is identified, the associated Kβ peak at 8.907 keV having approximately 20 per cent the intensity of the Kα should also be seen. The copper L line will also be present at 1.1 keV. Once all lines for that element are identified, the examiner should progress to the next most prominent peak, continuing until all peaks are identified. The order of peak identification is not critical as long as it is thorough and correct.

Throughout the peak identification process, the examiner must be aware of overlapping peaks and false peaks. Consideration of the operating conditions and the type of sample being examined is also helpful when sorting through peak identifications.

11.7.2.1 Overlapping peaks

When performing peak identifications, be aware that many elements have lines that overlap with one another. Sometimes the peak from one element will obscure the peak from another. Some of the most notorious overlaps encountered in paint analysis are titanium/barium (Ti/Ba), sulphur/lead/molybdenum (S/Pb/Mo), chromium/manganese (Cr/Mn), iron/manganese (Fe/Mn) and nickel/zinc (Ni/Zn). Most spectrometer software includes features that perform peak deconvolutions and will allow peaks to be stripped away from one element to see if there is any contribution from another. The observation of a broadened or non-Gaussian-shaped peak is an indication of overlapping peaks.

11.7.2.2 False peaks

The most common false peak is the escape peak. This peak is generated when a silicon (Si) $K\alpha$ X-ray is generated from the detector crystal and then escapes from the detector rather than being reabsorbed. When the Si X-ray escapes, it carries away its associated energy of 1.74 keV. This will cause a signal or pulse to be generated which is 1.74 keV less than the original incoming X-ray. Typically, a small peak, less than 2 per cent of the major peak, is noted. These escape peaks can be observed for any peak having an energy greater than the excitation energy of an Si atom (1.84 keV). Most EDS peak identification software includes the ability to identify and even strip away escape peaks.

Another type of false peak routinely encountered is a pulse pile-up or sum peak that occurs when the incoming count rate is so high that the detector cannot correctly identify a single incoming X-ray event. One of the most common examples of this occurs when high count rates are used in the examination of a metallic paint layer. When two incoming aluminium (Al) $K\alpha$ X-rays (1.48 keV) strike the detector, a sum peak with an energy of 2.96 keV is recorded. This peak energy is consistent with argon. Obviously, an examiner would not expect to detect argon gas in a paint layer. Sum peaks may occur with any combination of X-ray energies. In a sample with high Cu and Fe levels, a sum peak of the Cu $K\alpha$ (8.04 keV) and the Fe $K\alpha$ (6.40 keV) may occur at 14.44 keV. If sum peaks are present, acquiring a new spectrum using a lower count rate should make these peaks disappear. A titanium sum peak can be seen in Figure 11.7.

Stray radiation peaks may also be observed. Knowledge of the SEM chamber geometry and the position of the EDS detector assists in evaluating the source(s) of stray radiation. The X-ray detector observes radiation occurring in a solid cone area projecting from the face of the detector. Backscattered electrons or X-rays may strike other objects within the SEM chamber causing additional X-rays to be generated. If these X-rays reach the detector, they may be interpreted incorrectly as having come from the sample. An example of this could be a Cu strip placed on the SEM stub along with the sample. The Cu strip is used as a quality control step to ensure that the system is properly functioning prior to the examination of the paint samples. However, during examination of the paint sample, electrons or X-rays generated from the sample may strike the Cu causing the emission of Cu X-rays. While stray radiation peaks are difficult to identify as stray radiation peaks, changing the position of the sample stub relative to the detector may cause these peaks to

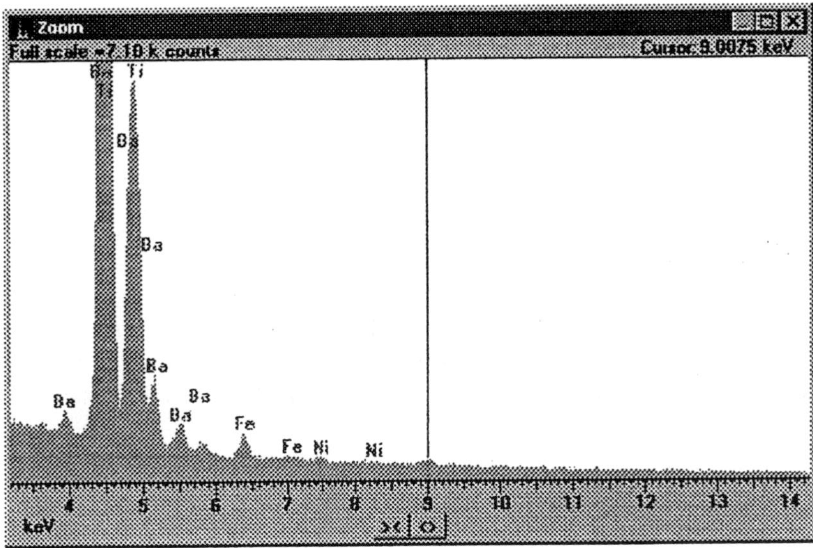

Figure 11.7 The cursor position at 9.007 keV marks a sum peak from two Ti Kα X-rays (4.510 keV)

increase, diminish or disappear. The examiner needs to know what elements may originate from the SEM chamber itself (e.g. molybdenum from the pole piece) and what stray radiation interference may be caused by other samples in the chamber.

11.7.3 Comparison of qualitative results

Once the elements have been identified for both the questioned and known samples, direct comparisons can begin. There are many occasions where simple qualitative analysis can provide useful information in forensic paint comparisons. If the direct comparison of questioned and known samples reveals an element whose presence can be repeatedly demonstrated in one sample and not the other, this is considered a significant difference (Figure 11.8).

Sometimes it is not the presence or absence of an element that will distinguish between two samples, but the distribution of the elements within the sample. A technique called X-ray dot mapping allows the examiner to look at the physical distribution of chosen elements within the sample. This is achieved by the placement of a coloured dot in the area of origin for each element of interest as the characteristic X-rays are detected by the EDS system. Two paint samples may have the same chemical formulations and elemental content, but a physical difference such as pigment settling within a layer might be evidenced by this technique.

11.7.4 Comparison of quantitative results

In order to obtain reliable quantitative results using EDS, clean, flat, smooth samples and appropriate standards must be used. Appropriate quantitative standards for paint would be

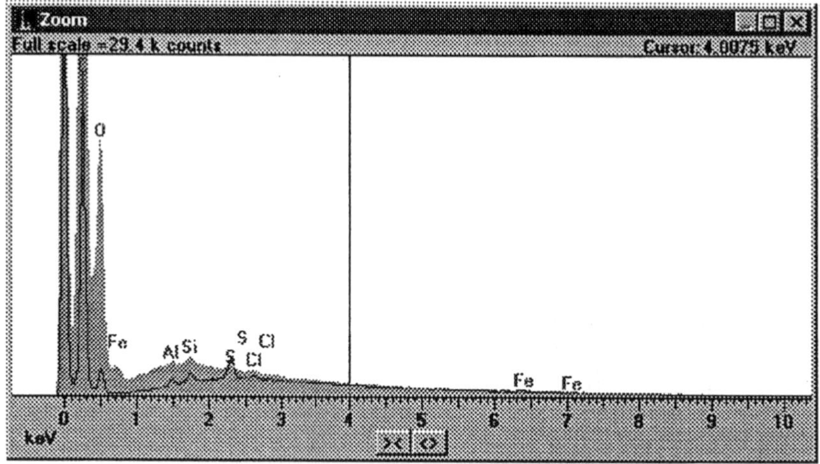

Figure 11.8 These spectra represent two clearcoats examined at 20 kV. Note the Fe L line (0.704 keV) present in the solid spectrum and the S K line (2.307 keV) present in the line spectrum. The presence of either of these elements in only one of the clearcoats is a significant difference

other paints containing known quantities of each element represented in the questioned sample. Without appropriate standards to recreate the interaction phenomena that take place within the sample, it is impossible to accurately translate X-rays detected into exact quantities of elements present. Some of these phenomena include: differences in X-ray behaviour which are dependent on the atomic number of the element; the number of X-rays absorbed before escaping the sample; and the number of X-rays resulting from secondary fluorescence within the sample. Secondary fluorescence is the production of one X-ray through excitation of an atom by another X-ray. Most software programs will allow the generation of quantitative or semi-quantitative data for each element. Several mathematical correction programs are used by different EDS manufacturers. The two most commonly used are ZAF and phi-rho-z. These basic models take into account the average atomic number of the sample, the effects of absorption, and the effects of fluorescence. There are different models of ZAF and pi-rho-z with corrections modified to fit the sample size and geometry (i.e. bulk sample or small particles). It is important to know which correction model a particular instrument uses. Most EDS software will also provide quantitative results without using any standards at all. When using standardless calculations, Goldstein *et al.* [7] recommend that elements are characterised as either trace (less than 1 weight per cent), minor (1–10 weight per cent) or major (greater than 10 weight per cent) rather than by attempting to report their numerical quantities.

Considering the variety of factors involved, the most useful practice when performing paint comparisons is to compare relative peak heights or areas under the peaks. These values can be generated in a numerical format. The area of each peak is simply a representation of how many X-rays were detected from a particular element; it is not directly proportional to the quantity of that element in that matrix. It is appropriate to evaluate the size of the peaks (Figure 11.9) when comparing questioned and known samples without attaching a quantitative value (weight per cent) to that element.

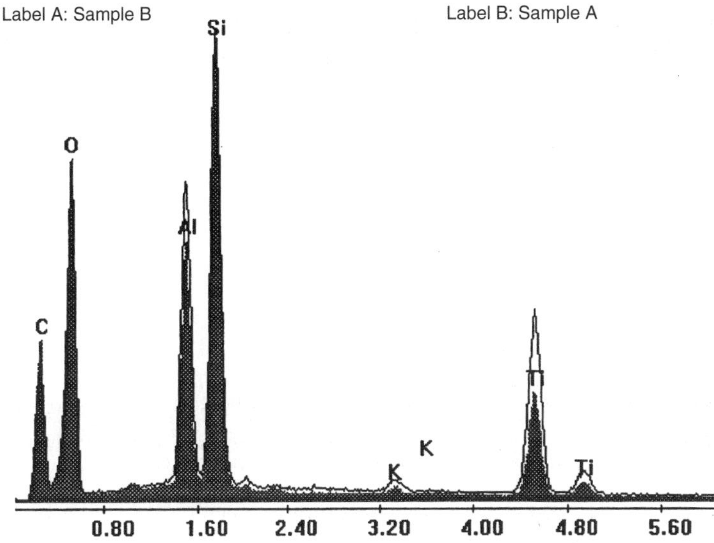

Figure 11.9 Based on information from the manufacturer, Samples A and B have different quantities of TiO_2 in the paint: Sample A (solid spectrum) has 6.0% TiO_2 and Sample B (line spectrum) has 10.7% TiO_2

When comparing paint samples by SEM/EDS, it is important to remember that element peak heights within a layer may vary from area to area. This can be caused by topography or lack of homogeneity within the layer arising from incomplete mixing or pigment settling. It can also be seen when an area of a sample is analysed that is too small to encompass an average of all the components in the layer. For example, a first analysis might be performed on top of a particularly large pigment particle. The next might be between the particles. Differences can also arise from contamination that has not been scraped or polished off the surface of the sample. Still other differences can arise from tiny particles of one layer being left behind on another layer, especially when using the stair-step preparation technique. It is important to ensure that a clean and well-prepared sample is used.

11.7.5 Component identification

It is sometimes advantageous to use the spot analysis mode of the SEM to focus on a particular particle or inclusion within a paint sample. The elemental information obtained from that isolated area or particle may aid in its identification.

As noted earlier, the complementary nature of IR and SEM/EDS serves as a check and balance system on data interpretation. For example, if SEM/EDS analysis of an automotive primer revealed a small magnesium peak accompanied by a larger aluminium peak and an intense silicon peak, this would typically be interpreted as an indication of a clay extender. However, careful review of the infrared spectrum of that same primer may demonstrate the presence of a small amount of talc along with the clay. Conversely, because of differences in the limits of detection of the two techniques, the presence of a low level of barium sulphate extender may go undetected by IR and yet the presence of barium and sulphur may be noted in the X-ray spectrum. These types of logical deductions are complicated by the fact that not all elements detected in the SEM/EDS analysis of a coating arise from inorganic pigments or extenders. Elements from organic constituents in the binder or additive portion of a coating will also be revealed.

11.7.6 Reference materials

Familiarity with the types of materials used as pigments and extenders in the coatings industry is an obvious prerequisite to their identification in a sample. This information can be gained from a variety of sources. Samples of pigments and extenders commonly used in automotive and architectural coatings can be acquired from their producers. These neat samples can be analysed and their spectra stored for future comparison with questioned pigments or extenders *in situ*. The classic printed reference in the field is *The Forensic Examination of Paints and Pigments* by Crown [12]. Although dated, this reference still stands as the only compendium listing common pigments and extenders, their elemental indicators and their usage. More current pigment information can be gleaned from short publications in the Federation Series on Coatings Technology [13–15] and from writings by Morgans [1], Poth [16] and Ryland [17]. Publications that demonstrate the ability to decipher which extenders, inorganic pigments and even organic pigments are present in a coating include those by Ryland [17], McCrone [18], Thornton [19] and Suzuki [20–25].

11.8 Other methods of elemental analysis

Forensic paint examiners sometimes employ methods of elemental analysis other than SEM/EDS. The choice of method(s) must take into consideration the strengths and weaknesses of each technique. Table 11.5 compares a few of these methods. It should be noted that extended-range FTIR (e.g. 4000 to 220 cm^{-1}) is also very useful for the identification of inorganic components.

11.8.1 X-ray fluorescence spectrometry

X-ray fluorescence (XRF) spectrometry is a technique similar to SEM/EDS and is also capable of non-destructive analysis of a small paint sample. The primary difference in the techniques is the use of an X-ray beam rather than an electron beam as an excitation source. XRF is often mistaken as a suitable replacement for SEM/EDS, but in fact provides information that is complementary to SEM/EDS.

XRF is more sensitive to higher atomic weight elements than SEM/EDS and less sensitive to the lower atomic weight elements that are commonly found in extender pigments. This makes XRF a good technique to examine paint components such as drier metals in alkyd enamels (e.g. Zr, Co, Mn) [22], low-level inorganic colouring pigments or organic colouring pigment metal complexes (e.g. Cu in phthalocyanine blue). The use of an X-ray beam eliminates the need for samples to be made electrically conductive by coating, and also provides a more energetic source (e.g. 50 kV or higher). The exposure to higher excitation potentials results in the detection of higher-energy K or L lines which makes the identification of heavy elements easier or even possible.

Table 11.5 Comparison of SEM/EDS, WDX and XRF

	EDS	WDX	XRF
Information gained	Elemental composition	Elemental composition	Elemental composition
Detection limit	0.1 wt%	0.01 wt%	0.01 wt%
Destructive technique	No	High beam current damage	No
Minimum analytical diameter	3 to 8 μm	5 to 10 μm	Poor (bulk sample technique)
X-ray energy resolution at Mn Kα	~135 eV	~2 to 20 eV	~160 eV

When dealing with multilayered coating systems, the ability to analyse small samples and to select single layers and isolated areas of a sample for analysis is crucial. XRF cannot rival SEM/EDS in this area; however, design advances made during the 1990s have brought it closer. Some modern XRF instruments offer focused X-ray beams as small as 30 μm in diameter. Older instruments had large beam diameters that could be collimated down to only 1 mm in diameter. Collimation of the beam rather than focusing reduces its intensity and lengthens collection times. Even though smaller beams provide better spatial resolution, X-ray beam penetration into adjacent layers is still a problem. An X-ray beam will penetrate further into a sample than a comparable electron beam. When using methods that tend to give a compilation of elements in all or many layers of a paint chip, care must be taken that the thickness of layers of the questioned and known samples are like one another. Varying elemental ratios may be detected in the two spectra as a result of slight layer-thickness variations as opposed to differing elemental concentrations in the respective layers. The possibility that two chips having different layer compositions could give the same bulk analysis results must always be considered.

11.8.2 Wavelength dispersive X-ray analysis (WDX)

A wavelength dispersive detector is mounted in an SEM chamber in a manner similar to the EDS detector. As the name implies, WDX detects X-rays according to their wavelength rather than their energy. The energy of X-rays emitted from a sample is inversely proportional to their wavelength. To detect the X-rays, a WDX spectrometer uses a series of crystals of known d-spacing. The crystals of known d-spacing work in concert with a detector which is moving through an arc within the detector. When an X-ray enters the WDX spectrometer, it is diffracted by the crystal following Bragg's equation ($\lambda = 2d\sin\theta$). When the detector is located at the appropriate angle relative to the crystal, the X-ray of wavelength λ is detected.

The resolution and sensitivity of WDX is better than EDS because WDX detects a narrow range of X-ray wavelengths at one time. WDX will allow the identification of X-ray lines that overlap in EDS such as Ba and Ti (Figure 11.10). However, in order to generate a detectable signal, WDX requires a much higher beam current than EDS. This higher beam current will cause damage to paint samples. In addition, the time required to complete a scan in the range of interest might be of the order of several minutes as opposed

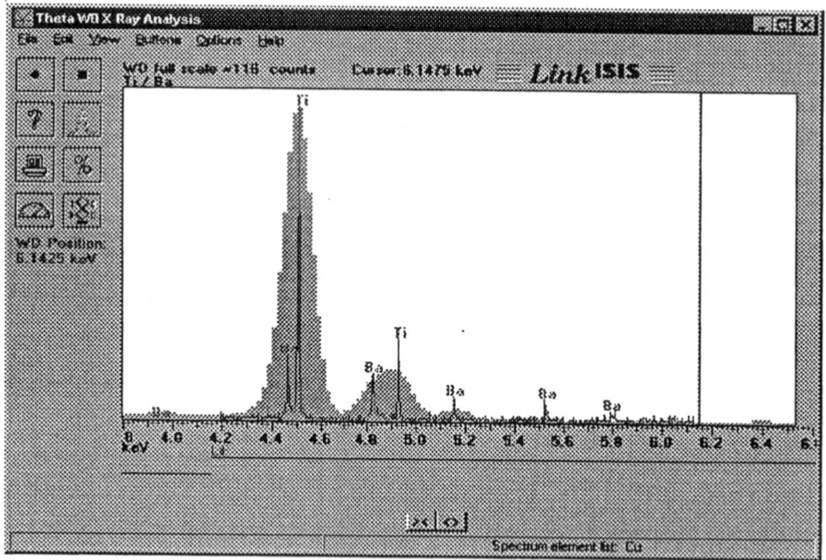

Figure 11.10 Barium L lines and titanium K lines are overlapped and unresolved by EDS in the shaded spectrum. In the solid line WDX spectrum, these elements are clearly observed as separate, distinct peaks

to several seconds for EDS. WDX is a technique that can allow better resolution of the X-ray lines and, therefore, more specific interpretation of the qualitative elemental data.

11.8.3 X-ray diffraction

X-ray diffraction (XRD) is another method occasionally used by forensic paint examiners. This technique analyses the crystalline structure of the material rather than its elemental content and has the ability to provide definitive identification of inorganic components. Like WDX, XRD theory is based on Bragg's equation ($\lambda = 2d \sin \theta$). In this technique, X-rays of known wavelength (λ) strike a sample and the detector measures the intensity of diffracted X-rays at each angle through an arc. Bragg's equation is solved for the d-spacing and that information is used to identify the crystalline material in the sample.

Paint samples examined by X-ray diffraction are typically small fragments that are mounted on the tip of a drawn glass capillary tube using a petroleum jelly or other non-crystalline adhesive. Relatively speaking, typical XRD samples are much larger than those examined by EDS. In addition, the time required to perform an XRD examination is measured in hours. The notable application of XRD to forensic paint examination is the differentiation of the two types of titanium dioxide (rutile and anatase).

11.9 Discrimination

The primary goal of a forensic paint comparison is to utilise an analytical scheme that will provide the highest level of discrimination between similar coatings. A typical examination begins with the morphological characterisation of the paint using a stereomicroscope. If a coating system has a complex layer structure, the examiner may decide that the use of instrumental analysis will not provide further discrimination. That scenario is far from the norm and most paint examinations necessitate the inclusion of some type of elemental analysis. How much additional value is provided by incorporating SEM/EDS into an analytical scheme is not, however, clear-cut. The importance of using elemental analysis in a paint comparison is recognised worldwide, as noted in the Forensic Paint Analysis and Comparison Guidelines written by the Paint Subgroup of the Scientific Working Group on Materials Analysis (SWGMAT) [26]. Several discrimination studies have been performed that will assist an examiner in making the types of judgements necessary to maximise discrimination using the least amount of time and effort [27–31].

11.10 Summary

The use of SEM/EDS has proven to be valuable for the elemental analysis of paints and coatings. The technique is prevalent in crime laboratories throughout the world. The success of EDS can be attributed to its suitability for use on small multilayered samples, its non-destructive nature and the speed and sensitivity of the technique.

As the pigment industry evolves, the emphasis on colour pigment production may continue to shift from a chemical phenomenon to a more physical/optical one. It is possible that one day the variety of elemental constituents found in colouring pigments may be but a fraction of its current abundance. Paint examiners of the future will have to adapt their analytical schemes to the products at hand. However, for today and several years to come, SEM/EDS is and will continue to be a crucial component of a forensic paint examination.

11.11 Acknowledgements

Many thanks to Beth Horton, Lynn Black and Sue Zurschmiede for their careful review of this chapter. A special note of thanks is also given to the members of the paint subgroup of SWGMAT. Led by Scott Ryland, they have demonstrated a constant willingness to share information and great tenacity while performing their arduous task of document writing. Their efforts will benefit the forensic paint community worldwide.

11.12 References

1. Morgans W. M. (1990). *Outlines of Paint Technology*, pp. 9–133. Halsted Press, New York.
2. Mystic – the look of car colors to come. http://www.basf.com/newsinfo/archive/1996/mystic1.html
3. Mystic color car paint – how does it work? http://www.basf.com/newsinfo/archive/1996/mystic2.html
4. A coat of many colors. http://www.dupont.com/corp/products/dupontmag/97/chromalusion.html
5. Wanlass M. (1997). Holographic pigments add new dimension to paint. *Paint and Coatings Industry* **August**:48–49.
6. Scott V. D., Love G., Read S. J. B. (1995). *Quantitative Electron-Probe Microanalysis*. Ellis Horwood, New York.
7. Goldstein J. I., Newburt D., Echlin P., Joy D., Romig A. D., Lyman C., Fiori C., Lifshin E. (1992). *Scanning Electron Microscopy and X-Ray Microanalysis*. Plenum Press, New York.

8. Postek M. T., Howard K., Johnson A., McMichael K. (1980). *Scanning Electron Microscopy, A Student's Handbook.* Michael T. Postek, Jr. and Ladd Research Industries, Inc.
9. Mathiew C. (1999). The limitations of X-ray microanalysis in the high pressure SEM. *The Americas Microscopy and Analysis* **July**:11–13.
10. Derrick M. R. (1995). Infrared microspectroscopy in the analysis of cultural artifacts. In *Practical Guide to Infrared Microspectroscopy* (ed. H. Humecki), pp. 294–298. Marcel Dekker, New York.
11. Gardiner L. R. (1981). The homogeneity of modern household paints using the scanning electron microscope-energy dispersive X-ray analyser (SEM-EDXA). Home Office Central Research Establishment (HOCRE) Report No. 408.
12. Crown D. A. (1968). *The Forensic Examination of Paints and Pigments.* Charles C. Thomas, Illinois.
13. Braun J. H. (1993). *Introduction to Pigments.* Federation of Societies for Coatings Technology, Pennsylvania.
14. Lewis P. A. (1988). *Organic Pigments.* Federation of Societies for Coatings Technology, Pennsylvania.
15. Smith A. (1988). *Inorganic Primer Pigments.* Federation of Societies for Coatings Technology, Pennsylvania.
16. Poth U. (1995). Topcoats for the automotive industry. In *Automotive Paints and Coatings* (ed. G. Fettis), pp. 127–130. Weinheim, New York.
17. Ryland S. G. (1995). Infrared microspectroscopy of forensic paint evidence. In *Practical Guide to Infrared Microspectroscopy* (ed. H. Humecki), pp. 163–243. Marcel Dekker, New York.
18. McCrone W. C. (1993). The scanning electron microscope supplemented by the polarized light microscope and vice versa, *Scanning Microscopy* **7**:1–4.
19. Thornton J. I. (1982). Forensic paint examinations. In *Forensic Science Handbook* (ed. R. Saferstein), pp. 529–571. Prentice Hall, New Jersey.
20. Suzuki E. M. (1996). Infrared spectra of U.S. automobile original topcoats (1974–1989): I. Differentiation and identification based on acrylonitrile and ferrocyanide C=N stretching absorptions. *Journal of Forensic Sciences* **41**:376–392.
21. Suzuki E. M. (1996). Infrared spectra of U.S. automobile original topcoats (1974–1989): II. Identification of some topcoat pigments using an extended range (4000–220 cm^{-1}) Fourier transform spectrometer. *Journal of Forensic Sciences* **41**:393–406.
22. Suzuki E. M. (1997). Infrared spectra of U.S. automobile original topcoats (1974–1989): III. In situ identification of some organic pigments used in yellow, orange, red, and brown nonmetallic and brown metallic finishes – benzimidazolones. *Journal of Forensic Sciences* **42**:619–648.
23. Suzuki E. M., Marshall W. P. (1998). Infrared spectra of U.S. automobile original topcoats (1974–1989): IV. Identification of some organic pigments used in red and brown nonmetallic and metallic monocoats – quinacridones. *Journal of Forensic Sciences* **43**:514–542.
24. Suzuki E. M. (1999). Infrared spectra of U.S. automobile original topcoats (1974–1989): V. Identification of organic pigments used in red nonmetallic and brown nonmetallic and metallic monocoats – DPP Red BO and Thioindigo Bordeaux. *Journal of Forensic Sciences* **44**:297–313.
25. Suzuki E. M. (1999). Infrared spectra of U.S. automobile original topcoats (1974–1989): VI. Identification and analysis of Yellow 3R, Isoindoline Yellow, Anthrapyrimidine Yellow, and miscellaneous Yellows. *Journal of Forensic Sciences* **44**:1151–1174.
26. Scientific Working Group on Materials Analysis (1999). Forensic paint analysis and comparison guidelines. *Forensic Science Communications* **1**:http://www.fbi.gov/programs/lab/fsc/backissu/july1999/painta.htm
27. Massonnet G. (1995). Comparison of X-ray fluorescence and X-ray diffraction techniques for the forensic analysis of automotive paint. *Crime Laboratory Digest* **22**:321–326.
28. May R. W., Porter J. (1975). An evaluation of common methods of paint analysis. *Journal of Forensic Sciences* **15**:137–146.
29. Ryland S. G., Kopec R. J. (1979). The evidential value of automobile paint chips. *Journal of Forensic Sciences* **24**:140–147.
30. Gothard J. A. (1976). Evaluation of automobile paint flakes as evidence. *Journal of Forensic Sciences* **21**:636–641.
31. Tippet C. F., Emerson V. J., Fereday M. J., Lawton F., Richardson A., Jones L. T., Lampert S. M. (1968). The evidential value of the comparison of paint flakes from sources other than vehicles. *Journal of the Forensic Science Society* **8**:61–65.

12

The interpretation of paint evidence

SHEILA WILLIS, JOHN MCCULLOUGH and SEÁN MCDERMOTT

12.1 Introduction

Paint, by which we mean any decorative or protective coating applied to a substrate, is one of the most widely encountered materials dealt with by forensic scientists working in the field of trace evidence. As a means of establishing association between people, places or things it has considerable potential as a source of evidence. Assuming that paint transfer has taken place, that paint is recovered and sent to the forensic laboratory for comparison, and assuming that whatever tests are performed cannot discriminate between the samples, the scientist has traditionally stated that the fragments under consideration match, or are indistinguishable, or some such phrase. As stated by Stoecklein in relation to traffic accidents: "If two multilayered paint fragments with the original layer sequence . . . are undifferentiable after using state-of-the-art examination techniques, it cannot be concluded with certainty that these samples came from the same source. . . . However, the probability is very low that a second automobile with identical morphological and chemical characteristics in its paint layer structure was in the vicinity of the crime scene at the time of the accident. Therefore, the evidential value of such results is nevertheless very significant". [1] The inevitable follow-on questions to this kind of statement, therefore, are 'So what? How significant?', questions which are legitimately asked by investigators, fellow scientists and, ultimately, by a court of law.

The scientist has come up with a set of analytical results but now has to put them in context, in other words to interpret them. The eventual function of the expert witness is to assist the court in an area of expertise not available to the layman, so interpretation of laboratory findings is just as much part of the scientist's work as the analytical and comparative work leading to the production of the laboratory report. As forensic science has evolved, so has the level of interpretation included in scientists' reports.

In the past, interpretation of paint evidence has largely been based on intuitive feelings, on experience and on the application of common sense: the so-called 'classical' approach. In more recent times the wider acceptance in the forensic community of Bayes' theorem and the statistical background which underpins it has led to a growth in interest in applying this approach to a variety of materials, including blood and glass. Bayes' theorem is not new (its originator, the Reverend Thomas Bayes, died in 1761) but it has only relatively recently been applied in the field of forensic science. More recent work [2] has extended this application of Bayes' theorem to vehicle paint.

In this chapter we present the classical approach to paint evidence interpretation and show how the Bayesian methodology can add value to the intuitive reasoning of the classical approach. Both household and vehicle paints are considered, although at present the Bayesian model is more applicable to vehicle paint as more of the necessary population survey work has been done in this area in order to substantiate the conclusions which Bayes' theorem can generate. Implicit in the Bayesian approach is the need to consider alternative scenarios, not just the one put forward by the prosecution side. Its application in the area of forensic science demonstrates the even-handed approach required by the scientist, whose aim should be to find the truth, not to support one side over the other.

12.2 Paint transfer mechanisms

The presence of foreign paint on a surface is generally recognised as being of evidential significance, indicating to a greater or lesser degree of certainty that contact of some kind had occurred between the surfaces in question. All subsequent laboratory comparison and analysis of paint assumes that transfer of paint is a possibility in the case in question, but how does transfer arise in the first place?

For paint transfer to occur, there are two related requirements: proximity and action of some kind. Paint transfer typically arises from direct contact between an object and a painted surface, although other transfer mechanisms, such as splashing or spraying onto nearby surfaces, are also possible.

Direct contact on its own is a necessary but not a sufficient condition for transfer to arise between dry painted surfaces. There also has to be sufficient force applied to remove paint from one or both surfaces. This can be caused by impact or by abrasion and the amount of material transferred will depend on the degree of force applied, the condition of the paint (sound, flaking, etc.), the duration of contact and the nature of the object in contact with the painted surface.

12.2.1 Wet paint

In the case of wet paint, transfer can occur by means of rubbing against a recently painted surface (for instance, in the course of a burglary) or by splashing or dripping (for example, in the case of a criminal damage incident). In such a case the suspect's clothing or shoes may have splashes of paint which can be compared with a known paint sample from a container or from the painted surface itself.

Although transferred wet paint generally has less evidential value than paint with a layer structure, when dry it can function as an individualising feature which adds significance to matching paint flakes. An example of this might be the comparison of vehicle paints with droplets of over-spray paint of a different colour on the surface or between the layers, providing further proof of association of the two samples.

A further source of wet paint is the anti-climb paint which is used to deter burglars. This paint is designed not to dry and provides reduced grip on areas like drainpipes where burglars might try to gain access to premises. This paint may be found on a suspect's hands and clothing.

12.2.2 Dry paint

Most paints examined by forensic scientists are dry flakes with one or more layers. These can be transferred in several ways:

- by flaking from a surface to which the paint is only loosely attached
- by forcing from a surface with a tool
- by impact with another surface.

Loose paint flakes may be transferred to clothing or other items by direct contact with the surface.

In most cases, however, the paint is firmly attached to the substrate and transfer requires additional force. Paint can be transferred to or from tools (e.g. screwdrivers, case-openers or jemmy bars) used in opening windows or doors during burglaries. It should also be noted that toolmarks may also be left on the painted surface by the tool. The possibility of a physical match should not be ignored: one of the authors of this chapter examined a painted iron bar used to break into cars and found a physical match between a paint flake recovered from the floor of one of the cars and the edge of the flaking painted surface of the iron bar.

Traffic accidents are a major source of transferred paint. Paint flakes are usually transferred between vehicles during impact, flakes may be lost from a vehicle's surface and left at an accident scene or they may be transferred to an accident victim's clothing. In addition, dry paint can be impacted or smeared onto another surface, suggesting forceful rather than gentle contact.

12.2.3 Persistence and secondary transfer

In most cases the question of secondary transfer of paint does not arise: paint may be left at a traffic accident scene, embedded in a victim's clothing or smeared on a tool. Wet paint transferred to an item will bond to the surface and so will persist for a much longer period than particulate matter such as paint fragments. Persistence is only an issue if paint is being used to establish whether a link exists between a suspect and a paint source.

While there have been surveys of the number of paint flakes found on clothing, no studies on the persistence of paint flakes on clothing have been published in the literature. The study of Pearson *et al.* [3] related to glass and paint recovered from pockets and turn-ups, not to clothing surfaces. Lau *et al.* [4] found that 86 per cent of outer garments examined had no paint on their surface and that most of the garments which had paint on them, only had one fragment present.

In terms of transfer and persistence the closest analogue is the study of glass particles on clothing. The number of glass fragments recovered from clothing immediately after a window is broken can be well in excess of 100. Most glass fragments are lost in the first half-hour, with a significantly smaller number (both in quantity and size) recovered after 8 hours [5].

Experiments involving glass on clothing show that the possibility of indirect transfer of significant numbers of glass fragments is low: for example, with 25 or more glass fragments on a seat, only one fragment was found to transfer to the trousers of the next person to sit there [6]. The amount of paint found on clothing is generally small in comparison to the amount of glass recoverable [4] and, in common with other extraneous material, one would expect it to be lost in an exponentially decreasing manner over time. If we assume the transfer and persistence characteristics of paint and glass are of the same order of magnitude, the possibility of secondary transfer of paint is lower than that of glass.

12.3 Household paints

Household paints, meaning all surface coverings other than vehicle paints, are considered in a different category when assessing the significance of their matching. Household paint is traditionally considered by forensic scientists to have the potential to provide strong evidence of association. The value of association is based intuitively on the wide variation and types of paint available.

In a 1968 'state-of-the-art' monograph entitled *The Forensic Examination of Paints and Pigments* [7], Crown outlined the analytical techniques available for the examination of paints but made no reference to the significance of matching two samples of paint, following comparison. Most of the work of the forensic scientist involves the comparison of two or more samples.

The significance of 'matching' samples is obviously dependent on the discriminating power of the test used. In paint examinations, more than other types of trace evidence, the discriminating power of the test will vary with the nature of the components of the paint. For example, Pfefferli [8] cites energy dispersive X-ray (EDX) analysis as a useful method in routine paint analysis but acknowledges that in the case of white household paint, especially single-layered paint chips, the results of EDX analysis are of low discriminating power since these paints are rather similar in elemental composition. In a similar way samples may exhibit the same resin composition when examined by infrared spectroscopy but be distinguishable by another technique because of the presence of differing pigments. Also it is acknowledged that colour is a significant discriminating feature yet care must be taken to ensure that metameric colour matches are eliminated. Laing *et al.* [9] recommended the use of microspectrophotometry to discriminate objectively on the basis of colour between sets of household paints. From the above it can be seen that analytical schemes for the examination of paint will vary from sample to sample.

The means by which paint samples are generated in a forensic context are outlined above. The comparison of smears of wet paint results in a low level of discrimination as it is generally not possible to distinguish such paint from any source of that paint manufactured at the same time. The level of significance of a match should then be judged against the volume of such a paint produced. This information is not generally available.

The examination and comparison of dried paint flakes is more usual for the forensic scientist than the examination of wet paint. The physical characteristics of such samples will be partly dependent on the conditions under which the paint film from which the flake originated was formed. Thus it may be possible to distinguish two samples of paint from the same batch because one is more weathered or degraded than another. It follows that matching dried flakes can be more significant than matching wet samples. In both instances, however, the significance of matching is dependent on the commonness of the paint, with white and black paint in general being more difficult to discriminate than coloured paints.

It is widely accepted that multiple layers provide strong evidence of association [10, 11]. An attempt to quantify this statement is difficult. The significance of matching multiple layers is easy to illustrate if we consider the number of options of colours that are available. Gallop and Stockdale [12] report that in daylight, the unaided human eye can distinguish 10,000,000 different colour surfaces. One paint manufacturer claims to be in a position to create up to 50,000 colours based on a selected range of paint types. In the light of these points, a top layer can, at a conservative estimate, be one of 100 distinguishable colours, and a second and third layer likewise. If the arrangement of layers is an independent event the combination can be calculated to have a one in a million chance of occurrence. This approach was outlined by Walls [11].

However, in most instances the situation is less clear-cut. A top-layer red paint is likely to have a similar colour underlayer, making the combination a dependent event. Nevertheless, there is value in the general argument as it is likely that even similarly coloured undercoats are distinguishable and are one of a very wide number of options. Hudson *et al.* [13] were sufficiently convinced of the value of matching paint flakes that they established a paint index to enable them to compare paint flakes recovered from the clothing of suspects taken into police custody in relation to 'break-in' cases. They considered that the flakes are often unique in appearance as they may be built up from coats of paint which may have a wide range of colours and thicknesses and so can be used to connect a particular garment with the scene of the crime.

Saferstein [10] acknowledges that layer structure is important for evaluating the significance of paint evidence, but he goes on to say that there are at present no books or journals that have compiled the information of frequency with which particular combinations of colours are observed to occur so that the forensic scientist is left to his or her own experience and knowledge when making decisions on concluding whether or not paints have a common origin. This is of little consolation to the forensic scientist who is still guided by Kirk and Kingston [14] when they say "statistical analysis provides the criminalist with a basis for his opinion and an evaluation of the likelihood that his testimony reflects the truth, rather than his personal belief or bias".

Tippett *et al.* [15] were among the first to attempt to apply statistical analysis to the problem of assessing the value of matching household paint flakes. They examined 2000 paint samples from buildings and found that over 98 per cent were differentiated by microscopy and solvent tests. An analysis of the results showed that it was possible to calculate the chance against two random samples agreeing, both when only microscopy and solvent tests had been carried out and when more sophisticated methods were used.

Use was made of these findings by one of the authors of this chapter who gave evidence of matching paint flakes in the trial of two suspects apprehended in connection with the explosion on the boat 'Shadow V' which caused the death of Lord Mountbatten and others in 1979. One of the suspects had a number of paint flakes on his clothing, some of which consisted of three layers of green paint, which matched some of the known paint from the boat. A smear of paint matched another type of paint from the boat. The value of these matches was assessed by referring to Tippett *et al.*, that the chance against two random paint samples agreeing in all the examinations carried out is of the order of a million to one.

Much of the recent literature on the interpretation of trace evidence relies on the use of Bayes' theorem to assess the significance of matching samples. To achieve this, surveys are typically used to determine the frequency of occurrence of the material in the environment or on the substrate from which it was recovered. The effectiveness of analytical methods to distinguish samples from different sources continues to be tested. The literature has many references to the effectiveness of analytical methods but there is little on frequency of occurrence surveys and no information on the use of Bayes' theorem to assess the value of finding household paints in various situations.

Castle *et al.* [16] carried out a survey of case openers. They noted wide variation in physical parameters as well as in paint formulations. Pearson *et al.* [3] carried out one of the first frequency of occurrence surveys when they examined 100 suits from a dry-cleaners and recovered a total of 1077 paint fragments from 97 per cent of the suits. More recently, Lau *et al.* [4] carried out a survey of the number of paint flakes on the shoes and outer clothing of high-school students in Canada. One or more paint flakes were found on 14 per cent of the jackets in this study. Yellow paint flakes were the most common type recovered. In one instance a pink fluorescent paint was removed from upper- and lower-body garments of individuals from the same school. The colour was not typical of the more common paint colours encountered. This was interpreted as indicating that the students from this school had come into contact with the same fluorescent pink-painted source in the school. Lau *et al.* cautioned care to ascribing significance to the commonness of paint colours.

The results of this survey differed from those carried out previously [3, 17]. The number of paint fragments recovered was lower than previously recorded. This difference highlights the continuing need for such surveys. Lau *et al.* [4] conclude that the mere finding of paint on clothing and footwear in their region can be assigned some level of significance. However, the discrepancy in the findings between this and other surveys makes it difficult to extrapolate the results from region to region. Nevertheless, it is clear that information such as this is necessary to enable Bayes' theorem to be applied to household paint evidence as it has been to other types of trace evidence.

12.4 Vehicle paint

In contrast to household paint, the examination of automobile paint and the assessment of its evidential value has been the basis of many studies over the years. Holden [18] stated that the evidential value of paint depended on such factors as type of paint, unusual pigments and the probability of another car of the same colour/model being at the scene at the same time as the suspect vehicle.

Tippett [19] counted almost 20,000 cars in an effort to know the distribution of car models of a particular colour and to check this distribution statistically. His conclusion was that the results obtained fitted a Poisson distribution. Tippett also looked at flow rates of traffic and where flow rates are low the chance of seeing two or more of a particular coloured model falls rapidly. Tippett states that with a two-way transfer of 'similar' paint flakes combined with a low flow rate and low frequency of the coloured model, the scientific witness can give a firm opinion that these two cars have been in contact.

Gothard [20] looked at automobile paint flakes as evidence in a different way. That report stated that the layer sequence of paint flakes is the most significant point of comparison particularly because of the variety of ways in which cars can be refinished. The report further states that a large number of layers agreeing with regard to colour, thickness and layer sequence can be taken as proof of common origin without further examination. The study was capable of differentiating all but four out of 500 samples. These four were two pairs from vehicles of similar make, model and colour. All refinished vehicles were easily differentiated.

Ryland and Kopec [21] conducted a similar study and examined the distribution of vehicles by topcoat colour, year of manufacture and vehicle make. This study also looked at the layer distribution in the samples studied. The conclusion of this report was that 94 per cent of the samples were differentiated by microscopic examination and solvent reactivity tests. Of the remaining 6 per cent that were undifferentiated, none of the paint chips had more than three layers. They further concluded that the probability of two matching paint chips originating from different sources is extremely remote when they have numerous layers (six or more) consistent in colour, tint, type of finish, layer thickness and reaction to solvents.

Ryland *et al.* [22] examined the frequency of occurrence of topcoat colours in the eastern United States. This survey examined vehicles in transit and also vehicles parked in public parking lots. Fourteen colours were used as a classification system and these were further divided by the use of the terms light/medium/dark. Discrimination between metallic and non-metallic was possible for the stationary vehicles but not for the moving vehicles. These authors note that the distribution values will still not truly reflect the high power of discrimination offered by a careful microscopic comparison of two similar automotive paints. The authors attempt to relate their result to the presentation of evidence in a court of law. They highlight the difficult responsibility of properly interpreting the meaning of paint evidence. They state that this responsibility should not be avoided as it is the expert's true reason for being in court. They state that the use of the term 'could have originated from' does not fully reflect the evidential value of paint results.

Buckle *et al.* [23] conducted a similar survey in Canada. A further study of a similar nature was conducted by Volpé *et al.* [24] in the province of New Brunswick. Their conclusion was that for comparisons containing physically and chemically indistinguishable original factory paint systems, the forensic conclusion 'probably originated from the same source' is applicable.

Lawton *et al.* [25] undertook a survey to elicit from other laboratories what conclusions were arrived at from various types of evidence. In their report the authors considered a range of case types and a descriptive interpretation was invited. The authors expressed their disappointment at the poor response. The report also stated that of all forensic examinations, paint evidence is one of the hardest to quantify.

Another report [26] surveyed paint examiners on a series of hypothetical paint transfer scenarios. The respondents were requested to use a scale of conclusions ranging from slight support to conclusive. The results of the survey are displayed in Table 12.1. In general there is good agreement on the choice of conclusion appropriate to each scenario. For any one scenario more than 75 per cent of the responses are covered by two adjacent points on the scale (e.g. for Scenario D more than 85 per cent of the participants used 'strong support' or 'very strong support' and for Scenario H more than 86 per cent of the participants used 'very strong support' or 'conclusive').

In the light of the above it is worthwhile looking at the use of language by forensic scientists in the interpretation of evidence.

Table 12.1 Conclusions used by paint examiners for paint exchange scenarios

Scenario	Conclusion (%)				
	Slight support	Support	Strong support	Very strong support	Conclusive
A	**77.4**	19.4	1.6	1.6	0
B	15.3	**64.5**	15.3	4.8	0
C	4.0	**68.5**	23.4	4.0	0
D	0.8	8.9	**51.6**	34.7	4.0
E	0.8	21.8	**57.3**	19.4	0.8
F	0.8	8.1	35.5	**46.8**	8.9
G	0.8	19.4	**44.4**	31.5	4.0
H	0.8	1.6	10.5	**51.6**	35.5

A, one layer transferred in one direction;
B, one layer transferred in each direction;
C, multilayer manufacturer's finish transferred in one direction;
D, multilayer manufacturer's finish transferred in each direction;
E, multilayer manufacturer's finish transferred in one direction and one layer transferred in the other direction;
F, multilayer non-manufacturer's finish transferred in one direction and one layer transferred in the other direction;
G, multilayer non-manufacturer's finish transferred in one direction;
H, multilayer non-manufacturer's finish transferred in each direction.

Rudram [27] examined the formulation of conclusions by forensic scientists. This paper reviews the literature in the area and many authors such as Craddock *et al.* [28] and Satterthwaite and Lambert [29] recommend the use of a well-publicised standard scale of words which reflect a Bayesian approach to the interpretation of evidence. Rudram points out that scientists present oral evidence in only a small minority of cases so a report will often be read aloud by someone else, therefore conclusions should be clear without being overstated. He concludes that putting final conclusions into context should reduce the level of misunderstanding of scientific reports. A published scale goes a long way towards openness and the author suggests that the use of a numerical scale of, for example, 1 (little evidence) to 10 (certain) should be avoided.

A recent survey [2] gathered information on colour, type of finish, make and year of vehicles in the Republic of Ireland and also established what portion of the vehicle population had evidence of being involved in a collision of some sort. This involved examining each vehicle for traces of foreign paint or for the presence of damage to the surface which would give rise to paint transfer. This study represents an estimation of the number of vehicles of a certain colour, finish, make and year on the roads there. More significantly, it gives an indication of the number of vehicles in the population which have been involved in some type of collision where paint was transferred to or from the vehicle. The authors used the results of this survey in a Bayesian statistical approach to get an estimate of the likelihood ratio for various paint transfer scenarios. The results of that survey showed the most common colour to be medium red solid. However, because of the ability to further subdivide this colour by microscopy the largest group was considered to be white at 12.7 per cent. The most common vehicle type was Toyota at 12.5 per cent. The results of the paint *transfer to* or *transfer from* vehicles aspect of the survey show that a significant number of vehicles (17.1 per cent) have evidence of having been the source of transferred paint while a smaller number of vehicles (9.4 per cent) have evidence of having foreign paint transferred to them.

12.5 A Bayesian approach

A detailed background and explanation of Bayes' theorem and its application to forensic science and specifically to transfer evidence is available [30]. However, a brief introduction is appropriate.

In the adversarial system of justice, we can visualise two competing hypotheses: C, the defendant committed the crime, and $\overline{C}$, the defendant did not commit the crime (i.e. someone else committed the crime). Bayes' theorem shows us the effect the scientific evidence has on the odds that C (the defendant committed the crime) is true. Bayes' theorem states that the odds on C after the scientific evidence (posterior odds) are simply the odds before the scientific evidence (prior odds) multiplied by a factor called the likelihood ratio (LR). This likelihood ratio is

$$\frac{\text{Probability of the evidence if C is true}}{\text{Probability of the evidence if } \overline{\text{C}} \text{ is true}} = \frac{P\langle F \mid Ci\rangle}{P\langle F \mid \overline{C}i\rangle}$$

where i denotes the background information and F represents the scientific findings.

Using Bayes' theorem in the manner published by Evett [31], the authors used information from the Irish survey to evaluate likelihood ratios for various scenarios. The work described by them was carried out in the absence of an appreciation for further published work on the hierarchy of propositions [32–36]. The approach used by the authors intermingles source and activity propositions.

12.5.1 Single layer of paint transferred from suspect vehicle to injured party's vehicle

In this instance paint is transferred to the injured party's vehicle and a suspect vehicle is later examined and its paint supplied as a possible source for the foreign paint on the injured party's vehicle. This is analogous to the transfers to the scene in Evett's examples. For the purpose of the exercise, the fact that there is damage to the paintwork of the suspect vehicle is ignored.

We are interested in establishing the likelihood ratio considering two possibilities

C: the paint on the injured party's vehicle originated from the suspect vehicle.
$\overline{C}$: the paint on the injured party's vehicle originated from a random vehicle.

Applying Bayes' theorem we find

Posterior odds $(O\langle C \mid Fi\rangle)$ = Likelihood ratio (LR) × Prior odds $(O\langle C \mid i\rangle)$

$$LR = \frac{P\langle F \mid Ci\rangle}{P\langle F \mid \overline{C}i\rangle} \qquad (1)$$

$P\langle F \mid Ci\rangle$ = Probability of the scientific findings if the suspect vehicle is responsible.
$P\langle F \mid \overline{C}i\rangle$ = Probability of the scientific findings if another vehicle is responsible.

Note: In the following, the background i is omitted for simplicity but the evidence will always be evaluated in the light of background information.

The scientific findings in this case are simple.

F_1: the foreign paint on the injured party's vehicle is white, for example.
F_2: the paint on the suspect vehicle is white.

The likelihood ratio now becomes:

$$\frac{P\langle F\,|\,C\rangle}{P\langle F\,|\,\overline{C}\rangle}=\frac{P\langle F_2\,|\,C\rangle}{P\langle F_2\,|\,\overline{C}\rangle}\times\frac{P\langle F_1\,|\,F_2C\rangle}{P\langle F_1\,|\,F_2\overline{C}\rangle} \tag{2}$$

$P\langle F_2\,|\,C\rangle$ = Probability that the suspect vehicle paint is white given that the suspect vehicle is responsible for the transfer.

$P\langle F_2\,|\,\overline{C}\rangle$ = Probability that the suspect vehicle paint is white given that another vehicle is responsible for the transfer.

$P\langle F_1\,|\,F_2C$ = Probability that the foreign paint on the injured party's vehicle is white given that the suspect vehicle is white and is responsible for the transfer.

$P\langle F_1\,|\,F_2\overline{C}\rangle$ = Probability that the foreign paint on the injured party's vehicle is white given that the suspect vehicle is white and another vehicle is responsible for the transfer.

$P\langle F_2\,|\,C\rangle = P\langle F_2\,|\,\overline{C}\rangle$ as the probability of F_2 (i.e. the paint on the suspect vehicle is white) is the same for scenarios C and $\overline{C}$. As the colour of the suspect vehicle is irrelevant if it is not responsible for the transfer of the paint then $P\langle F_1\,|\,F_2\overline{C}\rangle$ becomes $P\langle F_1\,|\,\overline{C}\rangle$.

Equation (2) now simplifies to

$$\frac{P\langle F\,|\,C\rangle}{P\langle F\,|\,\overline{C}\rangle}=\frac{P\langle F_1\,|\,F_2C\rangle}{P\langle F_1\,|\,\overline{C}\rangle} \tag{3}$$

The numerator $P\langle F_1\,|\,F_2C\rangle = 1$, i.e. the probability that the paint on the injured party's vehicle would be the same as the suspect vehicle if the suspect vehicle left the paint = 1.

In the case of the denominator we must evaluate the scientific findings in the light of the foreign paint originating from any other vehicle, i.e.

$P\langle F_1\,|\,\overline{C}\rangle$ = frequency of that particular colour. Let this frequency be denoted by f_c.

Now

$$\frac{P\langle F\,|\,C\rangle}{P\langle F\,|\,\overline{C}\rangle}=\frac{1}{f_c} \tag{4}$$

Substituting the frequency value for white solid paint value from the survey (12.7 per cent), the likelihood ratio becomes:

$$\frac{P\langle F\,|\,C\rangle}{P\langle F\,|\,\overline{C}\rangle}=\frac{1}{0.127}=7.9$$

This likelihood ratio obviously varies depending on colour. Substituting the value for the more unusual colours from this survey can result in likelihood ratios of approximately 1000.

12.5.2 Single layer of paint transferred from injured party to suspect vehicle

The distinction between transfer to and transfer from the injured party is only appropriate when an activity level proposition is addressed [33]. If we now look at the same type of transfer in the other direction the situation is more complicated. In this example the foreign paint found on a suspect vehicle matches the injured party's vehicle. Again no inference is taken from the presence of damage to the injured party's vehicle.

The possibilities to be examined are:

C: the paint on the suspect vehicle originated from the injured party's vehicle.
$\overline{\text{C}}$: the paint on the suspect vehicle originated from a random source.

The scientific findings consist of:

F_1: the injured party's vehicle is white.
F_2: the foreign paint on the suspect vehicle is a matching white.

So the likelihood ratio

$$\text{LR} = \frac{P\langle F \mid C\rangle}{P\langle F \mid \overline{C}\rangle} = \frac{P\langle F_1 \mid C\rangle}{P\langle F_1 \mid \overline{C}\rangle} \times \frac{P\langle F_2 \mid F_1 C\rangle}{P\langle F_2 \mid F_1 \overline{C}\rangle} \tag{5}$$

As in scenario 12.5.1

$$P\langle F_1 \mid C\rangle = P\langle F_1 \mid \overline{C}\rangle$$

In the $\overline{\text{C}}$ scenarios, i.e. a source other than the injured party's vehicle gave rise to the paint on the suspect vehicle, then the foreign paint found on the suspect vehicle is independent of the injured party's vehicle so $P\langle F_2 \mid F_2 \overline{C}\rangle$ becomes $P\langle F_2 \mid \overline{C}\rangle$.

Equation (5) is simplified to

$$\frac{P\langle F \mid C\rangle}{P\langle F \mid \overline{C}\rangle} = \frac{P\langle F_2 \mid F_1 C\rangle}{P\langle F_2 \mid \overline{C}\rangle} \tag{6}$$

We now need information from the survey to evaluate the probabilities.

We first consider the numerator, i.e. the C situation. Let b signify the probability that a random vehicle will have foreign paint on it, q signify the probability that the foreign paint would match that from the injured party's vehicle, and t denote the likelihood of paint transferring in the course of this accident and consisting of one topcoat layer.

In this instance the white paint can be on the suspect vehicle for one of two reasons:

(i) no paint was transferred and there was already white paint on the vehicle, i.e.

$$\text{Probability} = (1 - t)bq$$

(ii) it was transferred from the injured party's vehicle, i.e.

$$\text{Probability} = t(1 - b)$$

Now by adding the two probabilities at (i) and (ii) we get the numerator

$$P\langle F_2 \mid F_1 C\rangle = (1 - t)bq + t(1 - b)$$

We now consider the denominator, i.e. the $\overline{\text{C}}$ situation:

$$P\langle F_2 \mid \overline{C}\rangle = bq$$

so the likelihood ratio becomes

$$\frac{P\langle F\,|\,C\rangle}{P\langle F\,|\,\overline{C}\rangle} = (1-t) + \frac{t(1-b)}{bq} \tag{7}$$

If the likelihood ratio is considerably greater than 1 then Equation (7) is further simplified to

$$\frac{P\langle F_2\,|\,F_1C\rangle}{P\langle F_2\,|\,\overline{C}\rangle} = \frac{t(1-b)}{bq} \tag{8}$$

A value of 0.8 for t is suggested by the authors on the basis of experience, i.e. we estimate that paint consisting of at least a top layer is transferred in 80 per cent of collisions investigated:

$t \approx 0.8$
$b \approx 0.094$ from the survey (9.4 per cent of vehicles have evidence of foreign paint)
$q \approx f_c = 0.127$ (for white solid as in scenario 12.5.1)

so

$$\frac{P\langle F\,|\,C\rangle}{P\langle F\,|\,\overline{C}\rangle} = 61$$

This ratio is obviously also greatly affected by the value for the frequency of colour but it can be seen to be more significant than the suspect to injured party transfer scenario by a factor of approximately 10.

12.5.3 Exchange of single layer paints

In their paper [2] the authors examined the exchange of single layered paint as independent events. A more appropriate approach would be to consider a conditional transfer probability as outlined by Champod and Taroni [37].

12.5.4 Transfer of multilayer manufacturer's finish from suspect vehicle to injured party

Using the logic in scenario 12.5.1

$$\frac{P\langle F\,|\,C\rangle}{P\langle F\,|\,\overline{C}\rangle} = \frac{P\langle F_1\,|\,F_2C\rangle}{P\langle F_1\,|\,\overline{C}\rangle}$$

Again

$$P\langle F_1\,|\,F_2C\rangle = 1$$

but in this instance

$$P\langle F\,|\,\overline{C}\rangle = (\text{frequency of colour } (f_c)) \times (\text{frequency of manufacturer } (f_m))$$

Note: Colour and manufacturer are treated as being independent because of the broad classification used under the heading 'colour'.

So

$$\frac{P\langle F \mid C\rangle}{P\langle F \mid \overline{C}\rangle} = \frac{1}{f f_m} \quad (9)$$

Choosing white solid colour and the most common manufacturer from our survey (i.e. Toyota at 12.5 per cent)

$$\frac{1}{f f_m} = \frac{1}{(0.127)(0.125)} = 63$$

This is approximately a tenfold increase in the likelihood ratio over the same situation when only the top layer is transferred.

It must be remembered that this figure is derived from the most commonly occurring colour and the most commonly occurring manufacturer.

12.5.5 Multilayer manufacturer's finish transferred from injured party to suspect vehicle

Once again, the distinction between transfer to and transfer from the injured party is only appropriate when an activity level proposition is addressed [33]. In this instance we can use Equation (8), i.e.

$$\frac{P\langle F \mid C\rangle}{P\langle F \mid \overline{C}\rangle} = \frac{t(1-b)}{bq}$$ and substitute the appropriate values

Let

t = the likelihood of a multilayer manufacturer's paint transferring in the course of this accident. The authors set that figure conservatively at 0.4, i.e. 50 per cent of cases where paint is transferred. (So t = 50 per cent of 0.8; see scenario 12.5.2.)
b = probability that a random vehicle will have foreign multilayer paint on it. In their survey the authors found foreign paint on 9.4 per cent of vehicles. In their opinion none of these seems to be multilayer though it is acknowledged that this is difficult to assess in the field. They suggest 1 per cent as a conservative value.
q = probability that the foreign paint would match the top coat of the injured party's vehicle and share the same manufacturer's layer structure, i.e. $f_c f_m$.

Substituting the values for white solid Toyotas from the survey

$$\frac{P\langle F \mid C\rangle}{P\langle F \mid \overline{C}\rangle} = \frac{(0.4)(0.99)}{(0.127)(0.125)(0.01)} = 2494$$

12.5.6 Exchange of multilayer manufacturer's finish

The same logic applies as in scenario 12.5.3 above.

The above work shows that significant values for likelihood ratio are obtained for the most common paint types. Much higher values are expected for paints which have a lower frequency of occurrence.

It must be stressed that this work was undertaken without the benefit of an understanding of the distinctions between source and activity level propositions [32–36].

The interpretation of paint evidence in the context of such propositions remains a challenge to be tackled.

12.6 Conclusion

The literature over the past thirty years implies that paint is one of the more convincing sources of associative evidence. This was based on the variation within paints, on the analytical methods available to discriminate them and on the combination of layers present in flakes. Nothing has been published to contradict these claims. However, a review of recent literature suggests that paint as a type of trace evidence is getting little attention.

In other types of trace evidence Bayes' theorem has been extensively used to assess the value of various findings. In the case of vehicle paint attempts have been made to use the same approach. For household paints, some surveys have been carried out in relation to their frequency of occurrence. However, no attempts have been published to extend the value of these surveys by using Bayes' theorem to check the significance of these findings. Meanwhile household paints continue to diversify and flakes of paint continue to be encountered as trace evidence in cases.

In the present environment of probability theory being rightly used as the logic underpinning scientific inference, the lack of suitable studies in the area of paint may mean that this significant evidence type does not get its due recognition. The interpretation of transferred paint findings remains a challenge for forensic scientists.

12.7 References

1. Stoecklein W. (1995). Forensic analysis of automotive paints at the Bundeskriminalamt: the evidential value of automotive paints. *Crime Laboratory Digest* **22**:98.
2. McDermott S. D., Willis S. M., McCullough J. P. (1999). The evidential value of paint. Part II: A Bayesian approach. *Journal of Forensic Sciences* **44**:263–269.
3. Pearson E. F., May R. W., Dabbs M. G. D. (1971). Glass and paint fragments found in men's outer clothing – a report of a survey. *Journal of Forensic Sciences* **16**:283–302.
4. Lau L., Beveridge A. D., Callowhill B. C., Conners N., Foster K., Groves R. J., Ohashi K. N., Sumner A. M., Wong H. (1997). The frequency of occurrence of paint and glass on the clothing of high school students. *Canadian Society of Forensic Science Journal* **30**:233–240.
5. Hicks T., Vanina R., Margot P. (1996). Transfer and persistence of glass fragments on garments. *Science and Justice* **36**:101–107.
6. Holcroft G. A., Shearer B. (1993). Personal communication.
7. Crown D. A. (1968). *The Forensic Examination of Paints and Pigments*. Charles C. Thomas, Springfield, Illinois.
8. Pfefferli P. W. (1984). Evaluation of the evidential value of white household paint by energy dispersive X-ray analysis. PhD thesis, University of Lausanne.
9. Laing D. K., Dudley R. J., Home J. M., Isaacs M. D. J. (1982). The discrimination of small fragments of household gloss paint by microspectrophotometry. *Forensic Science International* **20**:191–200.
10. Saferstein R. (1995). *Criminalistics – An Introduction to Forensic Science*, p. 230. Prentice Hall, Upper Saddle River, New Jersey, USA.
11. Walls H. J. (1974). *Forensic Science – An Introduction to Scientific Crime Detection*, pp. 26–40. Sweet and Maxwell, London.

12. Gallop A., Stockdale R. (1998). Trace and contact evidence. In *Crime Scene to Court* (ed. P. White), pp. 47–72. Royal Society of Chemistry, Cambridge, UK.
13. Hudson G. D., Andahl R. O., Butcher S. J. (1977). The paint index – the colour classification and use of a collection of paint samples taken from scenes of crime. *Journal of the Forensic Science Society* **17**:27–32.
14. Kirk P. K., Kingston C. R. (1964). Evidential evaluation and problems in general criminalistics. *Journal of Forensic Sciences* **9**:437.
15. Tippett C. F., Emerson V. J., Fereday M. J., Lawton F., Richardson A., Jones L. T., Lampert S. M. (1968). The evidential value of the comparison of paint flakes from sources other than vehicles. *Journal of the Forensic Science Society* **8**:61–65.
16. Castle D. A., Curry C. J., Russell L. W. (1984). A survey of case openers. *Forensic Science International* **24**:285–294.
17. McQuillan J., Edgar K. (1992). A survey of the distribution of glass on clothing. *Journal of the Forensic Science Society* **32**:333–348.
18. Holden I. G. (1962). The evaluation of scientific evidence in relation to road accidents. *Medicine, Science and the Law* **3**:541–545.
19. Tippett C. F. (1964). Car distribution statistics and the hit-and-run driver. *Medicine, Science and the Law* **4**:91–97.
20. Gothard J. A. (1976). Evaluation of automobile paint flakes as evidence. *Journal of Forensic Sciences* **21**:636–641.
21. Ryland S. G., Kopec R. J. (1979). The evidential value of automobile paint chips. *Journal of Forensic Sciences* **24**:140–147.
22. Ryland S. G., Kopec R. J., Somerville P. N. (1981). The evidential value of automobile paint. Part II: The frequency of occurrence of topcoat colours. *Journal of Forensic Sciences* **26**:64–74.
23. Buckle J., Fung T., Ohashi K. (1987). Automotive topcoat colours: occurrence frequencies in Canada. *Canadian Society of Forensic Science Journal* **20**:45–56.
24. Volpé G. G., Stone H. S., Rioux J. M., Murphy K. J. (1988). Vehicle topcoat colour and manufacturer: frequency distribution and evidential significance. *Canadian Society of Forensic Science Journal* **21**:11–18.
25. Lawton M. E., Buckleton J. S., Walsh K. A. J. (1988). An international survey of the reporting of hypothetical cases. *Journal of the Forensic Science Society* **28**:243–252.
26. McDermott S. D., Willis S. M. (1997). A survey of the evidential value of paint transfer evidence. *Journal of Forensic Sciences* **42**:1012–1018.
27. Rudram D. A. (1996). Interpretation of scientific evidence. *Science and Justice* **36**:133–138.
28. Craddock J. G., Lamb P., Moffat A. C. (1989). Problems of written communication: understanding and misunderstanding forensic scientists' statements. Personal communication.
29. Satterthwaite J., Lambert J. (1989). Interpreting the interpretations: a survey to assess the effectiveness of conclusions in statements written by forensic scientists. Personal communication.
30. Aitken C. G. G., Stoney D. A. (1991). *The Use of Statistics in Forensic Science*. Ellis Horwood Ltd., Chichester, West Sussex, UK.
31. Evett I. W. (1990). The theory of interpreting scientific transfer evidence. In *Forensic Science Progress* (eds A. Maehly, R. L. Williams), Vol. 4, pp. 141–179. Springer-Verlag, Berlin, Germany.
32. Cook R. *et al.* (1998). A model for case assessment and interpretation. *Science and Justice* **38**:151–156.
33. Cook R. *et al.* (1988). A hierarchy of propositions: deciding which level to address in casework. *Science and Justice* **38**:231–239.
34. Cook R. *et al.* (1999). Case pre-assessment and review in a two-way transfer case. *Science and Justice* **39**:103–111.
35. Evett I. W. *et al.* (2000). More on the hierarchy of propositions: exploring the distinction between explanations and propositions. *Science and Justice* **40**:3–10.
36. Evett I. W. *et al.* (2000). The impact of the principles of evidence interpretation on the structure and content of statements. *Science and Justice* **40**:233–239.
37. Champod C. and Taroni F. (1999). Interpretation of fibres evidence – the Bayesian approach. In *Forensic Examination of Fibres* (2nd edn) (eds Robertson J., Grieve M.) pp. 379–398. Taylor & Francis, London.

Index